U.S. AIR FORCE POCKET SURVIVAL HANDBOOK

THE PORTABLE AND ESSENTIAL GUIDE TO STAYING ALIVE

United States Air Force

Edited by
Jay McCullough

Skyhorse Publishing

Skyhorse Publishing books may be purchased in bulk at special discounts for sales promotion, corporate gifts, fund-raising, or educational purposes. Special editions can also be created to specifications. For details, contact the Special Sales Department, Skyhorse Publishing, 307 West 36th Street, 11th Floor, New York, NY 10018 or info@skyhorsepublishing.com.

Skyhorse® and Skyhorse Publishing® are registered trademarks of Skyhorse Publishing, Inc.®, a Delaware corporation.

www.skyhorsepublishing.com

10 9 8 7 6

Library of Congress Cataloging-in-Publication Data is available on file.

ISBN: 978-1-62087-104-1

Printed in the United States of America

DEPARTMENT OF THE AIR FORCE
Headquarters U.S. Air Force
Washington DC 20330-5000

AF REGULATION 64-4
VOLUME I

Search and Rescue

SURVIVAL TRAINING

This regulation describes the various environmental conditions affecting human survival, and describes individual activities necessary to enable that survival. This regulation is for instructor and student use in formal and USAF survival and survival continuation training. This regulation also applies to U.S. Air Force Reserve and Air National Guard units and members. Sources used to compile this regulation are listed in the bibliography, attachment 1.

Page

Page

Part One

THE ELEMENTS OF SURVIVING

Chapter 1

MISSION

1-1. Introduction. An ejection sequence, a bailout, or crash landing ends one mission for the crew but starts another—to successfully return from a survival situation. Are they prepared? Can they handle the new mission, not knowing what it entails? Unfortunately, many aircrew members are not fully aware of their new mission or are not fully prepared to carry it out. All instructors teaching aircrew survival must prepare the aircrew member to face and successfully complete this new mission.

1-2. Aircrew Mission. The moment an aircrew member leaves the aircraft and encounters a survival situation, the assigned mission is to: "return to friendly control without giving aid or comfort to the enemy, to return early and in good physical and mental condition."

a. On first impressions, "friendly control" seems to relate to a combat situation. Even in peacetime, however, the environment may be quite hostile. Imagine parachuting into the arctic when it's minus 40° F. Would an aircrew member consider this "friendly?" No. If the aircraft is forced to crash-land in the desert where temperatures may soar above 120°F, would this be agreeable? Hardly. The possibilities for encountering hostile conditions affecting human survival are endless. Crewmembers who egress an aircraft may confront situations difficult to endure.

b. The second segment of the mission, "without giving aid or comfort to the enemy," is directly related to a combat environment. This part of the mission may be most effectively fulfilled by following the moral guide—the Code of Conduct. Remember, however, that the Code of Conduct is useful to a survivor at all times and in all situations. Moral obligations apply to the peacetime situation as well as to the wartime situation.

c. The final phase of the mission is "to return early and in good physical and mental condition." A key factor in successful completion of this part of the mission may be the *will to survive*. This will is present, in varying degrees, in all human beings. Although successful survival is based on many factors, those who maintain this important attribute will increase their chance of success.

1-3. Goals. Categorizing this mission into organizational components, the three goals or duties of a survivor are to maintain life, maintain honor, and return. Survival training instructors and formal survival training courses provide training in the skills, knowledge, and attitudes necessary for an aircrew member to successfully perform the fundamental survival duties.

1-4. Survival. Surviving is extremely stressful and difficult. The survivor may be constantly faced with hazardous and difficult situations. The stresses, hardships, and hazards (typical of a survival episode) are caused by the cumulative effects of existing conditions. (See Chapter 2 pertaining to conditions affecting survival.) Maintaining life and honor and returning, regardless of the conditions, may make surviving difficult or unpleasant. The survivor's mission forms the basis for identifying and organizing the major needs of a survivor. (See survivor's needs in Chapter 3.)

1-5. Decisions. The decisions survivors make and the actions taken in order to survive determine their prognosis for surviving.

1-6. Elements. The three primary elements of the survivor's mission are: the conditions affecting survival, the survivor's needs, and the means for surviving.

Chapter 2

CONDITIONS AFFECTING SURVIVAL

2-1. Introduction. Five basic conditions affect every survival situation (Figure 2-1). These conditions may vary in importance or degree of influence from one situation to another and from individual to individual. At the onset, these conditions can be considered to be neutral—being neither for nor against the survivor—and should be looked upon as neither an advantage nor a disadvantage. The aircrew member may succumb to their effects—or use them to best advantage. These conditions exist in each survival episode, and they will have great bearing on the survivor's every need, decision, and action.

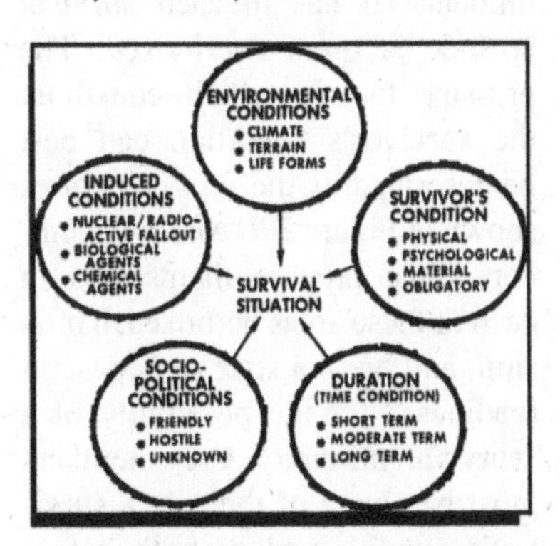

Figure 2-1. Five Basic Conditions.

2-2. Environmental Conditions. Climate, terrain, and life forms are the basic components of all environments. These components can present special problems for the survivor. Each component can be used to the survivor's advantage. Knowledge of these conditions may very well contribute to the success of the survival mission.

a. Climate. Temperature, moisture, and wind are the basic climatic elements. Extreme cold or hot temperatures, complicated by moisture (rain, humidity, dew, snow, etc.) or lack of moisture, and the possibility of wind, may have a life threatening impact on the survivor's needs, decisions, and actions. The primary concern, resulting from the effects of climate, is the need for personal protection. Climatic conditions also have a significant impact on other aspects of survival (for example, the availability of water and food, the need and ability to travel, recovery capabilities, physical and psychological problems, etc.).

b. Terrain. Mountains, prairies, hills, and lowlands are only a few examples of the infinite variety of landforms which describe "terrain." Each of the landforms have a different effect on a survivor's needs, decisions, and actions. A survivor may find a combination of several terrain forms in a given situation. The existing terrain will affect the survivor's needs and activities in such areas as travel, recovery, sustenance, and, to a lesser extent, personal protection. Depending on its form, terrain may afford security and concealment for an evader; cause travel to be easy or difficult; provide protection from cold, heat, moisture, wind, or make surviving a seemingly impossible task (Figure 2-2).

SURVIVOR'S CONDITION
◯ PHYSICAL
◯ PSYCHOLOGICAL
◯ MATERIAL
◯ OBLIGATORY

Figure 2-2. Survivor's Condition.

c. Life Forms. For survival and survival training purposes, there are two basic life-forms (other than human)—plant life and animal life. NOTE: The special relationship and effects of people on the survival episode are covered separately. Geographic areas are often identified in terms of the abundance of life (or lack thereof). For example, the barren arctic or desert, primary (or secondary) forests, the tropical rain forest, the polar ice cap, etc., all produce images regarding the quantities of life forms. These examples can have special meaning not only in terms of the hazards or needs they create, but also in how a survivor can use available life forms.

(1) Plant Life. There are hundreds of thousands of different types and species of plant life. In some instances, geographic areas are identified by the dominant types of plant life within that area. Examples of this are savannas, tundra, deciduous forests, etc. Some species of plant life can be used advantageously by a survivor— if not for the food or the water, then for improvising camouflage, shelter, or providing for other needs.

(2) Animal Life. Reptiles, amphibians, birds, fish, insects, and mammals are life-forms that directly affect a survivor. These creatures affect the survivor by posing hazards (which must be taken into consideraron), or by satisfying needs.

2-3. The Survivor's Condition.

The survivor's condition and the influence it has in each survival episode is often overlooked. The primary factors which constitute the survivor's condition can best be described by the four categories shown in Figure 2-2. Aircrew members must prepare themselves in each of these areas before each mission, and be in a state of "constant readiness" for the possibility of a "survival mission." Crewmembers must be aware of the role a survivor's condition plays both before and during the survival episode.

a. Physical. The physical condition and the fitness level of the survivor are major factors affecting survivability. Aircrew members who are physically fit will be better prepared to face survival episodes than those who are not. Further, a survivor's physical condition (injured or uninjured) during the initial phase of a survival episode will be a direct result of circumstances surrounding the ejection, bailout, parachute landing, or crash landing. In short, high levels of physical fitness and good post-egress physical condition will enhance a survivor's ability to cope with such diverse variables as: (1) temperature extremes, (2) rest or lack of it, (3) water availability, (4) food availability, and (5) extended survival episodes. In the last instance, physical weakness may increase as a result of nutritional deficiencies, disease, etc.

b. Psychological. Survivors' psychological state greatly influences their ability to successfully return from a survival situation.

(1) Psychological effectiveness in a survival episode (including captivity) results from effectively coping with the following factors:

(a) Initial shock—Finding oneself in a survival situation following the stress of ejection, bailout, or crash landing.

(b) Pain—Naturally occurring or induced by coercive manipulation.

(c) Hunger—Naturally occurring or induced by coercive manipulation.

(d) Thirst—Naturally occurring or induced by coercive manipulation.

(e) Cold or Heat—Naturally occurring or induced by coercive manipulation.

(f) Frustration—Naturally occurring or induced by coercive manipulation.

(g) Fatigue (including sleep deprivation)—Naturally occurring or induced by coercive manipulation.

(h) Isolation—Includes forced (captivity) and the extended duration of any episode.

(i) Insecurity—Induced by anxiety and self- doubts.

(j) Loss of self-esteem—Most often induced by coercive manipulation.

(k) Loss of self-determination—Most often induced by coercive manipulation.

(l) Depression—Mental "lows."

(2) A survivor may experience emotional reactions during a survival episode due to the previously stated factors, previous (life) experiences (including training), and the survivor's psychological tendencies. Emotional reactions commonly occurring in survival (including captivity) situations are:

(a) Boredom—sometimes combined with loneliness.

(b) Loneliness

(c) Impatience

(d) Dependency

(e) Humiliation

(f) Resentment

(g) Anger—sometimes included as a subelement of hate.

(h) Hate

(i) Anxiety

(j) Fear—often included as a part of panic or anxiety.

(k) Panic

(3) Psychologically survival episodes may be divided into "crisis" phases and "coping" phases. The initial crisis period will occur at the onset of the survival situation. During this initial period, "thinking" as well as "emotional control" may be disorganized. Judgment is impaired, and behavior may be irrational (possibly to the point of panic). Once the initial crisis is under control, the coping phase begins and the survivor is able to respond positively to the situation. Crisis periods may well recur, especially during extended situations (captivity). A survivor must strive to control if avoidance is impossible.

(4) The most important psychological tool that will affect the outcome of a survival situation is the will to survive. Without it, the survivor is surely doomed to failure—a strong will is the best assurance of survival.

c. Material. At the beginning of a survival episode, the clothing and equipment in the aircrew member's possession, the contents of available survival kits, and salvageable resources from the parachute or aircraft are the sum total of the survivor's material assets. Adequate premission preparations are required (must be stressed during training). Once the survival episode has started, special attention must be given to the care, use, and storage of all materials to ensure they continue to be serviceable and available. Items of clothing and equipment should be selectively augmented with improvised items.

(1) Clothing appropriate to anticipated environmental conditions (on the ground) should be worn or carried as aircraft space and mission permit.

(2) The equipment available to a survivor affects all decisions, needs, and actions. The survivor's ability to improvise may provide ways to meet some needs.

d. Legal and Moral Obligations. A survivor has both legal and moral obligations or responsibilities. Whether in peacetime or combat, the survivor's responsibilities as a member of the military service continues. Legal obligations are expressly identified in the Geneva Conventions, Uniform Code of Military Justice (UCMJ), and Air Force directives and policies. Moral obligations are expressed in the Code of Conduct. (Figure2-3.)

(1) Other responsibilities influence behavior during survival episodes and influence the *will to survive.* Examples include feelings of obligation or responsibilities to family, self, and(or) spiritual beliefs.

(2) A survivor's individual perception of responsibilities influence

U.S. FIGHTING MAN'S

CODE OF CONDUCT

I

I am an American fighting man. I serve in the forces which guard my country and our way of life. I am prepared to give my life in their defense.

II

I will never surrender of my own free will. If in command, I will never surrender my men while they still have the means to resist.

III

If I am captured, I will continue to resist by all means available. I will make every effort to escape and aid others to escape. I will accept neither parole nor special favors from the enemy.

IV

If I become a prisoner of war, I will keep faith with my fellow prisoners. I will give no information or take part in any action which might be harmful to my comrades. If I am senior, I will take command. If not, I will obey the lawful orders of those appointed over me and will back them up in every way.

V

When questioned, should I become a prisoner of war, I am required to give name, rank, service number, and date of birth. I will evade answering further questions to the utmost of my ability. I will make no oral or written statements disloyal to my country and its allies or harmful to their cause.

VI

I will never forget that I am an American fighting man, responsible for my actions, and dedicated to the principles which made my country free. I will trust in my God and in the United States of America.

Figure 2-3. Code of Conduct.

survival needs, and affect the psychological state of the individual both during and after the survival episode. These perceptions will be reconciled either consciously through rational thought or subconsciously through attitude changes. Training specifically structured to foster and maintain positive attitudes provides a key asset to survival.

2-4. Duration—The Time Condition. The duration of the survival episode has a major effect upon the aircrew member's needs. Every decision and action will be driven in part by an assessment of when recovery or return is probable. Air superiority, rescue capabilities, the distances involved, climatic conditions, the ability to locate the survivor, or captivity are major factors which directly influence the duration (time condition) of the survival episode. A survivor can never be certain that rescue is imminent.

2-5. Sociopolitical Condition. The people a survivor contacts, their social customs, cultural heritage, and political attitudes will affect the survivor's status. Warfare is one type of sociopolitical condition, and people of different cultures are another. Due to these sociopolitical differences, the interpersonal relationship between the survivor and any people with whom contact is established is crucial to surviving. To a survivor, the attitude of the people contacted will be friendly, hostile, or unknown.

a. Friendly People. The survivor who comes into contact with friendly people, or at least those willing (to some degree) to provide aid, is indeed fortunate. Immediate return to home, family, or home station, however, may be delayed. When in direct association with even the friendliest of people, it is essential to maintain their friendship. These people may be of a completely different culture in which a commonplace American habit may be a gross and serious insult. In other instances, the friendly people may be active insurgents in their country and constantly in fear of discovery. Every survivor action, in these instances, must be appropriate and acceptable to ensure continued assistance.

b. Hostile People. A state of war need not exist for a survivor to encounter hostility in people. With few exceptions, any contact with hostile people must be avoided. If captured, regardless of the political or social reasons, the survivor must make all efforts to adhere to the Code of Conduct and the legal obligations of the UCMJ, the Geneva Conventions, and U.S. AF policy.

c. Unknown People. The survivor should consider all factors before contacting unknown people. Some primitive cultures and closed societies still exist in which outsiders are considered a threat. In other areas of the world, differing political and social attitudes can place a survivor "at risk" in contacting unknown people.

2-6. Induced Conditions. Any form of warlike activity results in "induced conditions." Three comparatively new induced conditions may occur during combat operations. Nuclear warfare and the resultant residual radiation, biological warfare, and chemical warfare (NBC) create life-threatening conditions from which a survivor needs immediate protection. The longevity of NBC conditions further complicates a survivor's other needs, decisions, and actions.

Chapter 3

THE SURVIVOR'S NEEDS

3-1. Introduction. The three fundamental goals of a survivor—to maintain life, maintain honor, and return—may be further divided into eight basic needs. In a noncombatant situation, these needs include: personal protection, sustenance, health, travel, and communications (signaling for recovery). During combat, additional needs must be fulfilled. They are: evasion, resistance if captured, and escape if captured. Meeting the individual's needs during the survival episode is essential to achieving the survivor's fundamental goals (Figure 3-1).

Figure 3-1. Survior's Needs.

3-2. Maintaining Life. Three elementary needs of a survivor in any situation which are categorized as the integral components of maintaining life are: personal protection, sustenance, and health.

a. Personal Protection. The human body is comparatively fragile. Without protection, the effects of environmental conditions (climate, terrain, and life forms) and of induced conditions (radiological, biological agents, and chemical agents) may be fatal. The survivor's primary defenses against the effects of the environment are clothing, equipment, shelter, and fire. Additionally, clothing, equipment, and shelter are the primary defenses against some of the effects of induced conditions (Figure 3-2).

(1) The need for adequate clothing and its proper care and use cannot be overemphasized. The human body's tolerance for temperature extremes is very limited. However, its ability to regulate heating and cooling is extraordinary. The availability of clothing and its proper use is extremely important to a survivor in using these abilities of the body. Clothing also provides excellent protection against the external effects of alpha and beta radiation, and may serve as a shield against the external effects of some chemical or biological agents.

(2) Survival equipment is designed to aid survivors throughout their episode. It must be cared for to maintain its effectiveness. Items found in a survival kit or

Figure 3-2. Personal Protection

aircraft can be used to help satisfy the eight basic needs. Quite often, however, a survivor must improvise to overcome an equipment shortage or deficiency.

(3) The survivor's need for shelter is twofold—as a place to rest and for protection from the effects of the environmental and(or) induced conditions (NBC). The duration of the survival episode will have some effect on shelter choice. In areas that are warm and dry, the survivor's need is easily satisfied using natural resting places. In cold climates, the criticality of shelter can be measured in minutes, and rest is of little immediate concern. Similarly, in areas of residual radiation, the criticality of shelter may also be measured in minutes (Figure 3-3).

(4) Fire serves many survivor needs: purifying water, cooking and preserving food, signaling, and providing a source of heat to warm the body and dry clothing (Figure 3-4).

b. Sustenance. Survivors need food and water to maintain normal body functions and to provide strength, energy, and endurance to overcome the physical stresses of survival.

(1) Water. The survivor must be constantly aware of the body's continuing need for water (Figure 3-5).

(2) Food. During the first hours of a survival situation, the need for food receives little attention. During the first 2 or 3 days, hunger becomes a nagging aggravation which a survivor can overcome. The first major food crisis occurs when the loss of energy,

Figure 3-3. Shelters.

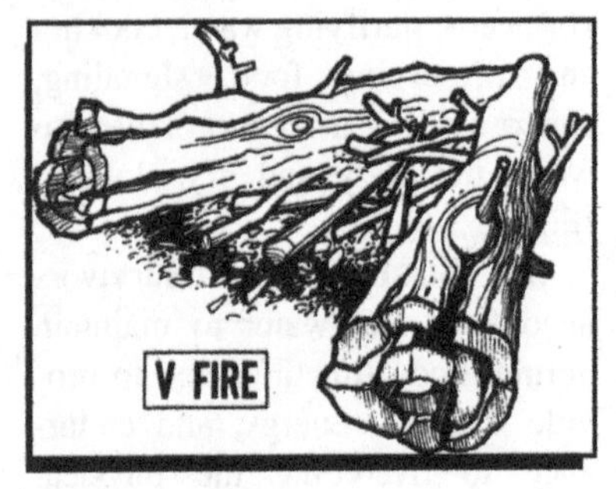

Figure 3-4. Fire.

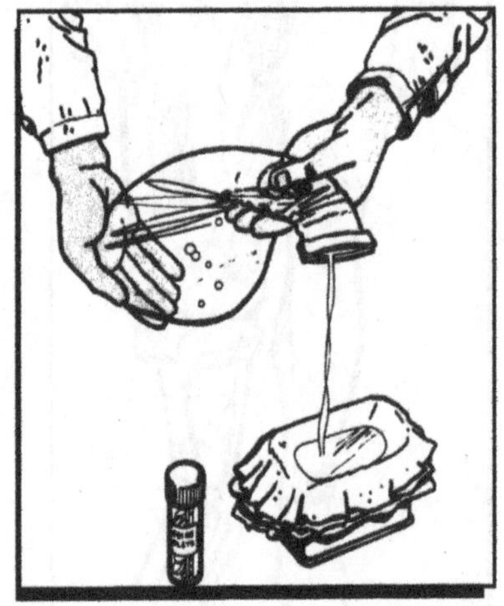

Figure 3-5. Water.

stamina, and strength begin to affect the survivor's physical capabilities. The second major food crisis is more insidious. A marked increase in irritability and other attitudes may occur as the starvation process continues. Early and continuous attention must be given to obtaining and using any and all available food. Most people have food preferences. The natural tendency to avoid certain types of food is a major problem which must be overcome early in the survival situation. The starvation process ultimately overcomes all food aversions. The successful survivor overcomes these aversions before physical or psychological deterioration sets in (Figure 3-6).

Figure 3-6. Food.

c. Health (Physical and Psychological). The survivor must be the doctor, nurse, corpsman, psychologist, and cheerleader. Self-aid is the survivor's sole recourse.

(1) Prevention. The need for preventive medicine and safety cannot be overemphasized. Attention to sanitation and personal hygiene is a major factor in preventing physical, morale, and attitudinal problems.

(a) The need for cleanliness in the treatment of injuries and illness is self-evident. The prisoner of war (PW) who used maggots to eat away rotting flesh caused by infection is a dramatic example. Prevention is much more preferred than such drastic procedures.

(b) Safety must be foremost in the mind of the survivor; carelessness is caused by ignorance and(or) poor judgment or bad luck. One miscalculation with a knife or ax can result in self-inflicted injury or death.

(2) Self-Aid:

(a) Injuries frequently occur during ejection, bailout, parachute landing, or ditching. Other post-egress factors may also cause injury. In the event of injury, the survivor's existence may depend on the ability to perform self-aid. In many instances, common first aid procedures will suffice; in others, more primitive techniques will be required (Figure 3-7).

(b) Illness and the need to treat it is more commonly associated with long-term situations such as an extended evasion episode or captivity. When preventive techniques have failed, the survivor must treat symptoms of disease in the absence of professional medical care.

Figure 3-7. Self-Aid.

(3) Psychological Health. Perhaps the survivor's greatest need is the need for emotional stability and a positive, optimistic attitude. An individual's ability to cope with psychological stresses will enhance successful survival. Optimism, determination, dedication, and humor, as well as many other psychological attributes, are all helpful for a survivor to overcome psychological stresses (Figure 3-8).

Figure 3-8. Health and Morale.

3-3. Maintaining Honor. Three elementary needs which a survivor may experience during combat survival situations are categorized as integral components of maintaining honor. These three elementary needs are: (a) avoiding capture or evading, (b) resisting (if captured), and (c) escaping (if captured).

a. Avoiding Capture. Evasion will be one of the most difficult and hazardous situations a survivor will face. However difficult and hazardous evasion may be, captivity is always worse. During an evasion episode, the survivor has two fundamental tasks. The first is to use concealment techniques. The second is to use evasion movement techniques. The effective use of camouflage is common to both of these activities.

(1) Hiding oneself and all signs of presence are the evader's greatest needs. Experience indicates that the survivor who uses effective concealment techniques has a better chance of evading capture. Capture results most frequently when the evader is moving.

(2) The evader's need to move depends on a variety of needs such as recovery, food, water, better shelter, etc. Evasion movement is more successful when proven techniques are used.

b. Resisting. The PW's need to resist is self-evident. This need is both a moral and a legal obligation. Resistance is much more than refusing to divulge some bit of classified information. Fundamentally, resisting is two distinctly separate behaviors expected of the prisoner:

(1) Complying with legal and authorized requirements only.

(2) Disrupting enemy activity through resisting, subtle harrassment, and tying up enemy guards who could be used on the front lines.

c. Escaping (When Possible and Authorized). Escape is neither easy nor without danger. The Code of Conduct states a survivor should make every effort to escape and aid others to escape.

Figure 3-9. Recovery.

3-4. Returning. The need to return is satisfied by successful completion of one or both of the basic tasks confronting the survivor: aiding with recovery and traveling (on land or water).

a. Aiding With Recovery. For survivors or evaders to effectively aid in recovery, they must be able to make their position and the situation on the ground known. This is done either electronically, visually, or both (Figure 3-9).

(1) Electronic signaling covers a wide spectrum of techniques. As problems such as security and safety during combat become significant factors, procedures for using electronic signaling to facilitate recovery become increasingly complex.

(2) Visual signaling is primarily the technique for attracting attention and pinpointing an exact location for rescuers. Simple messages or information may also be transmitted with visual signals.

b. Travel On Land. A survivor may need to move on land for a variety of reasons, ranging from going for water to attempting to walk out of the situation. In any survival episode, the survivor must weigh the need to travel against capabilities and(or) safety (Figure 3-10). Factors to consider may include:

(1) The ability to walk or traverse existing terrain. In a non-survival situation, a twisted or sprained ankle is an inconvenience accompanied by some temporary pain and restricted activity. A sur-

Figure 3-10. Travel.

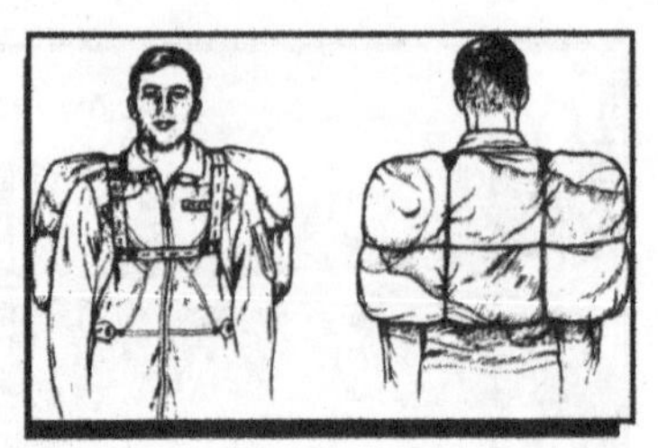

Figure 3-11. Burden Carrying.

vivor who loses the mobility, due to injury, to obtain food, water, and shelter, can face death. There is a safe and effective way to travel across almost any type of terrain.

(2) The need to transport personal possessions (burden carrying). There are numerous documented instances of survivors abandoning equipment and clothing simply because carrying was a bother. Later, the abandoned materials were not available when needed to save life, limb, or aid in rescue. Burden carrying need not be difficult or physically stressful. There are many simple ways for a survivor to carry the necessities of life (Figure 3-11).

(3) The ability to determine present position. Maps, compasses, star charts, Weems plotters, etc., permit accurate determination of position during extended travel. Yet, the knowledgeable, skillful, and alert survivor can do well without a full complement of these aids. Constant awareness, logic, and training in nature's clues to navigation may allow a survivor to determine general location even in the absence of detailed navigation aids.

(4) Restrictions or limitations to select and maintain a course of travel. The tools used in determining position are the tools used to maintain a course of travel. A straight line course to a destination is usually the simplest, but may not always be the best course for travel. Travel courses may need to be varied for diverse reasons, such as to get food or water, to enhance covert travels, or to avoid hazardous or impassable obstacles or terrain. Careful planning and route selection before and during travel is essential.

c. Travel On Water. Two differing circumstances may require survivors to travel on water. First, those who crash-land or parachute into the open sea are confronted with one type of situation. Second, survivors who find a river or stream which leads in a desirable direction are faced with a different situation. In each instance, however, a common element is to stay afloat.

(1) The survivor's initial problems on the open sea are often directly related to the winds and size of the waves. Simply getting into a liferaft and staying there are often very difficult tasks. On the open sea, the winds and ocean currents have a significant effect on the direction of travel. As the survivor comes closer to shore, the direction in which the tide is flowing also becomes a factor. There are some techniques a survivor can use to aid with stabilizing the raft, controlling the direction and rate of travel, and increasing safety.

(2) Survivors using rivers or streams for travel face both hazards and advantages as compared to overland travel. First, floating with the current is far less difficult than traveling over land. An abundance of food and water are usually readily available. Even in densely forested areas, effective signaling sites are generally available along streambeds. A survivor must use care and caution to avoid drowning, the most serious hazard associated with river travel.

Part Two

PSYCHOLOGICAL ASPECTS OF SURVIVAL

Chapter 4

CONTRIBUTING FACTORS

4-1. Introduction. Aircrew members in a survival situation must recognize that coping with the psychological aspects of survival are at least as important as handling the environmental factors. In virtually any survival episode, the aircrew will be in an environment that can support human life. The survivors' problems will be compounded because they never really expected to bail out or crash-land in the jungle, over the ocean, or anywhere else. No matter how well prepared, aircrews probably will never completely convince themselves that "it can happen to them." However, the records show it can happen. Before aircrew members learn about the physical aspects of survival, they must first understand that psychological problems may occur and that solutions to those problems must be found if the survival episode is to reach a successful conclusion (Figure 4-1).

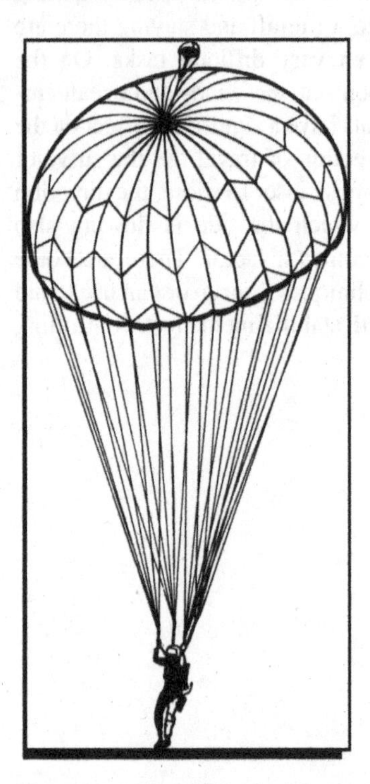

Figure 4-1. Psychological Aspects.

4-2. Survival Stresses:

a. The emotional aspects associated with survival must be completely understood just as survival conditions and equipment are understood. An important factor bearing on success or failure in a survival episode is the individual's psychological state. Maintaining an even, positive psychological state or outlook depends on the individual's ability to cope with many factors. Some include:

(1) Understanding how various physiological and emotional

signs, feelings, and expressions affect one's bodily needs and mental attitude.

(2) Managing physical and emotional reactions to stressful situations.

(3) Knowing individual tolerance limits, both psychological and physical.

(4) Exerting a positive influence on companions.

b. Nature has endowed everyone with biological mechanisms which aid in adapting to stress. The bodily changes resulting from fear and anger, for example, tend to increase alertness and provide extra energy to either run away or fight. These and other mechanisms can hinder a person under survival conditions. For instance, a survivor in a raft could cast aside reason and drink sea water to quench a thrist; or, evaders in enemy territory, driven by hunger pangs, could expose themselves to capture when searching for food. These examples illustrate how "normal" reactions to stress could create problems for a survivor.

c. Two of the gravest threats to successful survival are concessions to comfort and apathy. Both threats represent attitudes which must be avoided. To survive, a person must focus planning and effort on fundamental needs.

(1) Many people consider comfort their greatest need. Yet, comfort is not essential to human survival. Survivors must value life more than comfort, and be willing to tolerate heat, hunger, dirt, itching, pain, and any other discomfort. Recognizing discomfort as temporary will help survivors concentrate on effective action.

(2) As the will to keep trying lessens, drowsiness, mental numbness, and indifference will result in apathy. This apathy usually builds on slowly, but ultimately takes over and leaves a survivor helpless. Physical factors can contribute to apathy. Exhaustion due to prolonged exposure to the elements, loss of body fluids (dehydration), fatigue, weakness, or injury are all conditions that can contribute to apathy. Proper planning and sound decisions can help a survivor avoid these conditions. Finally, survivors must watch for signs of apathy in companions and help prevent it. The first signs are resignation, quietness, lack of communication, loss of appetite, and withdrawal from the group. Preventive measures could include maintaining group morale by planning, activity, and getting the organized participation of all members.

d. Many common stresses cause reactions which can be recognized and dealt with appropriately in survival situations. A survivor must understand that stresses and reactions often occur at the same time. Although survivors will face many stresses, the following common stresses will occur in virtually all survival episodes: pain, thirst, cold and heat, hunger, frustration, fatigue, sleep deprivation, isolation, insecurity, loss of self-esteem, loss of self-determination, and depression.

4-3. Pain:

a. Pain, like fever, is a warning signal calling attention to an injury or damage to some part of the body. Pain is discomforting but is not, in itself, harmful or dangerous. Pain can be controlled, and in an extremely grave situation, survival must take priority over giving in to pain (Figure 4-2).

b. The biological function of pain is to protect an injured part by warning the individual to rest it or avoid using it. In a survival situation, the normal pain warnings may have to be ignored in order to meet more critical needs. People have been known to complete a fight with a fractured hand, to run on a fractured or sprained ankle, to land an aircraft despite severely burned hands, and to ignore pain during periods of intense concentration and determined effort. Concentration and intense effort can actually stop or reduce feelings of pain. Sometimes this concentration may be all that is needed to survive.

Figure 4-2. Pain.

c. A survivor must understand the following facts about pain:

(1) Despite pain, a survivor can move in order to live.

(2) Pain can be reduced by:

(a) Understanding its source and nature.

(b) Recognizing pain as a discomfort to be tolerated.

(c) Concentrating on necessities like thinking, planning, and keeping busy.

(d) Developing confidence and self-respect. When personal goals are maintaining life, honor, and returning, and these goals are valued highly enough, a survivor can tolerate almost anything.

4-4. Thirst and Dehydration:

a. The lack of water and its accompanying problems of thirst and dehydration are among the most critical problems facing survivors. Thirst, like fear and pain, can be tolerated if the will to carry on, supported by calm, purposeful activity, is strong. Although thirst indicates the body's need for water, it does not indicate how much water is needed. If a person drinks only enough to satisfy thirst, it is still possible to slowly dehydrate. Prevention of thirst and the more debilitating dehydration is possible if survivors drink plenty of water any time it is available, and especially when eating (Figure 4-3).

b. When the body's water balance is not maintained, thirst and discomfort result. Ultimately, a water imbalance will result in dehydration. The need for water may be increased if the person:

(1) Has a fever.

(2) Is fearful.

(3) Perspires unnecessarily.

(4) Rations water rather than sweat.

c. Dehydration decreases the body's efficiency or ability to function. Minor degrees of dehydration may not noticeably affect a survivor's performance, but as it becomes more severe, body functioning will become increasingly impaired. Slight dehydration and thirst can also cause irrational behavior. One survivor described it:

> "The next thing I remember was being awakened by an unforgettable sensation of thirst. I began to move about aimlessly and finally found a pool of water."
>
> "We finally found water. In the water were two dead deer with horns locked. We went down to the water and drank away. It was the best damned drink of water I ever had in my life. I didn't taste the stench of the deer at all."

While prevention is the best way to avoid dehydration, virtually any degree of dehydration is reversible simply by drinking water.

Figure 4-3. Thirst.

5. Cold and Heat. The average normal body temperature for a person is 98.6°F. Victims have survived a body temperature as low as 20 °F below normal, but consciousness is clouded and thinking numbed at a much smaller drop. An increase of 6 to 8 degrees above normal for any prolonged period may prove fatal. Any deviation from normal temperature, even as little as 1 or 2 degrees, reduces efficiency.

a. Cold is a serious stress since even in mild degrees it lowers efficiency. Extreme cold numbs the mind and dulls the will to do anything except get warm again. Cold numbs the body by lowering the flow of blood to the extremities, and results in sleepiness. Survivors have endured prolonged cold and dampness through exercise, proper hygiene procedures, shelter, and food. Wearing proper clothing and having the proper climatic survival equipment when flying in cold weather areas is essential to enhance survivability (Figure 4-4).

(1) One survivor described cold and its effect:

> "Because of the cold water, my energy was going rapidly and all I could do was to hook my left arm over one side of the raft, hang on, and watch the low flying planes as they buzzed me . . . As time progressed, the numbing increased . . . and even seemed to impair my thinking."

(2) Another survivor remembered survival training and acted accordingly:

> "About this time, my feet began getting cold. I remembered part of the briefing I had received about feet freezing so I immediately took action. I thought about my shoes, and with my jack knife, cut off the bottom of my Mark II immersion suit and put them over my shoes. My feet immediately felt warmer and the rubber feet of the immersion suit kept the soles of my shoes dry."

Figure 4-4. Cold.

b. Just as "numbness" is the principal symptom of cold, "weakness" is the principle symptom of heat. Most people can adjust to high temperatures, whether in the hold of a ship or in a harvest field on the Kansas prairie. It may take from 2 days to a week before circulation, breathing, heart action, and sweat glands are all adjusted to a hot climate. Heat stress also accentuates dehydration, which was discussed earlier. In addition to the problem of water, there are many other sources of discomfort and impaired efficiency which are directly attributable to heat or to the environmental conditions in hot climates. Extreme temperature changes, from extremely hot days to very cold nights, are experienced in desert and plains areas. Proper use of clothing and shelters can decrease the adverse effects of such extremes (Figure 4-5).

c. Bright sun has a tremendous effect on eyes and any exposed skin. Direct sunlight or rays reflecting off the terrain require dark glasses or improvised eye protectors. Previous suntanning provides little protection; protective clothing is important.

Figure 4-5. Heat.

d. Blowing wind, in hot summer, has been reported to get on some survivors' nerves. Wind can constitute an additional source of discomfort and difficulty in desert areas when it carries particles of sand and dirt. Protection against sand and dirt can be provided by tying a cloth around the head after cutting slits for vision.

e. Acute fear has been experienced among survivors in sandstorms and snowstorms. This fear results from both the terrific impact of the storm itself and its obliteration of landmarks showing direction of travel. Finding or improving shelter for protection from the storm itself is important.

f. Loss of moisture, drying of the mouth and mucous membranes, and accelerating dehydration can be caused by breathing through the mouth and talking. Survivors must learn to keep their mouths shut in desert winds as well as in cold weather.

g. Mirages and illusions of many kinds are common in desert areas. These illusions not only distort visual perception but sometimes account for serious incidents. In the desert, distances are usually greater than they appear and, under certain conditions, mirages obstruct accurate vision. Inverted reflections are a common occurrence.

4-6. Hunger. A considerable amount of edible material (which survivors may not initially regard as food) may be available under survival conditions. Hunger and semistarvation are more commonly experienced among survivors than thirst and dehydration. Research has revealed no evidence of permanent damage nor any decrease in mental efficiency from short periods of total fasting (Figure 4-6).

a. The prolonged and rigorous Minnesota semistarvation studies during World War II revealed the following behavioral changes:

(1) Dominance of the hunger drive over other drives.

(2) Lack of spontaneous activity.

(3) Tired and weak feeling.

(4) Inability to do physical tasks.

(5) Dislike of being touched or caressed in any way.

Figure 4-6. Hunger.

(6) Quick susceptibility to cold.

(7) Dullness of all emotional responses (fear, shame, love, etc.)

(8) Lack of interest in others—apathy.

(9) Dullness and boredom.

(10) Limited patience and self-control.

(11) Lack of a sense of humor.

(12) Moodiness—reaction of resignation.

b. Frequently, in the excitement of some survival, evasion, and escape episodes, hunger is forgotten. Survivors have gone for considerable lengths of time without food or awareness of hunger pains. An early effort should be made to procure and consume food to reduce the stresses brought on by food deprivation. Both the physical and psychological effects described are reversed when food and a protective environment are restored. Return to normal is slow and the time necessary for the return increases with the severity of starvation. If food deprivation is complete and only water is ingested, the pangs of hunger disappear in a few days, but even then the mood changes of depression and irritability occur. The individual tendency is still to search for food to prevent starvation and such efforts might continue as long as strength and self-control permit. When the food supply is limited, even strong friendships are threatened.

c. Food aversion may result in hunger. Adverse group opinion may discourage those who might try foods unfamiliar to them. In some groups, the barrier would be broken by someone eating the particular food rather than starving. The solitary individual has only personal prejudices to overcome and will often try strange foods.

d. Controlling hunger during survival episodes is relatively easy if the survivor can adjust to discomfort and adapt to primitive conditions. This man would rather survive than be fussy:

> "Some men would almost starve before eating the food. There was a soup made of lamb's head with the lamb's eyes floating around in it. . . . When there was a new prisoner, I would try to find a seat next to him so I could eat the food he refused."

4-7. Frustration. Frustration occurs when one's efforts are stopped, either by obstacles blocking progress toward a goal or by not having a realistic goal. It can also occur if the feeling of self-worth or self-respect is lost (Figure 4-7).

a. A wide range of obstacles, both environmental and internal, can lead to frustration. Frustrating conditions often create anger, accompanied by a tendency to attack and remove the obstacles to goals.

Figure 4-7. Frustration.

b. Frustration must be controlled by channeling energies into a positive and worthwhile obtainable goal. The survivor should complete the easier tasks before attempting more challenging ones. This will not only instill self-confidence, but also relieve frustration.

4-8. Fatigue. In a survival episode, a survivor must continually cope with fatigue and avoid the accompanying strain and loss of efficiency. A survivor must be aware of the dangers of overexertion. In many cases, a survivor may already be experiencing strain and reduced efficiency as a result of other stresses such as heat or cold, dehydration, hunger, or fear. A survivor must judge capacity to walk, carry, lift, or do necessary work, and plan and act accordingly. During an emergency, considerable exertion may be necessary to cope with the situation. If an individual understands fatigue and the attitudes and feelings generated by various kinds of effort, that individual should be able to call on available reserves of energy when they are needed (Figure 4-8).

Figure 4-8. Fatigue

a. A survivor must avoid complete exhaustion which may lead to physical and psychological changes. A survivor should be able to distinguish between exhaustion and being uncomfortably tired. Although a person should avoid working to complete exhaustion, in emergencies certain tasks must be done in spite of fatigue.

(1) Rest is a basic factor for recovery from fatigue and is also important in resisting further fatigue. It is essential that the rest (following fatiguing effort) be sufficient to permit complete recovery; otherwise, the residual fatigue will accumulate and require longer periods of rest to recover from subsequent effort. During the early stages of fatigue, proper rest provides a rapid recovery. This is true of muscular fatigue as well as mental fatigue. Sleep is the most complete form of rest available and is basic to recovery from fatigue.

(2) Short rest breaks during extended stress periods can improve total output. There are five ways in which rest breaks are beneficial:

(a) They provide opportunities for partial recovery from fatigue.

(b) They help reduce energy expenditure.

(c) They increase efficiency by enabling a person to take maximum advantage of planned rest.

(d) They relieve boredom by breaking up the uniformity and monotony of the task.

(e) They increase morale and motivation.

(3) Survivors should rest before output shows a definite decline. If rest breaks are longer, fewer may be required. When efforts are highly strenuous or monotonous, rest breaks should be more frequent. Rest breaks providing relaxation are the most effective. In mental work, mild exercise may be more relaxing. When work is monotonous, changes of activity, conversation, and humor are effective relaxants. In deciding on the amount and frequency of rest periods, the loss of efficiency resulting from longer hours of effort must be weighed against the absolute requirements of the survival situation.

(4) Fatigue can be reduced by working "smarter." A survivor can do this in two practical ways:

(a) Adjust the pace of the effort. Balance the load, the rate, and the time period. For example, walking at a normal rate is a more economical effort than fast walking.

(b) Adjust the technique of work. The way in which work is done has a great bearing on reducing fatigue. Economy of effort is most important. Rhythmic movements suited to the task are best.

(5) Mutual group support, cooperation, and competent leadership are important factors in maintaining group morale and efficiency, thereby reducing stress and fatigue. A survivor usually feels tired and weary before the physiological limit is reached. In addition, other stresses experienced at the same time; such as cold, hunger, fear, or despair, can intensify fatigue. The feeling of fatigue involves not only the physical reaction to effort, but also subtle changes in attitudes and motivation. Remember, a person has reserves of energy to cope with an important emergency even when feeling very tired.

b. As in the case of other stresses, even a moderate amount of fatigue reduces efficiency. To control fatigue, it is wise to observe a program of periodic rest. Because the main objective—to establish contact with friendly forces—survivors may overestimate their strength and risk exhaustion. On the other hand, neither an isolated individual nor a group leader should underestimate the capacity of the individual or the group on the basis of fatigue. The only sound basis for judgment must be gained from training and past experience. In training, a person should form an opinion of individual capacity based on actual experience. Likewise, a group leader must form an opinion of the capacities of fellow aircrew members. This group didn't think:

> "By nightfall, we were completely bushed. . . . We decided to wrap ourselves in the 'chute instead of making a shelter. We were too tired even to build a fire. We just cut some pine boughs, rolled ourselves in the nylon and went to sleep. . .and so, of course, it rained, and not lightly. We stood it until we were soaked, and then we struggled out and made a shelter. Since it was pitch dark, we didn't get the sags out of the canopy, so the water didn't all run off. Just a hell of a lot of it came through. Our hip and leg joints ached as though we had acute rheumatism. Being wet and cold accentuated the pain. We changed positions every 10 minutes, after gritting our teeth to stay put that long."

4-9. Sleep Deprivation. The effects of sleep loss are closely related to those of fatigue. Sleeping at unaccustomed times, sleeping under strange circumstances (in a strange place, in noise, in light, or in other distractions), or missing part or all of the accustomed amount of sleep will cause a person to react with feelings of weariness, irritability, emotion tension, and some loss of efficiency. The extent of an individual's reaction depends on the amount of disturbance and on other stress factors which may be present at the same time (Figure 4-9).

a. Strong motivation is one of the principal factors in helping to compensate for the impairing effects of sleep loss. Superior physic and mental conditioning, opportunities to rest, food and water, and companions help in enduring sleep deprivation. If a person is in reasonably good physical and mental condition, sleep deprivation can be endured 5 days or more without damage, although efficiency during the latter stages may be poor. A person must learn to get as much sleep and rest as possible. Restorative effects of sleep are felt even after "catnaps." In some instances, survivors may need to stay awake. Activity, movement, conversation, eating, and drinking are some of the ways a person can stimulate the body to stay awake.

Figure 4-9. Sleep Deprivation.

b. When one is deprived of sleep, sleepiness usually comes in waves. A person may suddenly be sleepy immediately after a period of feeling wide awake. If this can be controlled, the feeling will soon pass and the person will be wide awake again until the next wave appears. As the duration of sleep deprivation increases, these periods between waves of sleepiness become shorter. The need to sleep may be so strong in some people after a long period of deprivation that they become desperate and do careless or dangerous things in order to escape this stress.

4-10. Isolation. Loneliness, helplessness, and despair experienced by survivors when they are isolated are among the most severe survival stresses. People often take their associations with family, friends, military collegues, and others for granted. But survivors soon begin to miss the daily interaction with other people. However, these, like the other stresses already discussed, can be conquered. Isolation can be controlled and overcome by knowledge, understanding, deliberate countermeasures, and a determined will to resist it (Figure 4-10).

Figure 4-10. Isolation.

4-11. Insecurity. Insecurity is the survivor's feeling of helplessness or inadequacy resulting from varied stresses and anxieties. These anxieties may be caused by uncertainty regarding individual goals, abilities, and the future in a survival situation. Feelings of insecurity may have widely different effects on the survivor's behavior. A survivor should establish challenging but attainable goals. The better a survivor feels about individual abilities to achieve goals and adequately meet personal needs, the less insecure the survivor will feel.

4-12. Loss of Self-Esteem. Self-esteem is the state or quality of having personal self-respect and pride. Lack of (or loss of) self-esteem in a survivor may bring on depression and a change in perspective and goals. A loss of self-esteem may occur in individuals in captivity. Humiliation and other factors brought on by the captor may cause them to doubt their own worth. Humiliation comes from the feeling of losing pride or self-respect by being disgraced or dishonored, and is associated with the loss of self-esteem. Prisoners must maintain their pride and not become ashamed either because they are PWs or because of the things that happen to them as a result of being a PW. The survivor who "loses face" (both personally and with the enemy) becomes more vulnerable to captor exploitation attempts. To solve this problem, survivors should try to maintain proper perspective about both the situation and themselves. Their feelings of self-worth may be bolstered if they recall the implied commitment in the Code of Conduct—PWs will not be forgotten (Figure 4-11).

4-13. Loss of Self-Determination. A self-determined person is relatively free from external controls or influences over his or her actions. In everyday society, these "controls and influences" are the laws and customs of our society and of the self-imposed elements of our personalities. In a survival situation, the "controls and influences" can be very different. Survivors may feel as if events, circumstances, and (in some cases) other people, are in control of the situation. Some factors which may cause individuals to feel they have

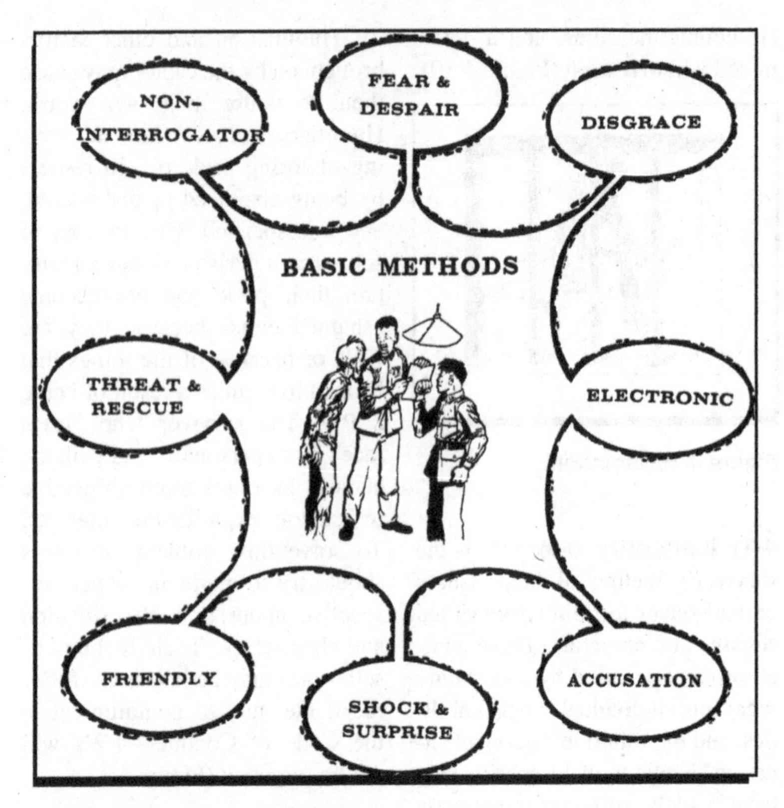

Figure 4-11. Loss of Self-Esteem.

lost the power of self-determination are a harsh captor, captivity, bad weather, or rescue forces that make time or movement demands. This lack of self-determination is more perceived than actual. Survivors must decide how unpleasant factors will be allowed to affect their mental state. They must have the self-confidence, fostered by experience and training, to live with their feelings and decisions, and to accept responsibility for both the way they feel and how they let those feelings affect them.

4-14. Depression. As a survivor, depression is the biggest psychological problem that has to be conquered. It should be acknowledged that everyone has mental "highs" as well as mental "lows." People experiencing long periods of sadness or other negative feelings are suffering from depression. A normal mood associated with the sadness, grief, disappointment, or loneliness that everyone experiences at times is also described as depression. Most of the emotional changes in mood are temporary and do not become chronic. Depressed survivors may feel fearful, guilty, or helpless. They may lose interest in the basic needs of life. Many cases of depression also involve pain, fatigue, loss of appetite, or other physical ailments. Some depressed survivors try to injure or kill themselves (Figure 4-12).

a. Psychiatrists have several theories as to the cause of depression.

Figure 4-12. Depression.

Some feel a person who, in everyday life and under normal conditions, experiences many periods of depression would probably have a difficult time in a survival situation. The main reason depression is a most difficult problem is that it can affect a wide range of psychological responses. The factors can become mutually reinforcing. For example, fatigue may lead to a feeling of depression. Depression may increase the feeling of fatigue, and this, in turn, leads to deeper depression and so on.

b. Depression usually begins after a survivor has met the basic needs for sustaining life, such as water, shelter, and food. Once the survivor's basic needs are met, there is often too much time for that person to dwell on the past, the present predicament, and on future problems. The survivor must be aware of the necessity to keep the mind and body active to eliminate the feeling of depression. One way to keep busy (daily) is by checking and improving shelters, signals, and food supply.

Chapter 5
EMOTIONAL REACTIONS

5-1. Introduction. Survivors may depend more upon their emotional reactions to a situation than upon calm, careful analysis of potential danger—the enemy, the weather, the terrain, the nature of the in-flight emergency, etc. Whether they will panic from fear, or use it as a stimulant for greater sharpness, is more dependent on the survivor's reactions to the situation than on the situation itself. Although there are many reactions to stress, the following are the most common and will be discussed in detail: fear, anxiety, panic, hate, resentment, anger, impatience, dependency, loneliness, boredom, and hopelessness.

5-2. Fear. Fear can SAVE A LIFE—or it can COST ONE. Some people are at their best when they are scared. Many downed fliers faced with survival emergencies have been surprised at how well they remembered their training, how quickly they could think and react, and what strength they had. The experience gave them a new confidence in themselves. On the other hand, some people become paralyzed when faced with the simplest survival situation. Some of them have been able "to snap themselves out of it" before it was too late. In other cases, a fellow aircrew member was on hand to assist them. However, others have not been so fortunate. They are not listed among the survivors (Figure 5-1).

Figure 5-1. Fear.

a. How a person will react to fear depends more upon the individual than it does upon the situation. This has been demonstrated both in actual survival situations and in laboratory experiments. It isn't always the physically strong or the happy-go-lucky people who handle fear most effectively. Timid and anxious people have met emergencies with remarkable coolness and strength.

b. Anyone who faces life-threatening emergencies experiences fear. Fear is conscious when it results from a recognized situation (such as an immediate prospect of bailout) or when experienced as apprehension of impending disaster. Fear also occurs at a subconscious level and creates feelings of uneasiness, general discomfort, worry, or depression. Fear may vary widely in intensity, duration, and frequency of occurrence, and affect behavior across the spectrum from mild uneasiness to complete disorganization and panic. People have many fears; some are learned through personal experiences and others are deliberately taught to them. Fear in children is directed through negative learning, as they are taught to be afraid of the dark, of animals, of noise, or of teachers. These fears may control behavior, and a survivor may react to feelings and imagination rather than to the problem causing fear.

c. When fantasy distorts a moderate danger into a major catastrophe, or vice versa, behavior can become abnormal. There is a general tendency to underestimate and this leads to reckless, foolhardy behavior. The principal means of fighting fear (in this case) is to pretend that it does not exist. There are no sharp lines between recklessness and bravery. It is necessary to check behavior constantly to maintain proper control.

d. One or more of the following signs or symptoms may occur in those who are afraid. However, they may also appear in circumstances other than fear.

(1) Quickening of pulse; trembling.

(2) Dilation of pupils.

(3) Increased muscular tension and fatigue.

(4) Perspiration of palms of hands, soles of feet, and armpits.

(5) Dryness of mouth and throat; higher pitch of voice; stammering.

(6) Feeling of "butterflies in the stomach," emptiness of the stomach, faintness, and nausea.

e. Accompanying these physical symptoms are the following common psychological symptoms:

(1) Irritability; increased hostility.

(2) Talkativeness in early stages, leading finally to speechlessness.

(3) Confusion, forgetfulness, and inability to concentrate.

(4) Feelings of unreality, flight, panic, or stupor.

f. Throughout military history, many people have coped successfully with the most strenuous odds. In adapting to fear, they have found support in previous training and experience. There is no limit to human control of fear. Survivors must take action to control fear. They cannot run away from fear. Appropriate actions should be to:

(1) Understand fear.

(2) Admit that it exists.

(3) Accept fear as reality.

g. Training can help survivors recognize what individual reactions may be. Using prior training, survivors should learn to think, plan, and act logically, even when afraid.

h. To effectively cope with fear, a survivor must:

(1) Develop confidence. Use training opportunities; increase capabilities by keeping physically and mentally fit; know what equipment is available and how to use it; learn as much as possible about all aspects of survival.

(2) Be prepared. Accept the possibility that "it can happen to me." Be properly equipped and clothed at all times; have a plan ready. Hope for the best, but be prepared to cope with the worst.

(3) Keep informed. Listen carefully and pay attention to all briefings. Know when danger threatens and be prepared if it comes; increase knowledge of survival environments to reduce the "unknown."

(4) Keep busy at all times. Prevent hunger, thirst, fatigue, idleness, and ignorance about the situation, since these increase fear.

(5) Know how fellow crewmembers react to fear. Learn to work together in emergencies—to live, work, plan, and help each other as a team.

(6) Practice religion. Don't be ashamed of having spiritual faith.

(7) Cultivate "good" survival attitudes. Keep the mind on a main goal and keep everything else in perspective. Learn to tolerate discomfort. Don't exert energy to satisfy minor desires which may conflict with the overall goal—to survive.

(8) Cultivate mutual support. The greatest support under severe stress may come from a tightly knit group. Teamwork reduces fear while making the efforts of every person more effective.

(9) Exercise leadership. The most important test of leadership and perhaps its greatest value lies in the stress situation.

(10) Practice discipline. Attitudes and habits of discipline developed in training carry over into other situations. A disciplined group has a better chance of survival than an undisciplined group.

(11) Lead by example. Calm behavior and demonstration of control are contagious. Both reduce fear and inspire courage.

i. Every person has goals and desires. The greatest values exercise the greatest influence. Because of strong religious, moral, or patriotic values, people have been known to face torture and death calmly rather than reveal information or compromise a principle. Fear can kill or it can save lives. It is a normal reaction to danger. By understanding and controlling fear through training, knowledge, and effective group action, fear can be overcome.

5-3. Anxiety:

Anxiety is a universal human reaction. Its presence can be felt when changes occur which affect an individual's safety, plans, or methods of living. It is generally felt when individuals perceive something bad is about to happen. A common description of anxiety is "butterflies in the stomach." Anxiety creates feelings of uneasiness, general discomfort, worry, or depression. Anxiety and fear differ mainly in intensity. Anxiety is a milder reaction and the specific cause(s) may not be readily apparent, whereas fear is a strong reaction to a specific, known cause. Common characteristics of anxiety are: fear of the future, indecision, feelings of helplessness, resentment (Figure 5-2).

Figure 5-2. Anxiety.

To overcome anxiety, the individual must take positive action by adopting a simple plan. It is essential to keep your mind off of your injuries and do something constructive. For example, one PW began to try and teach English to the Chinese and to learn Chinese from them.

5-4. Panic. In the face of danger, a person may panic or "freeze" and cease to function in an organized manner. A person experiencing panic may have no conscious control over individual actions. Uncontrollable, irrational behavior is common in emergency situations. Anybody can panic, but some people go to pieces more easily than others. Panic is brought on by a sudden overwhelming fear and can often spread quickly through a group of people. Every effort must be made to bolster morale and calm the panic with leadership and discipline. Panic has the same signs as fear and should be controlled in the same manner as fear. This survivor allowed pain to panic him.

> "His parachute caught in the tree, and he found himself suspended about five feet above the ground. . . one leg strap was released while he balanced in this aerial position and he immediately slipped toward the ground.
>
> In doing so, his left leg caught in the webbing and he was suspended by one leg with his head down. Unfortunately, the pilot's head touched an ant hill and biting ants immediately swarmed over him. Apparently, in desperation, the flier pulled his gun and fired five rounds into the webbing holding

his foot. When he did not succeed in breaking the harness by shooting at it, he placed the last shot in his head and thus took his own life. It was obvious from the discoverer's report that if the pilot had even tried to turn around or to swing himself from his inverted position, he could have reached either the aerial roots or the latticed trunk of the tree. With these branches, he should have been able to pull himself from the harness...The fact that his head was in a nest of stinging ants only added to his panic, which led to the action that took his life." (Figure 5-3)

Figure 5-3. Panic.

5-5. Hate. Hate—feelings of intense dislike, extreme aversion, or hostility—is a powerful emotion which can have both positive and negative effects on a survivor. An understanding of the emotion and its causes is the key to learning to control it. Hate is an acquired emotion rooted in a person's knowledge or perceptions. The accuracy or inaccuracy of the information is irrelevant to learning to hate.

a. Any person, any object, or anything that may be understood intellectually, such as political concepts or religious dogma, can promote feelings of hate. Feelings of hate (usually accompanied with a desire for vengence, revenge, or retribution) have sustained former prisoners of war through their harsh ordeals. If an individual loses perspective while under the influence of hate and reacts emotionally, rational solutions to problems may be overlooked, and the survivor may be endangered.

To effectively deal with this emotional reaction, the survivor must first examine the reasons why the feeling of hate is present. Once that has been determined, survivors should then decide what to do about those feelings. Whatever approach is selected, it should be as constructive as possible. Survivors must not allow hate to control them.

5-6. Resentment. Resentment is the experiencing of an emotional state of displeasure or indignation toward some act, remark, or person that has been regarded as causing personal insult or injury. Luck and fate may play a role in any survival situation. A hapless survivor may feel jealous resentment toward a fellow PW, travel partner, etc., if that other person is perceived to be enjoying a success or advantage not presently experienced by the observer. The survivor must understand that events cannot always go as expected. It is detrimental to morale and could affect survival chances if feelings of resentment over another's attainments become too strong. Imagined slights or insults are common. The survivor should try to maintain a sense of humor and perspective about ongoing events and realize that stress and lack of self-confidence play roles in bringing on feelings of resentment.

5-7. Anger. Anger is a strong feeling of displeasure and belligerence aroused by a real or supposed wrong. People become angry when they cannot fulfill a basic need or desire which seems important to them. When anger is not relieved, it may turn into a more enduring attitude of hostility, characterized by a desire to hurt or destroy the person or thing causing the frustration. When anger is intense, the survivor loses control over the situation, resulting in impulsive behavior which may be destructive in nature. Anger is a normal response which can serve a useful purpose when carefully controlled. If the situation warrants and there is no threat to survival, one could yell or scream, take a walk, do some vigorous exercise, or just get away from the source of the anger, even if only for a few minutes. Here is a man who couldn't hold it.

> "I tried patiently to operate it (radio) in every way I had been shown. Growing more angry and disappointed at its failure, I tore the aerial off, threw the cord away, beat the battery on the rocks, then threw the pieces all over the hillside. I was sure disappointed." (Figure5-4.)

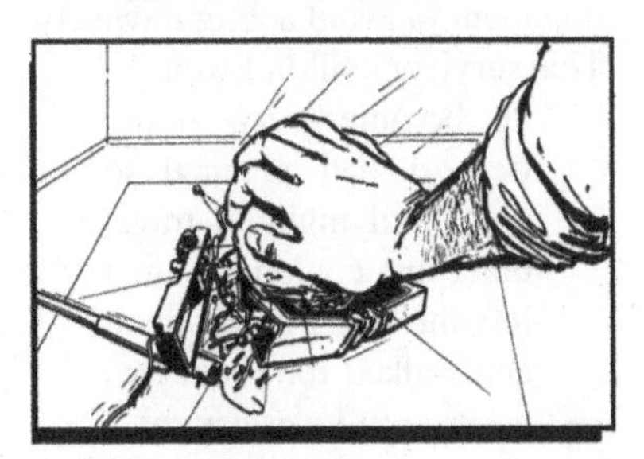

Figure 5-4. Anger.

5-8. Impatience:

The psychological stresses brought about by feelings of impatience can quickly manifest themselves in physical ways. Internally, the effects of impatience can cause changes in physical and mental well-being. Survivors who allow impatience to control their behavior may find that their efforts prove to be counterproductive and possibly dangerous. For example, evaders who don't have the ability or willingness to suppress annoyance when confronted with delay may expose themselves to capture or injury.

Potential survivors must understand they have to bear pain, misfortune, and annoyance without complaint. In the past, many survivors have displayed tremendous endurance, both mental and physical, in times of distress or misfortune. While not every survivor will be able to display such strength of character in all situations, each person should learn to recognize the things which may make them impatient to avoid acting unwisely. This survivor couldn't wait:

> "I became very impatient. I had planned to wait until night to travel but I just couldn't wait. I left the ditch about noon and walked for about two hours until I was caught."

5-9. Dependence. The captivity environment is the prime area where a survivor may experience feelings of dependency. The captor will try to develop in prisoners feelings of need, support, and trust for the captor. By regulating the availability of basic needs like food, water, clothing, social contact, and medical care, captors show their power and control over the prisoners' fate. Through emphasis on the prisoners' inability to meet their own basic needs, captors seek to establish strong feelings of prisoner dependency. This dependency can make prisoners extremely vulnerable to captor exploitation—a major captor objective. PW recognition of this captor tactic is key to countering it. Survivors must understand that, despite captor controls, they do control their own lives. Meeting even one physical or mental need can provide a PW with a "victory" and provide the foundation for continued resistance against exploitation (Figure 5-5).

Figure 5-5. Dependence.

5-10. Loneliness. Loneliness can be very debilitating during a survival episode. Some people learn to control and manipulate their environment and become more self-sufficient while adapting to changes. Others rely on protective persons, routines, and familiarity of surroundings to function and obtain satisfaction (Figure 5-6).

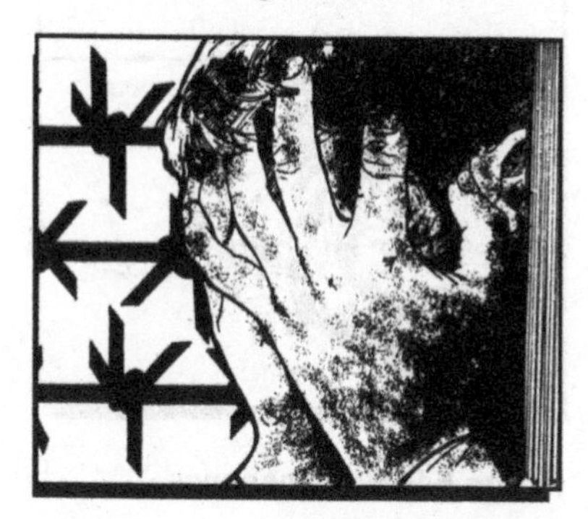

Figure 5-6. Loneliness.

a. The ability to combat feelings of loneliness during a survival episode must be developed long before the episode occurs. Self-confidence and self-sufficiency are key factors in coping with loneliness. People develop these attributes by developing and demonstrating competence in performing tasks. As the degree of competence increases, so does self-confidence and self-sufficiency. Military training, more specifically survival training, is designed to provide individuals with the competence and self-sufficiency to cope with and adapt to survival living.

b. In a survival situation, the countermeasure to conquer loneliness is to be active, to plan and think purposely. Development of self-sufficiency is the primary protection since all countermeasures in survival require the survivor to have the ability to practice self-control.

5-11. Boredom. Boredom and fatigue are related and frequently confused. Boredom is accompanied by a lack of interest and may include feelings of strain, anxiety, or depression, particularly when no relief is in sight and the person is frustrated. Relief from boredom must be based on correction of the two basic sources, repetitiveness and uniformity. Boredom can be relieved by a variation of methods—rotation of duties, broadening the scope of a particular task or job, taking rest breaks, or other techniques of diversification which may actually interfere with efficient performance of the job. The un-gratifying nature of a task can be counteracted by clearing up its meaning, objectives, and, in some cases, its relation to the total plan.

a. This survivor couldn't think of anything to do:

> "The underground representative took me to a house to wait for another member of the underground to pick me up. This was the worst part of the whole experience—this waiting. I just sat in the house and waited for two weeks. I thought I would go mad." (Figure 5-7)

Figure 5-7. Boredom.

b. This survivor invented something to do:

> "Not knowing what to do, I decided to kill all the bugs. There were a lot of spiders, the big ones that do not hurt humans, so I killed the flies and gave them to the spiders to eat."

5-12. Hopelessness. Hopelessness stems from negative feelings—regardless of actions taken, success is impossible, or the certainty that future events will turn out for the worst no matter what a person tries to do. Feelings of hopelessness can occur at virtually any time during a survival episode. Survivors have experienced loss of hope in trying to maintain health due to an inability to care for sickness, broken bones, or injuries; considering their chances of returning home alive; seeing their loved ones again; or believing in their physical or mental ability to deal with the situation; for example, evade long distances or not give information to an interrogator (Figure 5-8).

Figure 5-8. Hopelessness.

a. During situations where physical exhaustion or exposure to the elements affects the mind, a person may begin to lose hope. The term "give-up-itis" was coined in Korea to describe the feeling of "hopelessness." During captivity, deaths occurred for no apparent cause. These individuals actually willed themselves to die or at least did not will themselves to live. The original premise (in the minds of such people) is that they are going to die. To them, the situation seemed totally futile and they had passively abandoned themselves to fate. It was possible to follow the process step by step. The people who died withdrew themselves from the group, became despondent, then lay down and gave up. In some cases, death followed rapidly.

b. One way to treat hopelessness is to eliminate the cause of the stress. Rest, comfort, and morale-building activities can help eliminate this psychological problem. Another method used in Korea was to make the person so angry that the person wanted to get up and attack the tormentors. A positive attitude has a powerful influence on morale and combating the feeling of hopelessness.

c. Since many stress situations cannot be dealt with successfully by either withdrawal or direct attack, it may be necessary to work out a compromising solution. The action may entail changing a survivor's method of operation or accepting substitute goals.

d. Evaders faced with starvation may compromise with their conscience and steal "just this one time." They may ignore their food aversion and eat worms, bugs, or even human flesh. A related form of compromise is acceptance of substitute means to achieve the same goals.

5-13. Summary. All the psychological factors may be overcome by survivors if they can recognize the problem, work out alternative solutions, decide on an appropriate course of action, take action, and evaluate the results. Perhaps the most difficult step in this sequence is deciding on an appropriate course of action. Survivors may face either one or several psychological problems. These problems are quite dangerous and must be effectively controlled or countered for survival to continue.

Chapter 6

THE WILL TO SURVIVE

6-1. Introduction. The will to survive is defined as the desire to live despite seemingly insurmountable mental and(or) physical obstacles. The tools for survival are furnished by the military, the individual, and the environment. The training for survival comes from survival training publications, instruction, and the individual's own efforts. But tools and training are not enough without a *will to survive.* In fact, the records prove that "will" alone has been the deciding factor in many survival cases. While these accounts are not classic examples of "how to survive," they illustrate that a singleminded survivor with a powerful *will to survive* can overcome most hardships. There are cases where people have eaten their belts for nourishment, boiled water in their boots to drink as broth, or have eaten human flesh—though this certainly wasn't their cultural instinct.

a. One incident where the *will to survive* was the deciding factor between life and death involved a man stranded in the Arizona desert for 8 days without food and water. He traveled more than 150 miles during searing daytime temperatures, losing 25 percent of his body weight due to the lack of water (usually 10 percent loss causes death). His blood became so thick that the lacerations he received could not bleed until he had been rescued and received large quantities of water. When he started on that journey, something must have clicked in his mind telling him to live, regardless of any obstacles which might confront him. And live he did— on guts and will alone! (Figure 6-1)

Figure 6-1. Will to survive.

b. Let's flip a coin and check the other side of "will." Our location is the Canadian wilderness. A pilot ran into engine trouble and chose to deadstick his plane onto a frozen lake rather than punch out. He did a beautiful job and slid to a stop in the middle of the lake. He left the aircraft and examined it for damage. After surveying the area, he noticed a wooded shoreline only 200 yards away where food and shelter could be provided—he decided to go there. Approximately halfway there, he changed his mind and returned to the cockpit of his aircraft where he smoked a cigar, took out his pistol, and blew his brains out. Less than 24 hours later, a rescue team found him. Why did he give up? Why was he unable to survive? Why did he take his own life? On the other hand, why do people eat their belts or drink broth from their boots? No one really knows, but it's all related to the *will to survive.*

6-2. Overcoming Stress. The ability of the mind to overcome stress and hardship becomes most apparent when there appears to be little chance of a person surviving. When there appears to be no escape from the situation, the "will" enables a person to begin to win "the battle of the mind." This mental attitude can bridge the gap between the crisis period and the coping period.

6-3. Crisis Period:

The crisis period is the point at which the person realizes the gravity of the situation and understands that the problem will not go away. At this stage, action is needed. Most people will experience shock in this stage as a result of not being ready to face this new challenge. Most will recover control of their faculties, especially if they have been prepared through knowledge and training.

Shock during a crisis is normally a response to being overcome with anxiety. Thinking will be disorganized. At this stage, direction will be required because the individual is being controlled by the environment. The person's center of control is external. In a group survival episode, a natural leader may appear who will direct and reassure the others. But if the situation continues to control the individual or the group, the response may be panic, behavior may be irrational, and judgment is impaired. In a lone-survivor episode, the individual must gain control of the situation and respond constructively. In either case, survivors must evaluate the situation and develop a plan of action.

During the evaluation, the survivor must determine the most critical needs to improve the chance of living and being rescued.

6-4. The Coping Period. The coping period begins after the survivor recognizes the gravity of the situation and resolves to endure it rather than succumb. The survivor must tolerate the effects of physical and emotional stresses. These stresses can cause anxiety which becomes the greatest obstacle to self-control and solving problems. Coping with the situation requires considerable internal control. For example, the survivor must often subdue urgent desires to travel when that would be counterproductive and dangerous. A person must have patience to sit in an emergency action shelter while confronted with an empty stomach, aching muscles, numb toes, and suppressed feelings of depression and hopelessness. Those who fail to think constructively may panic. This could begin a series of mistakes which result in further exhaustion, injury, and sometimes death. Death comes not from hunger pains but from the inability to manage or control emotions and thought processes.

6-5. Attitude. The survivor's attitude is the most important element of the *will to survive*. With the proper attitude, almost anything is possible. The desire to live is sometimes based on the feelings toward another person and(or) thing. Love and hatred are two emotional extremes which have moved people to do exceptional things physically and mentally. The lack of a *will to survive* can sometimes be identified by the individual's motivation to meet essential survival needs, emotional control resulting in reckless, paniclike behavior, and self-esteem.

a. It is essential to strengthen the *will to survive* during an emergency. The first step is to avoid a tendency to panic or "fly off the handle." Sit down, relax, and analyze the situation rationally. Once thoughts are collected and thinking is clear, the next step is to make decisions. In normal living, people can avoid decisions and let others do their planning. But in a survival situation, this will seldom work. Failure to decide on a course of action is actually a decision for inaction. This lack of decision-making may even result in death. However, decisiveness must be tempered with flexibility and planning for unforeseen circumstances. As an example, an aircrew member down in an arctic nontactical situation decides to construct a shelter for protection from the elements. The planning and actions must allow sufficient flexibility so the aircrew can monitor the area for indications of rescuers and be prepared to make contact—visually, electronically, etc.—with potential rescuers.

b. Tolerance is the next topic of concern. A survivor or evader will have to deal with many physical and psychological discomforts, such as unfamiliar animals, insects, loneliness, and depression. Aircrew members are trained to tolerate uncomfortable situations. That training must be applied to deal with the stress of environments.

c. Survivors in both tactical and nontactical situations must face and overcome fears to strengthen the *will to survive*. These fears may be founded or unfounded, be generated by the survivor's uncertainty or lack of confidence, or be based on the proximity of enemy forces. Indeed, fear may be caused by a wide variety of real and imagined dangers. Despite the source of the fear, survivors must recognize fear and make a conscious effort to overcome it.

6-6. Optimism. One of a survivor's key assets is optimism—hope and faith. Survivors must maintain a positive, optimistic outlook on their circumstance and how well they are doing. Prayer or meditation can be helpful. How a survivor maintains optimism is not so important as its use.

6-7. Summary. Survivors do not choose or welcome their fate and would escape it if they could. They are trapped in a world of seemingly total domination—a world hostile to life and any sign of dignity or resistance. The survival mission is not an easy one, but it is one in which success can be achieved. This has been an introduction to the concepts and ideas that can help an aircrew member return. Having the *will to survive* is what it's all about!

Part Three

BASIC SURVIVAL MEDICINE

Chapter 7

SURVIVAL MEDICINE

7-1. Introduction:

a. Foremost, among the many things that can compromise a survivor's ability to return, are medical problems encountered during ejection, parachute descent, and(or) parachute landing. In the Southeast Asian conflict, some 30 percent of approximately 1,000 U.S. Air Force survivors, including 322 returned PWs, were injured by the time they disentangled themselves from their parachutes. The most frequently reported injuries were fractures, strains, sprains, and dislocations, as well as burns and other types of wounds (Figure 7-1).

Figure 7-1. Survival Medicine.

b. Injuries and illnesses peculiar to certain environments can reduce survival expectancy. In cold climates, and often in an open sea survival situation, exposure to extreme cold can produce serious tissue trauma, such as frostbite, or death from hypothermia. Exposure to heat in warm climates, and in certain areas on the open seas, can produce heat cramps, heat exhaustion, or life-threatening heatstroke.

c. Illnesses contracted during evasion or in a captivity environment can interfere with successful survival. Among these are gastrointestinal disorders, respiratory diseases, skin infections and infestations, malaria, typhus, cholera, etc.

d. A review of the survival experiences from World War II, Korea, and Southeast Asia indicates that, while U.S. military personnel generally knew how to administer first aid to others, there was a marked inability to administer self-aid. Further, only the most basic medical care had been taught to most military people. Lastly, it was repeatedly emphasized that even minor injuries or ailments, when ignored, became major problems in a survival situation. Thus, prompt attention to the most minor medical problem is essential in a survival episode. Applying principles of survival medicine should enable military members to maintain health and well-being in a hostile or nonhostile environment until rescued and returned to friendly control.

e. Information in this chapter and Chapter 8 is a basic reference to self-aid techniques used by PWs in captivity and techniques found in folk medicine. The information describes procedures which can maintain health in medically austere situations. It includes items used to prevent and treat injuries and illnesses. Because there is no "typical" survival situation, the approach to self-aid must be flexible, placing emphasis on using what is available to treat the injury or illness. Further, survivors recognize that medical treatment offered by people of other cultures may be far different from our own. For example, in the rural areas of Vietnam, a poultice of python meat was and is used to treat internal lower back pain. Such treatment may be repugnant to some U.S. military personnel; however, medical aid offered to survivors in non-U.S. cultures may be the best available in the given circumstance.

f. The procedures in this chapter and Chapter 8 must be viewed in the reality of a true survival situation. The results of treatment may be substandard compared with present medical standards. However, these procedures will not compromise professional medical care which becomes available following rescue. Moreover, in the context of a survival situation, they may represent the best available treatment to extend the individual's survival expectancy.

7-2. Procedures and Expedients. Survival medicine encompasses procedures and expedients that are:

a. Required and available for the preservation of health and the prevention, improvement, or treatment of injuries and illnesses encountered during survival.

b. Suitable for application by nonmedical personnel to themselves or comrades in the circumstances of the survival situation.

(1) Survival medicine is more than first aid in the conventional sense. It approaches final definitive treatment in that it is not dependent upon the availability of technical medical assistance within a reasonable period of time.

(2) To avoid duplication of information generally available, the basic principles of first aid will not be repeated, nor will the psychological factors affecting survival which were covered in Part Two.

7-3. Hygiene. In a survival situation, cleanliness is essential to prevent infection. Adequate personal cleanliness will not only protect against disease germs that are present in the individual's surroundings, but will also protect the group by reducing the spread of these germs (Figure 7-2).

Figure 7-2. Hygiene.

a. Washing, particularly the face, hands, and feet, reduces the chances of infection from small scratches and abrasions. A daily bath or shower with hot water and soap is ideal. If no tub or shower is available, the body should be cleaned with a cloth and soapy water, paying particular attention to the body creases (armpits, groin, etc.), face, ears, hands, and feet. After this type of "bath," the body should be rinsed thoroughly with clear water to remove all traces of soap which could cause irritation.

b. Soap, although an aid, is not essential to keeping clean. Ashes, sand, loamy soil, and other expedients may be used to clean the body and cooking utensils.

c. When water is in short supply, the survivor should take an "air bath." All clothing should be removed and the body simply exposed to the air. Exposure to sunshine is ideal, but even on an overcast day or indoors, a 2-hour exposure of the naked body to the air will refresh the body. Care should be taken to avoid sunburn when bathing in this manner. Exposure in the shade, shelter, sleeping bag, etc., will help if the weather conditions do not permit direct exposure.

d. Hair should be kept trimmed, preferably 2 inches or less in length, and the face should be clean-shaven. Hair provides a surface for the attachment of parasites and the growth of bacteria. Keeping the hair short and the face clean-shaven will provide less habitat for these organisms. At least once a week, the hair should be washed with soap and water. When water is in short supply, the hair should be combed or brushed thoroughly and covered to keep it clean. It should be inspected weekly for fleas, lice, and other parasites. When parasites are discovered, they should be removed.

e. The principal means of infecting food and open wounds is contact with unclean hands. Hands should be washed with soap and water, if available, after handling any material which is likely to carry germs. This is especially important after each visit to the latrine, when caring for the sick and injured, and before handling food, food utensils, or drinking water. The fingers should be kept out of the mouth and the fingernails kept closely trimmed and clean. A scratch from a long fingernail could develop into a serious infection.

7-4. Care of the Mouth and Teeth. Application of the following fundamentals of oral hygiene will prevent tooth decay and gum disease:

a. The mouth and teeth should be cleansed thoroughly with a toothbrush and dentifrice at least once each day. When a toothbrush is not available, a "chewing stick" can be fashioned from a twig. The twig is washed, then chewed on one end until it is frayed and brushlike. The teeth can then be brushed very thoroughly with the stick, taking care to clean all tooth surfaces. If necessary, a clean strip of cloth can be wrapped around the finger and rubbed on the teeth to wipe away food particles which have collected on them. When neither toothpaste nor toothpowder are available, salt, soap, or baking soda can be used as substitute dentifrices. Parachute inner core can be used by separating the filaments of the inner core and using this as a dental floss. Gargling with willow bark tea will help protect the teeth.

b. Food debris which has accumulated between the teeth should be removed by using dental floss or toothpicks. The latter can be fashioned from small twigs.

c. Gum tissues should be stimulated by rubbing them vigorously with a clean finger each day.

d. Use as much care cleaning dentures and other dental appliances, removable or fixed, as when cleaning natural teeth. Dentures and removable bridges should be removed and cleaned with a denture brush or "chew stick" at least once each day. The tissue under the dentures should be brushed or rubbed regularly for proper stimulation. Removable dental applicances should be removed at night or for a 2- to 3-hour period during the day.

7-5. Care of the Feet. Proper care of the feet is of utmost importance in a survival situation, especially if the survivor has to travel. Serious foot trouble can be prevented by observing the following simple rules:

a. The feet should be washed, dried thoroughly, and massaged each day. If water is in short supply, the feet should be "air cleaned" along with the rest of the body (Figure 7-3).

b. Toenails should be trimmed straight across to prevent the development of ingrown toenails.

c. Boots should be broken in before wearing them on any

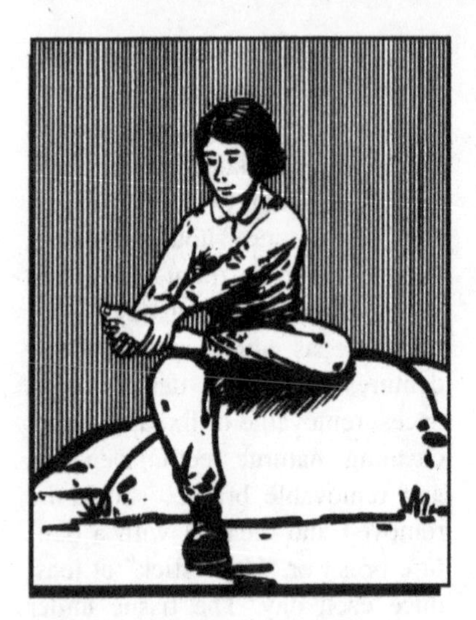

Figure 7-3. Care of the Feet.

mission. They should fit properly, neither so tight that they bind and cause pressure spots nor so loose that they permit the foot to slide forward and backward when walking. Insoles should be improvised to reduce any friction spots inside the shoes.

d. Socks should be large enough to allow the toes to move freely but not so loose that they wrinkle. Wool socks should be at least one size larger than cotton socks to allow for shrinkage. Socks with holes should be properly darned before they are worn. Wearing socks with holes or socks that are poorly repaired may cause blisters. Clots of wool on the inside and outside should be removed from wool socks because they may cause blisters. Socks should be changed and washed thoroughly with soap and water each day. Woolen socks should be washed in cool water to lessen shrinkage. In camp, freshly laundered socks should be stretched to facilitate drying by hanging in the sun or in an air current. While traveling, a damp pair of socks can be dried by placing them inside layers of clothing or hanging them on the outside of the pack. If socks become damp, they should be exchanged for dry ones at the first opportunity.

e. When traveling, the feet should be examined regularly to see if there are any red spots or blisters. If detected in the early stages of development, tender areas should be covered with adhesive tape to prevent blister formation.

7-6. Clothing and Bedding. Clothing and bedding become contaminated with any disease germs which may be present on the skin, in the stool, in the urine, or in secretions of the nose and throat. Therefore, keeping clothing and bedding as clean as possible will decrease the chances of skin infection and decrease the possibility of parasite infestation. Outer clothing should be washed with soap and water when it becomes soiled. Under clothing and socks should be changed daily. If water is in short supply, clothing should be "air cleaned." For air cleaning, the clothing is shaken out of doors, then aired and sunned for 2 hours. Clothing cleaned in this manner should be worn in rotation. Sleeping bags should be turned inside out, fluffed, and aired after each use. Bed linen should be changed at least once a week, and the blankets, pillows, and mattresses should be aired and sunned (Figure 7-4).

Figure 7-4. Bedding.

7-7. Rest. Rest is necessary for the survivor because it not only restores physical and mental vigor, but also promotes healing during an illness or after an injury.

a. In the initial stage of the survival episode, rest is particularly important. After those tasks requiring immediate attention are done, the survivor should inventory available resources, decide upon a plan of action, and even have a meal. This "planning session" will provide a rest period without the survivor having a feeling of "doing nothing."

b. If possible, regular rest periods should be planned in each day's activities. The amount of time allotted for rest will depend on a number of factors, including the survivor's physical condition, the presence of hostile forces, etc., but usually, 10 minutes each hour is sufficient. During these rest periods, the survivor should change either from physical activity to complete rest or from mental activity to physical activity as the case may be. The survivor must learn to become comfortable and to rest under less than ideal conditions.

7-8. Rules for Avoiding Illness. In a survival situation, whether short-term or long-term, the dangers of disease are multiplied. Application of the following simple guidelines regarding personal hygiene will enable the survivor to safeguard personal health and the health of others:

a. ALL water obtained from natural sources should be purified before consumption.

b. The ground in the camp area should not be soiled with urine or feces. Latrines should be used, if available. When no latrines are available, individuals should dig "cat holes" and cover their waste.

c. Fingers and other contaminated objects should never be put into the mouth. Hands should be washed before handling any food or drinking water, before using the fingers in the care of the mouth and teeth, before and after caring for the sick and injured, and after handling any material likely to carry disease germs.

d. After each meal, all eating utensils should be cleaned and disinfected in boiling water.

e. The mouth and teeth should be cleansed thoroughly at least once each day. Most dental problems associated with long-term survival episodes can be prevented by using a toothbrush and toothpaste to remove accumulated food debris. If necessary, devices for cleaning the teeth should be improvised.

f. Bites and insects can be avoided by keeping the body clean, by wearing proper protective clothing, and by using head nets, improvised bed nets, and insect repellants.

g. Wet clothing should be exchanged for dry clothing as soon as possible to avoid unnecessary body heat loss.

h. Personal items such as canteens, pipes, towels, toothbrushes, handkerchiefs, and shaving items should not be shared with others.

i. All food scraps, cans, and refuse should be removed from the camp area and buried.

j. If possible, a survivor should get 7 or 8 hours of sleep each night.

k. Aircrew members should keep all immunization "shots" current.

7-9. General Management of Injuries:

a. Bleeding. Control of bleeding is most important in survival situations where replacement transfusions are not possible. Immediate steps should be taken to stop the flow of blood, regardless of its source. The method used should be commensurate with the type and degree of bleeding. The tourniquet, when required and properly used, will save life. If improperly used, it may cost the life of the survivor. The basic characteristics of a tourniquet and the methods of its use are well covered in standard first-aid texts; however, certain points merit emphasis in the survival situation. A tourniquet should be used only after every alternate method has been attempted. If unable to get to medical aid within 2 hours, after 20 minutes, gradually loosen the tourniquet. If bleeding has stopped, remove the tourniquet; if bleeding continues, reapply and leave in place. The tourniquet should be applied as near the site of the bleeding as possible, between the wound and the heart, to reduce the amount of tissue lost.

b. Pain:

(1) Control of Pain. The control of pain accompanying disease or injury under survival situations is both difficult and essential. In addition to its morale-breaking discomfort, pain contributes to shock and makes the survivor more vulnerable to enemy influences. Ideally, pain should be eliminated by the removal of the cause. However, this is not always immediately possible, hence measures for the control of pain are beneficial.

(2) Position, Heat, and Cold. The part of the body that is hurting should be put at rest, or at least its

activity restricted as much as possible. The position selected should be the one giving the most comfort and be the easiest to maintain. Splints and bandages may be necessary to maintain the immobilization. Elevation of the injured part, with immobilization, is particularly beneficial in the throbbing type pain such as is typical of the "mashed" finger. Open wounds should be cleansed, foreign bodies removed, and a clean dressing applied to protect the wound from the air and chance contacts with environmental objects. Generally, the application of warmth reduces pain—toothache, bursitis, etc. However, in some conditions, application of cold has the same effect—strains and sprains. Warmth or cold is best applied by using water due to its high specific heat, and the survivor can try both to determine which is most beneficial.

(3) Pain Killers. Drugs are very effective in reducing pain; however, they are not likely to be available in the survival situation. Hence, the importance of the above "natural" procedures. Aspirin, APCs, and such tablets are primarily intended to combat the discomforts of colds and upper respiratory diseases, and, at best, will just take the edge off severe pain. They should be taken, however, if available. If no aspirin is available, there are some parts of vegetation that can be used. For example, most of the willows have been used for their pain-relieving and fever-lowering properties for hundreds of years. The fresh bark contains salicin, which probably decomposes into salicylic acid in the human body. Wintergreen, also known as checkerberry, was used by some Indians for body aches and pains. The leaves are made into a tea. The boiled bark of the magnolia tree helps relieve internal pains and fever, and has been known to stop dysentery. To be really effective in control of pain, stronger narcotic drugs such as codeine and morphine are required. During active hostilities, morphine may be available in aircraft and individual first-aid kits.

c. Shock:

(1) Circulatory Reaction. Shock in some degree accompanies all injuries to the body, and frequently it is the most serious consequence of the injury. In essence, shock is a circulatory reaction of the body (as a whole) to an injury (mechanical or emotional). While the changes to the circulatory system initially favor body resistance to the injury (by ensuring adequate blood supply to vital structures), they may progress to the point of circulatory failure and death. All aircrew members should be familiar with the signs and symptoms of shock so that the condition may be anticipated, recognized, and dealt with effectively. However, the best survival approach is to treat ALL moderate and severe injuries for shock. No harm will be done, and such treatment will speed recovery.

(2) Fluids. Normally, fluids administered by mouth are generally prohibited in the treatment of shock following severe injury. Such fluids are poorly absorbed when given by mouth, and they may interfere with later administration of anesthesia for surgery. In survival medicine, however, the situation is different in that the treatment being given is the final treatment. Survivors cannot be deprived of water for long periods just because they have been injured; in fact, their recovery depends upon adequate hydration. Small amounts of warm water, warm tea, or warm coffee given frequently early in shock are beneficial if the patient is conscious, can swallow, and has no internal injuries. In later shock, fluids by mouth are less effective as they are not absorbed from the intestines. Burns, particularly, require large amounts of water to replace fluid lost from injured areas. Alcohol should never be given to a person in shock or who may go into shock.

(3) Psychogenic Shock. Psychogenic shock is frequently noted during the period immediately following an emergency; for example, bailout. Psychogenic shock, which occurs even without injury, requires attention to limit it, both in degree and duration. The degree of this post-impact shock varies widely among individuals but its occurrence is almost universal. In reality, the survivor has passed through two major emergencies almost simultaneously; the aircraft incident leading to the survival situation, and the situation itself. Should the survivor be injured (and the majority of them are), a third emergency is added. It is not uncommon, then, that some psychogenic reaction with circulatory implications occurs. Resistance to this type of shock depends upon the individual's personality and the amount of training previously received. Treatment consists of stopping all activities (when possible), relaxing, evaluating the situation, and formulating a plan of action before the survival situation begins.

d. Fractures:

(1) Proper immobilization of fractures, dislocations, and sprains is even more important in survival medicine than in conventional first aid. Rather than merely making the patient comfortable during transport to eventual treatment, in survival medicine, the initial immobilization is part of the ultimate treatment. Immobilizing body parts to help control pain was discussed earlier. In addition, immobilization in proper position hastens healing of fractures and improves the ultimate functional result. In the survival situation, the immobilization must suffice for a relatively long period of time and permit the patient to maintain a fairly high degree of mobility. Materials for splinting and bandaging are

available in most survival situations, and proper techniques are detailed in most first-aid manuals.

(2) The reduction of fractures is normally beyond the scope of first aid; however, in the prolonged survival situation, the correction of bone deformities is necessary to hasten healing and obtain the greatest functional result. The best time for manipulation of a fracture is in the period immediately following the injury, before painful muscle spasms ensue. Traction is applied until overriding fragments of bone are brought into line, (check by the other limb) and the extremity is firmly immobilized. Frequently, it is advantageous to continue traction after reduction to ensure the proper alignment of the bones.

(3) As plaster casts are not available in the survival situation, improvising an immobilization device is necessary. This may be done by using several parallel, pliable willow branches, woven together with vines or parachute lines. Use care so that the extremity is not constricted when swelling follows the injury. In an escape and evasion situation, it may be necessary to preserve the mobility of the survivor after reduction of the fracture. This is difficult in fractures of the lower extremities, although tree limbs may be improvised as crutches. With companions, the use of improvised litters may be possible.

(4) Reduction of dislocated joints is done similar to that of fractures. Gentle, but firm traction is applied and the extremity is manipulated until it "snaps" back into place. If the survivor is alone, the problem is complicated but not impossible. Traction can still be applied by using gravity. The distal portion of the extremity is tied to (or wedged) into the fork of a tree or similar point of fixation. The weight of the body is then allowed to exert the necessary traction, with the joint being manipulated until the dislocation is reduced.

e. Infection:

(1) Infection is a serious threat to the survivor. The inevitable delay in definite medical treatment and the reality of the survival situation increases the chances of wound infection. Antibiotics may not be available in sufficient amounts in the survival situation. In survival medicine, one must place more emphasis on the prevention and control of infection by applying techniques used before the advent of antibiotics.

(2) Unfortunately, survivors have little control over the amount and type of infection introduced at the time of injury. However, they can help control the infection by wearing clean clothes. Use care to prevent additional infection into wounds. Wounds, regardless of the type or severity, should not be touched with dirty hands or objects. One exception to this rule is the

essential control of arterial bleeding. Clothing should be removed from wounds to avoid contamination surrounding skin areas.

(3) All wounds should be promptly cleansed. Water is the most universally available cleaning agent, and should be (preferably) sterile. At sea level, sterilize water, by placing it in a covered container and boiling it for 10 minutes. Above 3,000 feet, water should be boiled for 1 hour (in a covered container) to ensure adequate sterilization. The water will remain sterile and can be stored indefinitely as long as it is covered.

(a) Irrigate wounds rather than scrubbing to minimize additional damage to the tissue. Foreign material should be washed from the wound to remove sources of continued infection. The skin adjacent to wounds should be washed thoroughly before bandaging. When water is not available for cleaning wounds, the survivor should consider the use of urine. Urine may well be the most nearly sterile of all fluids available and, in some cultures, is preferred for cleaning wounds. Survivors should use urine from the midstream of the urine flow.

(b) While soap is not essential to clean wounds, a bar of medicated soap placed in a personal survival kit and used routinely would do much to prevent the infection of seemingly inconsequential injuries. External antiseptics are best used for cleaning abrasions, scratches, and the skin areas adjacent to lacerations. Used in deep, larger wounds, antiseptics produce further tissue damage.

(c) Nature also provides antiseptics which can be used for wound care. The American Mountain Ash is found from Newfoundland south to North Carolina and its inner bark has antiseptic properties. The red berries contain ascorbic acid and have been eaten to cure scurvy. The Sweet Gum bark is still officially recognized as being an antiseptic agent. Water from boiled Sweet Gum leaves can also be used as antiseptic for wounds.

f. The "Open Treatment" Method. This is the only safe way to manage survival wounds. No effort should be made to close open wounds by suturing or by other procedure. In fact, it may be necessary to open the wound even more to avoid entrapment or infection and to promote drainage. The term "open" does not mean that dressings should not be used. Good surgery requires that although wounds are not "closed," nerves, bone, and blood vessels should be covered with tissue. Such judgment may be beyond the capability of the aircrew member, but protection of vital structures will aid in the recovery and ultimate function. A notable exception to "open treatment" is the early closure of facial wounds which interfere with breathing, eating, or drinking. Wounds, left open, heal by formation of infection-resistant granulation tissue (proud flesh). This tissue is easily recognized by its moist red granular appearance, a good sign in any wound.

g. Dressings and Bandages. After cleansing, all wounds should be covered with a clean dressing. The dressing should be sterile; however, in the survival situation, any clean cloth will help to protect the wound from further infection. A proper bandage will anchor the dressing to the wound and afford further protection. Bandages should be snug enough to prevent slippage, yet not constrictive. Slight pressure will reduce discomfort in most wounds and help stop bleeding. Once in place, dressings should not be changed too frequently unless required. External soiling does not reduce the effectiveness of a dressing and pain and some tissue damage will accompany any removal. In addition, changing dressings increases the danger of infection.

h. Physiological "Logistics." Despite all precautions, some degree of infection is almost universal in survival wounds. This is the primary reason for the "open" treatment advocated above. The human body has a tremendous capacity for combating infections if it is permitted to do so. The importance of proper rest and nutrition to wound healing and control of infection has been mentioned. In addition, the "logistics" of the injured part should be improved. The injury should be immobilized in a position to favor adequate circulation, both to and from the wound. Avoid constrictive clothing or bandages. Applying heat to an infected wound further aids in mobilizing local body defense measures. Lukewarm saltwater soaks will help draw out infection and promote oozing of fluids from the wound, thereby removing toxic products. Poultices, made of clean clay, shredded bark of most trees, ground grass seed, etc., do the same thing.

i. Drainage. Adequate natural drainage of infected areas promotes healing. Generally, wicks or drains are unnecessary. On occasion, however, it may be better to remove an accumulation of pus (abscess) and insert light, loose packing to ensure continuous drainage. The knife or other instrument used in making the incision for drainage must be sterilized to avoid introducing other types of organisms. The best way to sterilize in the field is with heat, dry or moist.

j. Antibiotics. Antibiotics, when available, should be taken for the control of infection. Consensus is that the drug should be of the so-called "broad spectrum type;" that is, be effective against any micro-organisms rather than specific for just one or two types. The exact amount to be included in survival kits will vary with the drug and basic assumptions as to the number and types of infections to be expected. Remember that antibiotics are potency-dated items (shelf-life about 4 years), and including them in survival kits requires kit inspection and drug replacement with active medical stocks.

k. Debridement. (The surgical removal of lacerated, devitalized, or contaminated tissue.) The debridement of severe wounds may be necessary to minimize infection (particularly of the gas gangrene type) and to reduce septic (toxic) shock. In essence, debridement is the removal of foreign material and dead or dying tissue. The procedure requires skill and should only be done by nonmedical personnel in case of dire emergency. If required, follow these general rules. Dead skin must be cut away. Muscle may be trimmed back to a point where bleeding starts and gross discoloration ceases. Fat which is damaged tends to die and should be cut away. Bone and nerves should be conserved where possible and protected from further damage. Provide ample natural drainage for the potentially infected wound and delay final closure of the wound.

l. Burns:

(1) Burns, frequently encountered in aircraft accidents and subsequent survival episodes, pose serious problems. Burns cause severe pain, increase the probability of shock and infection, and offer an avenue for the loss of considerable body fluids and salts. Direct initial treatment toward relieving pain and preventing infection. Covering the wound with a clean dressing of any type reduces the pain and chance for infection. Further, such protection enhances the mobility of the patient and the capability for performing other vital survival functions. In burns about the face and neck, ensure the victim has an open airway. If necessary, cricothyroidotemy should be done before the patient develops extreme difficulties. Burns of the face and hands are particularly serious in a survival situation as they interfere with the capability of survivors to meet their own needs. Soaking certain barks (willow, oak, maple) in water soothes and protects burns by astringent action. This is a function of the acid content of the bark used.

(2) Maintenance of body fluids and salts is essential to recover from burns. The only way to administer fluids in a survival situation is by mouth; hence the casualty should ingest sufficient water early before the nausea and vomiting of toxicity intervenes. Consuming the eyes and blood (both cooked) of animals can help restore electrolyte levels if salt tablets are not available. NOTE: The survivor may also pack salt in personal survival kits to replace electrolytes (¼ teaspoon per quart of water).

m. Lacerations: Lacerations (cuts) are best left open due to the probability of infection. Clean thoroughly, remove foreign material, and apply a protective dressing.

Frequently, immobilization will hasten the healing of major lacerations. On occasion (tactical), it may be necessary to close (cover) the wound, despite the danger of infection, in order to control bleeding or increase the mobility of the patient. If a needle is available, thread may be procured from parachute lines, fabric, or clothing, and the wound closed by "suturing." If suturing is required, place the stitches individually and far enough apart to permit drainage of underlying parts. Do not worry about the cosmetic effect; just approximate the tissue. For scalp wounds, hair may be used to close after the wound is cleansed. Infection is less a danger in this area due to the rich blood supply.

n. Head Injuries. Injuries to the head pose additional problems related to brain damage as well as interfering with breathing and eating. Bleeding is more profuse in the face and head area, but infections have more difficulty in taking hold. This makes it somewhat safer to close such wounds earlier to maintain function. Cricothyroidotemy may be necessary if breathing becomes difficult due to obstruction of the upper airways. In the event of unconsciousness, watch the patient closely and keep him or her still. Even in the face of mild or impending shock, keep the head level or even slightly elevated if there is reason to expect brain damage. Do not give fluids or morphine to unconscious persons.

o. Abdominal Wounds. Wounds of the abdomen are particularly serious in the survival situation. Such wounds, without immediate and adequate surgery, have an extremely high mortality rate and render patients totally unable to care for themselves. If intestines are not extruded through the wound, a secure bandage should be applied to keep this from occurring. If intestine is extruded, do not replace it due to the almost certain threat of fatal peritonitis. Cover the extruded bowel with a large dressing and keep the dressing wet with any fluid that is fit to drink, or urine. The patient should lie on the back and avoid any motions that increase intra-abdominal pressure which might extrude more bowel. Keep the survivor in an immobile state or move on a litter. "Nature" will eventually take care of the problem; either through death or walling-off of the damaged area.

p. Chest Injuries. Injuries of the chest are common, painful and disabling. Severe bruises of the chest or fractures of the ribs require that the chest be immobilized to prevent large painful movements of the chest wall. The bandage is applied while the patient deeply exhales. In the survival situation, it may be necessary for survivors to wrap their own chest. This is more difficult but can be done by attaching one end of the long bandage (parachute material) to a tree or other fixed object, holding the other end in the hand, and slowly rolling body toward the tree, keeping enough counterpressure on the bandage to ensure a tight fit.

q. Sucking Chest Wounds. These wounds are easily recognized by the sucking noise and appearance of foam or bubbles in the wound. These wounds must be closed immediately before serious respiratory and circulatory complications occur. Ideally, the patient should attempt to exhale while holding the mouth and nose closed (Valsalva) as the wound is closed. This inflates the lungs and reduces the air trapped in the pleural cavity. Frequently, a taped, airtight dressing is all that is needed, but sometimes it is necessary to put in a stitch or two to make sure the wound is closed.

r. Eye Injuries. Eye injuries are quite serious in a survival situation due to pain and interference with other survival functions. The techniques for removing foreign bodies and for treating snow blindness are covered in standard first-aid manuals. More serious eye injuries involving disruption of the contents of the orbit may require that the lids of the affected eye be taped closed or covered to prevent infection.

s. Thorns and Splinters. Thorns and splinters are frequently encountered in survival situations. Reduce their danger by wearing gloves and proper footgear. Their prompt removal is quite important to prevent infection. Wounds made by these agents are quite deep compared to their width, which increases chances of infection by those organisms (such as tetanus) that grow best in the absence of oxygen. Removal of splinters is aided by the availability of a sharp instrument (needle or knife), needle nose pliers, or tweezers. Take care to get all of the foreign body out; sometimes it is best to open the wound sufficiently to properly cleanse it and to allow air to enter the wound. When cleaned, treat as any other wound.

t. Blisters and Abrasions. Care for blisters and abrasions promptly. Foot care is extremely important in the survival situation. If redness or pain is noted, the survivor should stop (if at all possible) to find and correct the cause. Frequently, a protective dressing or bandage and(or) adhesive will be sufficient to prevent a blister. If a blister occurs, do not remove the top. Apply a sterile (or clean) dressing. Small abrasions should receive attention to prevent infection. Using soap with a mild antiseptic will minimize the infection of small abrasions which may not come to the attention of the survivor.

u. Insect Bites. Bites of insects, leeches, ticks, chig-gers, etc., pose several hazards. Many of these organisms transmit diseases, and the bite itself is likely to become infected, especially if it itches and the survivor scratches it.

The body should be inspected frequently for ticks, leeches, etc., and these should be removed immediately. If appropriate and possible, the survivor should avoid infested areas. These parasites can best be removed by applying heat or other irritant to them to encourage a relaxation of their hold on the host. Then the entire organism may be gently detached from the skin, without leaving parts of the head imbedded. Treat such wounds as any other wound. Applying cold wet dressings will reduce itching, scratching, and swelling.

7-10. Illnesses. Many illnesses which are minor in a normal medical environment become major in a survival situation when the individual is alone without medications or medical care. Survivors should use standard methods (treat symptoms) to prevent expected diseases since treatment in a survival situation is so difficult. Key preventive methods are to maintain a current immunization record, maintain a proper diet, and exercise.

a. Food Poisoning. Food poisoning is a significant threat to survivors. Due to sporadic food availability, excess foods must be preserved and saved for future consumption. Methods for food preservation vary with the global area and situation. Bacterial contamination of food sources has historically caused much more difficulty in survival situations than the ingestion of so-called poisonous plants and animals. Similarly, dysentery or water-borne diseases can be controlled by proper sanitation and personal hygiene.

b. Treatment of Food Poisoning. Supportive treatment is best if the food poisoning is due to preformed toxin; staphylococcus, botulism, etc. (acute symptoms of nausea, vomiting, and diarrhea soon after ingestion of the contaminated food). Keep the patient quiet and lying down, and ensure the patient drinks substantial quantities of water. If the poisoning is due to ingestion of bacteria which grow within the body (delayed gradual onset of same symptoms), take antibiotics (if available). In both cases, symptoms may be alleviated by frequently eating small amounts of fine, clean charcoal. In PW situations, if chalk is available, reduce it to powder, and eat to coat and soothe the intestines. Proper sanitation and personal hygiene will help prevent spreading infection to others in the party or continuing reinfection of the patient.

Chapter 8

PW MEDICINE

8-1. Introduction:

Imprisoned PWs are, in the physical sense at least, under the control of their captors. Thus, the application of survival medicine principles will depend on the amount of medical service and supplies the captors can, and will, give to their prisoners. An enemy may both withhold supplies and confiscate survivor's supplies. Some potential enemies (even if they wanted to provide PW medical support) have such low standards of medical practice that their best efforts could jeopardize the recovery of the patient (Figure 8-1).

An interesting and important sociological problem arises in getting medical care for PWs. How far should prisoners go in their efforts to get adequate rations and medicines for themselves or those for whom they are responsible? The Code of Conduct is quite specific concerning consorting with the enemy. Individuals must use considerable judgment in deciding whether to forget the welfare of fellow prisoners in order to follow the letter of the Code. Even more questionable is the individual who will offer such a justification for personal actions. Again, these questions involve more than purely medical consideration. In combat, there are apt to be frequent situations in which medical considerations are outweighed by more important ones.

Figure 8-1. PW Medicine.

8-2. History:

As in past wars, there were professional medical personnel among the captives in North Vietnam; however, these personnel were not allowed to care for the sick and injured as in the past. Medical care and assistance from the captors were limited and generally below comparable standards of the United States. Yet 566 men returned, most in good physical and psychological condition, having relied to a large extent on their own ingenuity, knowledge, and common sense in treating wounds and diseases. They were able to recall childhood first aid, to learn by trial and error, and to use available resources. Despite their measures of success in this respect, many released personnel felt that with some prior training, considerable improvement in self-help techniques was possible even in the most primitive conditions.

To determine how the services could help and to assist future PWs to care for themselves if the situation required it, the Medical Section of the Air Force Intelligence Service, with the Surgeon General of the Air Force, sponsored a 5-day seminar to examine the pertinent medical experiences of captivity and to recommend appropriate additions and changes in training techniques. As a basis for seminar discussion, Air Training Command provided data on the major diseases, wounds, and ailments, and the treatment methods used by the captives in Southeast Asia. Transcripts (325) of debriefing material were screened for medical data. Significant disease categories were established for analysis simplification based on the freqency of the problems encountered.

PROBLEMS	MAJOR CATEGORIES ESTABLISHED
Dysentery	Trauma (lacerations, burns, fractures)
Fungus	Gastrointestinal problems
Dental problems	Communicable diseases
Intestinal problems	Nutritional diseases
Fractures	Dermatological ailments
Lacerations	Dental problems
Respiratory ailments	
Burns	

In examining these major categories, attention was focused on those medical problems considered significant by the prisoners themselves in evaluating their primitive practices (self-help).

8-3. Trauma:

Most of the prisoners began their captivity experience with precapture injuries—burns, wounds, fractures, and lacerations. Other injuries were the result of physical abuse while a prisoner. Most of these individuals, upon their return, expressed a need to know more about managing their injuries in captivity and also what to expect about the long-term effects of injuries. It was not evident to them that the practice of a few simple rules will generally lead to acceptable results in wound treatment, and that much can be done after repatriation to correct cosmetic and functional defects.

The groundwork for management of injuries should begin well before an individual enters the captivity or survival environment. The treatment of injuries in survival or captivity depends primarily on providing the body the best possible circumstances to "repair itself." It is vital, therefore, to have the body in the best possible physical condition before exposure to survival or captivity. This means good cardiovascular conditioning, good muscle strength and tone, and good nutritional status. Physiological and nutritional status will markedly influence the rate and degree of healing in response to injury. The opportunity for maintaining the best possible physical conditioning and nutritional status in captivity will be greatly reduced. (Once in a captivity or survival setting, it is important to do everything possible to maintain a good physiological and nutritional status.)

8-4. Gastrointestinal Problems:

a. Diarrhea. This was a common ailment in the prison environment, not only in Vietnam, but also in WW II and Korea. It plagued the forces of North Vietnam and the allied forces. This was the second most frequent malady afflicting the Viet Cong forces. The causative factors of this almost epidemic state were varied. A variety of infectious agents gaining access to the body by use of contaminated food and water certainly contributed to the problem. Equally important as causative agents were the low level of sanitation and hygiene practices within the camps. Psychogenic responses to unappetizing diet, nutritional disturbances, and viral manifestations also contributed.

(1) Captor Therapy. This consisted primarily of local or imported antidiarrheal agents, antibiotics, and vitamins. Appropriate diet therapy was instituted.

(2) Captive Self-Therapy. After instituting diet restrictions (solid food denial and increased liquid intake), afflicted personnel were administered "concoctions" of banana skins, charcoal, chalk, or tree bark tea.

(3) Treatment Evaluation. The accepted therapy for diarrhea focuses on the causative factors which in the captivity experience

were largely neglected. From a symptomatic perspective, the principles of self-treatment are simple to master: restrict intake to nonirritating foods (avoid vegetables and fruits), establish hygienic standards, increase fluid intake, and, when available, use antidiarrheal agents. The prisoners often resorted to a more exotic therapeutic regimen consisting of banana skins, charcoal, chalk, salt restriction, rice, or coffee. Charcoal, chalk, and the juice of tree barks have a scientific basis for their therapeutic success. Inasmuch as diarrhea was a source of concern and a disability for the North Vietnamese as well as the captives, therapy was often offered on request and was appropriate and successful.

(4) Conclusion. Diarrhea was frequent among PWs during captivity. Seldom fatal, it was disabling and a source of concern to those afflicted. Most captives were treated on demand and improved. This condition lends itself to some form of self-therapy through an understanding of its physiological derangements. The PW responded with intelligence, common sense, and a reasonably effective self-help regimen.

b. Dysentery. From a symptomatic perspective, dysentery is a severe form of diarrhea with passage of mucous and blood. Treatment and conclusions are similar to those for diarrhea.

c. Worms and Intestinal Parasites. Worms were extremely common among the captives. Twenty-eight percent of the released prisoners indicated worms as a significant medical problem during captivity. Worms often caused gastrointestinal problems similar to those resulting from a variety of other causes. The pin worm appears to have been the primary cause. This is not surprising, as its distribution is worldwide and the most common cause of helminthic infection of people in the United States. It requires no intermediate host; hence, infection is more rapidly acquired under poor hygienic conditions so commonly found in warm climates and conditions similar to the captivity environment. Seldom fatal, worms are significant, as they can lower the general resistance of the patient and may have an adverse affect on any intercurrent illness.

(1) Captor Therapy. This consisted of anti-helminthic agents (worm medicine) dispensed without regularity, but with satisfactory results.

(2) Captive Therapy. The nuisance and irritating aspects of worms led to severe rectal itching, insomnia, and restlessness. This motivated the prisoner to find some form of successful self-therapy. Prevention was a simple and readily obtainable goal. Shoes were worn when possible; hands

were washed after defecation; and fingernails were trimmed close and frequently. Peppers, popular throughout the centuries in medicine, contain certain substances chemically similar to morphine. They are effective as a counter-irritant for decreasing bowel activity. Other "house remedies" popular among the captives included drinking saltwater (a glass of water with 4 tablespoons of salt added), eating tobacco from cigarettes (chewing up to two or three cigarettes and swallowing them), and infrequently drinking various amounts of kerosene. All of these remedies have some degree of therapeutic effectiveness, but are not without danger and therefore deserve further comment. Saltwater alters the environment in the gastrointestinal tract and can cause diarrhea and vomiting. Too large an amount can have harmful effects on body fluid mechanisms and can lead to respiratory complications and death. Tobacco contains nicotine and historically was popular in the 19th century as an emetic expectorant and was used for the treatment of intestinal parasites. Nicotine is, however, one of the most toxic of all drugs and can cause death when more than 60 mg is ingested. A single cigarette contains about 30 mg of nicotine, so the captives who ate two or more cigarettes had been using a cure more dangerous than the disease. Kerosene is also toxic with 3 to 4 ounces capable of causing death. It is particularly destructive to the lungs and if through vomiting it were to make its way into the trachea and eventually the lungs, the complications would then again be far worse than the presence of worms.

(3) Treatment Evaluation. The antihelminthics therapy used by the captors was extremely effective. The problem during confinement was the nonavailability of such medication on demand. In addition, the inability to practice proper hygienic standards assured the continuation of, and reinfection with, worms.

(4) Conclusion. Worm infection in confinement is common and expected. It is seldom fatal, but contributes to general disability and mental depression due to its nuisance symptoms. Under certain circumstances, worms can assist in the spread of other diseases. The principle to follow in self-care is simple—use as high a hygienic standard as possible, and use medication causing bowel paristalsis and worm expulsion. Substances which interfere with the environment of the worms will aid in their expulsion. The toxic "house remedies" must be weighed against their possible complications.

8-5. Hepatitis. Infection of the liver was fairly common in some camps and present among the prison population throughout the captivity experience. Diagnosis was usually made on the basis of change of skin color to yellow (jaundice).

a. Captor Treatment. The Vietnamese seemed to have followed the standard therapy of rest, dietary management, and vitamin supplementation. They also displayed a heightened fear of the disease and avoided direct contact, when possible, with those afflicted.

b. Captive Therapy. For the most part, it parallels the therapy of the captors. This disease allows for little ingenuity or inventiveness of therapy.

c. Comments. Hepatitis is worldwide. Presumably most cases of hepatitis in captivity were viral in origin and easily disseminated to fellow prisoners. Conditions of poor sanitation and hygiene with close communal living foster its spread. Prevention through proper hygienic practices is the most effective tool. Equally important is an understanding of the disease characteristics. The majority of the cases recovered completely and less than 1 percent succumbed to this disease.

8-6. Nutritional Deficiencies. Symptoms attributed to malnutrition were frequent in the early years of confinement and continued up through 1969. The use of polished rice and the lack of fresh fruits and vegetables contributed to vitamin and protein deficiencies. From 1969 through 1973, food supplements were provided, and by release time, few obvious manifestations of diseases were present among those returning. The primary problems during the early years were vitamin deficiencies.

a. Vitamin B Deficiency (Beri-Beri). Presumably present among several PWs (especially those confined to the Briarpatch [Xom Ap Lo] about 15 miles west of Sontay), it was rarely diagnosed on return. Its primary manifestation was pain in the feet described by the captives as "like a minor frostbite that turned to shooting pains."

(1) Captor Therapy. Prisoners were treated with vitamin injections and increased caloric content.

(2) Captive Therapy. Increasing caloric intake by eating anything of value. No specific self-care program existed for this malady.

(3) Comments. Beri-beri is a nutritional disease resulting from a deficiency of vitamin B (Thiamine). It is widespread in the Orient and in tropical areas where polished rice is a basic dietary staple. Of the various forms of the disease, dry beri-beri would seem most important to the confinement condition. Early signs and symptoms of the disease include muscle weakness and atrophy, loss of vibratory sensation over parts of the extremities, numbness, and

tingling in the feet. From the comments of the PWs, it is difficult to formulate a diagnosis. Modern therapy consists of vitamin B or sources of the vitamin in food (such as green peas, cereal grains, and unpolished rice).

(4) Conclusion. In Vietnam, the possible early onset of the dry form of beri-beri was encountered. This is supported by the symptoms described and by the existence of dietary shortages of vitamin B and other nutritional deficiency.

b. Vitamin A Deficiency. There were several reported cases of decreased vision (primarily at night) attributed to vitamin A deficiency. This problem usually occurred during periods of punishment or politically provoked action when food was withheld as part of the discipline. The condition responded well to increased caloric intake and deserves little special mention. An understanding of the transient nature of this problem and its remedial response to therapy is important.

8-7. Communicable Diseases. Some communicable diseases were endemic in North Vietnam and certainly responsible for large scale disability among the personnel of the enemy forces. Plague, cholera, and malaria are frequent and a serious public health menace. Thanks to the immunization practices of the American forces, these diseases were of little concern to Americans during their captivity.

8-8. Skin Diseases:

a. Lesions. Dermatological lesions were common to the various prison experiences. Their importance lies not in their lethality (as they apparently did not cause any deaths), but for their irritant quality and the debilitating and grating effect on morale and mental health. Boils, fungi, heat rash, and insect bites appeared frequently and remained a problem throughout the captivity experience.

b. Boils and Blisters. A deep-seated infection usually involves the hair follicles and adjacent subcutaneous tissue, especially parts exposed to constant irritation.

(1) Captor Treatment. Prisoner complaints about the presence of boils usually brought about some action by the captors. Treatment varied considerably and obviously depended on the knowledge of medics, doctors treating their prisoners, the availability of medical supplies, and the current camp policy. For the most part, systemic antibiotics, sulfa, and tetracycline were administered. In other instances, the boils were lanced or excised and treated with topical astringents.

(2) Captive Treatment. As the medics normally responded to pleas about boils, self-treatment

was practiced primarily when there was distrust of captor techniques. Prisoners would attempt to lance the boil with any sharp instrument such as needle, wire, splinters, etc., and exude their contents by applying pressure. The area was then covered with toothpaste and, when available, iodine.

(3) Comments. As noted above, the boil is an infection of hair follicles. It is more frequent in warm weather and aggravated by sweat which provides ideal conditions for the bacteria. Boils seldom appear singularly. Once present, they are disseminated by fingers, clothing, and discharges from the nose, throat, and groin. Modern therapy consists of hot compresses to hasten localization, and then conservative incision and drainage. Topical antibiotics and systemic antibiotics are then used. Boils increase in frequency with a decrease in resistance as seen in malnutrition and exhaustion states in a tropical environment. This almost mimics the prison conditions.

(4) Conclusions. Self-help treatment is limited. Of importance here is sterility when handling the boils, cleanliness, exposure to sunlight, keeping the skin dry, and getting adequate nutrition. The disease is self-limiting and not fatal. The application of any material or medication with a detergent effect may be used (soaks in saline, soap, iodine, and topical antibiotics).

8-9. Fungal Infections. Fungal infections were also a common skin problem for those in Southeast Asia captivity. As with other skin lesions, they are significant for their noxious characteristics and weakening effect on morale and mental health. Superficial fungal infections of the skin are widespread throughout the world. Their frequency among PWs reflects the favorable circumstances of captivity for cultivating fungal infections.

a. Captor Therapy. Treatment consisted of medication described by many PWs as iodine and the occasional use of sulfa powder.

b. Captive Therapy. Treatment (often the result of memory of childhood experiences and trial and error observations) consisted of the removal of body hair (to prevent or improve symptoms in the case of heat rash), exposure to sunlight to dry out fungal lesions, and development of effective techniques to foster body cooling and to decrease heat generation. Considerable effort was directed at keeping the body clean.

c. Comments. Superficial Dermatoses (skin lesions) due to fungi were common. Their invasive powers are at best uniformly weak, and because of this, infections are limited to the superficial portions of the skin and seldom by themselves fatal. Modem therapy since 1958 has relied heavily on an oral antifungal agent effective against many superficial fungi. This drug is expensive and not available in many parts of the world. Several lotions and emulsions can be used with some success. Elemental iodine is widely used as a germicide and fungicide. It is an effective antiseptic and obviously found favor in North Vietnam because of its availability. Without professional therapy, self-help, although limited in scope, can be effective. The principle of wet soaks for dry lesions and dry soaks for wet lesions is a fairly reliable guide. The use of the sun as a drying agent can also be very effective.

d. Conclusions. Skin problems are common to the captivity environment. More importantly, extreme personal discomfort, accompanied by infection, was detrimental to the physical and mental well-being of the prisoner.

8-10. Dental Problems. These were common among all captives, not only during confinement, but also before capture. They were secondary to facial injury during egress, or caused by physical abuse during interrogation. Periodontitis (inflammation of tissue surrounding the tooth), pyorrhea (discharge of pus), and damage to teeth consistent with poor hygiene and "wear and tear" were also present.

a. Specific Complaints. Pain associated with the common toothache represented one of the most distressing problems faced by the PW. It affected the PW's nutrition and robbed the PW of the physical pleasure of eating (a highlight of isolated captivity). The inability of the PW to adequately deal with this problem caused persistent anxiety and decreased the ability to practice successful resistance techniques. In a few isolated instances, PWs actually considered collaboration with the captors in exchange for treatment and relief from tooth pain.

b. Captor Treatment. Treatment varied considerably and was no doubt influenced by political considerations. "Dentists" were infrequently available in camps before 1969. Cavities were filled, although usually inadequately, with subsequent loss of the filling. Use of local anesthesia also varied depending on the dentist providing care.

c. Captive Therapy. The PWs often chose to treat themselves rather than seek or accept prison dentistry when it was available. Abscesses were lanced with sharp instruments made locally out of wood, bamboo, or whatever was available. Brushing was excessive, again using whatever was available; chew sticks common to Asia were widely used. Aspirin (ASA), when available, was applied directly to the tooth or cavity.

d. Commentary. The self-help practices noted previously had many positive aspects. The basic principle of maintaining a well-planned cleaning program using fiber, brushes, or branches certainly contributed to the relatively low incidence of cavities and infection among the prisoners. The lancing of abscesses using bamboo sticks, although not a professional maneuver, has merit insofar as the pressure is relieved and the tendency to develop into cellulitis (widespread infection) decreased. The application of aspirin directly into the cavity should be discouraged as might the application of any other substance not directly produced for this purpose.

e. Conclusions. The most effective tool against dental complications in captivity is proper preventive dentistry. The present program of the three services, if adhered to, is adequate to ensure a high state of dental hygiene while captive.

8-11. Burns. Burns were an extremely frequent injury among PWs. Severity ranged from first through third degree and occurred frequently on hands and arms.

a. Captor Treatment. For the most part, burns were treated by captors by cleaning the burns and applying antiseptics and bandages. The results obtained were, by and large, inadequate, with frequent infections and long-term debilitation.

b. Captive Treatment. No specific treatment was developed among the PWs for burns. Reliance for some form of therapy was almost completely left to the captor.

c. Commentary. Burns are extremely painful and can severely interfere with the ability to escape or to survive in captivity. The basic principle here is prevention.

d. Prevention. Adequate protection of exposed surfaces while flying (flame-retardant suits, gloves, boots, and helmet with visor down) is the best preventive action.

8-12. Lacerations and Infections:

a. Treatment. Captor treatment for lacerations and infections reflected the medical standards in North Vietnam and their domestic priorities. Wound and infection treatment varied considerably from being adequate to substandard and malpracticed. Obviously, the availability of trained physicians, a changing political climate, and difficulty in obtaining sophisticated medical supplies and equipment dictated and influenced the quality of the care delivered. The prisoners could do little professionally with this type of injury. As with diseases, the maintenance of good nutritional standards, cleanliness, and "buddy self-care" were the basic treatments.

b. Comments. When soft tissue is split, torn, or cut, there are three primary concerns—bleeding, infection, and healing of the wound.

(1) Bleeding is the first concern and must be controlled as soon as possible. Most bleeding can be controlled by direct pressure on the wound and that should be the first treatment used. If that fails, the next line of defense would be the use of classic pressure points to stop hemorrhaging. And the last method for controlling hemorrhage would be the tourniquet. The tourniquet should be used only as a last resort. Even in more favorable circumstances where the tourniquet can be applied as a first aid measure and left in place until trained medical personnel remove it, the tourniquet may result in the loss of the limb. The tourniquet should be used only when all other measures have failed, and it is a life and death matter. To control bleeding by direct pressure on the wound, sufficient pressure must be exerted to stop the bleeding, and that pressure must be maintained long enough to "seal off" the bleeding surfaces. Alternate pressing and then releasing to see if the wound is still bleeding is not desirable. It is best to apply the pressure and keep it in place for up to 20 minutes. Oozing blood from a wound of an extremity can be slowed or stopped by elevating the wound above the level of the heart.

(2) The next concern is infection. In survival or captivity, consider all breaks in the skin due to mechanical trauma contaminated, and treat appropriately. Even superficial scratches should be cleaned with soap and water and treated with antiseptics, if available. Antiseptics should generally not be used in wounds which go beneath the skin's surface since they may produce tissue damage which will delay healing. Open wounds must be thoroughly cleansed with boiled water. Bits of debris such as clothing, plant materials, etc., should be rinsed out of wounds by pouring large amounts of water into the wounds and ensuring that even the deepest parts are clean. In a fresh wound where bleeding has been a problem, care must be taken not to irrigate so vigorously that clots are washed away and the bleeding resumes. Allow a period of an hour or so after the bleeding has been stopped before beginning irrigation with the boiled water. Begin gently at first, removing unhealthy tissue, increasing the vigor of the irrigation over a period of time. If the wound must be cleaned, use great care to avoid doing additional damage to the wound. The

wound should be left open to promote cleansing and drainage of infection. In captivity, frequently deep open wounds will become infested with maggots. The natural tendency is to remove these maggots, but actually, they do a good job of cleansing a wound by removing dead tissue. Maggots may, however, damage healthy tissue when the dead tissue is removed. So the maggots should be removed if they start to affect healthy tissue. Remember that it is imperative that the wound be left open and allowed to drain.

(3) An open wound will heal by a process known as secondary intention or granulation. During the healing phase, the wound should be kept as clean and dry as possible. For protection, the wound may be covered with clean dressings to absorb the drainage and to prevent additional trauma to the wound. These dressings may be loosely held in place with bandages (clean parachute material may be used for dressings and bandages). The bandages should not be tight enough to close the wound or to impair circulation. At the time of dressing change, boiled water may be used to gently rinse the wound. The wound may then be air dried and a clean dressing applied. (The old dressing may be boiled, dried, and reused.) Nutritional status is interrelated with the healing process, and it is important to consume all foods available to provide the best possible opportunity for healing.

c. Conclusions. Obviously the PW is at a distinct disadvantage in treating wounds, lacerations, and infections without modem medicine. Yet, knowledge of the basic principles previously mentioned, locally available equipment and resources, and optimism and common sense can help a survivor to maintain life.

8-13. Fractures and Sprains. Fractures and sprains often occurred during shootdown and(or) egress from the aircraft. They also occurred during evasion attempts.

a. Captor Treatment. As with other treatment, treatment of sprains and fractures varied considerably depending on the severity of the injury and the resources available for treatment. Even after immediate treatment or surgical procedures, there was little follow-up therapy. Prisoners were usually returned to camp to care for themselves or to rely on the help of fellow prisoners.

b. Captive Treatment. Captive therapy was primarily that of helping each other to exercise or immobilize the injured area, and in severe cases, to provide nursing care.

c. Comments:

(1) An acute nonpenetrating injury to a muscle or joint can best

be managed by applying cold as soon as possible after the injury. Icepacks or cold compresses should be used intermittently for up to 48 hours following the injury. This will minimize hemorrhage and disability. Be careful not to use snow or ice to the point where frostbite or cold injury occurs. As the injured part becomes numb, the ice should be removed to permit rewarming of the tissues. Then the ice can be reapplied. Following a period of 48 to 72 hours, the cold treatment can be replaced by warm packs to the affected part. A "sprain or strain" may involve a wide variety of damage ranging from a simple bruise to deep hemorrhage or actual tearing of muscle fibers, ligaments, or tendons. While it is difficult to establish specific guidelines for treatment in the absence of a specific diagnosis, in general, injuries of this type require some period of rest (immobilization) to allow healing. The period of rest is followed by a period of rehabilitation (massage and exercise) to restore function. For what appears to be a simple superficial muscle problem, a period of 5 to 10 days rest followed by a gradual progressive increase in exercise is desirable. Pain should be a limiting factor. If exercise produces significant pain, the exercise program should be reduced or discontinued. In captivity, it is probably safest to treat severe injuries to a major joint like a fracture with immobilization (splint, cast) for a period of 4 to 6 weeks before beginning movement of the joint.

(2) Bone fractures are of two general types, open and closed. The open fracture is associated with a break in the skin over the fracture site which may range all the way from a broken bone protruding through the skin to a simple puncture from a bone splinter. The general goals of fracture management are: restore the fracture to a functional alignment; immobilize the fracture to permit healing of the bone; and rehabilitation. Restoring or reducing the fracture simply means realigning the pieces of bones, putting the broken ends together as close to the original position as possible. The natural ability of the body to heal a broken bone is remarkable and it is not necessary that an extremity fracture be completely straight for satisfactory healing to occur. In general, however, it is better if the broken bone ends are approximated so that they do not override. Fractures are almost always associated with muscle spasms which become stronger with time. The force of these muscle spasms tends to cause the ends of the broken bones to override one another, so the fracture should be reduced as soon as possible. To overcome the muscle spasm, force

must be exerted to reestablish the length of the extremity. Once the ends of the bone are realigned, the force of the muscle spasm tends to hold the bones together. At this point, closed fractures are ready to be immobilized, but open fractures require treatment of the soft tissue injury in the manner outlined earlier. In other words, the wound must be cleansed and dressed, then the extremity should be immobilized. The immobilization preserves the alignment of the fracture and prevents movement of the fractured parts which would delay healing. For fractures of long bones of the body, it becomes important to immobilize the joints above and below the fracture site to prevent movement of the bone ends. In a fracture of the mid forearm, for example, both the wrist and the elbow should be immobilized. In immobilizing a joint, it should be fixed in a "neutral" or functional position. That is, neither completely straight nor completely flexed or bent, but in a position about midway between. In splinting a finger, for example, the finger should be curved to about the same position the finger would naturally assume at rest.

(3) A splint of any rigid material such as boards, branches, bamboo, metal boot insoles, or even tightly rolled newspaper may be almost as effective as plaster or mud casts. In conditions such as continuous exposure to wetness, the splint can be cared for more effectively than the plaster or mud cast. In cases where there is a soft tissue wound in close proximity to the fracture, the splint method of immobilization is more desirable than a closed cast because it permits change of dressing, cleaning, and monitoring of the soft tissue injury. The fracture site should be loosely wrapped with parachute cloth or soft plant fibers; then the splints can be tied in place extending at least the entire length of the broken bone and preferably fashioned in such a way as to immobilize the joint above and below the fracture site. The splints should not be fastened so tightly to the extremity that circulation is impaired. Since swelling is likely to occur, the bindings of the splint will have to be loosened periodically to prevent the shutting off of the blood supply.

(4) The time required for immobilization to ensure complete healing is very difficult to estimate. In captivity, it must be assumed that healing time will, in general, be prolonged. This means that for a fracture of the upper extremity of a "nonweight-bearing bone," immobilization might have to be maintained for 8 weeks or more to ensure complete healing. For a fracture of the lower extremity or a

"weight-bearing bone," it might require 10 or more weeks of immobilization.

(5) Following the period of immobilization and fracture healing, a program of rehabilitation is required to restore normal functioning. Muscle tone must be reestablished and the range of motion of immobilized joints must be restored. In cases where joints have been immobilized, the rehabilitation program should be started with "passive range of motion exercises." This means moving the joint through a range of motion without using the muscles which are normally used to move that joint. For example, if the left wrist has been immobilized, a person would begin the rehabilitation program by using the right hand to passively move the left wrist through a range of motion which can be tolerated without pain. When some freedom of motion of the joint has been achieved, the individual should begin actively increasing that range of motion using the muscles of the joint involved. Do not be overly forceful in the exercise program—use pain as a guideline—the exercise should not produce more than minimal discomfort. Over a period of time, the joint movement should get progressively greater until the full range of motion is restored. Also, exercises should be started to restore the tone and strength of muscles which have been immobilized. Again, pain should be the limiting point of the program and progression should not be so rapid as to produce more than a minimal amount of discomfort.

8-14. Summary. Common sense and basic understanding of the type of injuries are most helpful in avoiding complication and debilitation. Adequate nourishment and maintenance of physical condition will materially assist healing of burns, fractures, lacerations, and other injuries—the body will repair itself.

8-15. Conclusions. In the management of trauma and burns in captivity or survival, remember that the body will do the healing or repair, and the purpose of the "treater" is to provide the body with the best possible atmosphere to conduct that self-repair. Some general principles are:

a. Be in the best possible physical, emotional, and nutritional status before being exposed to the potential survival or captivity setting.

b. Minimize the risk of injury at the time of survival or captivity by following appropriate safety procedures and properly using protective equipment.

c. Maintain the best possible nutritional status while in captivity or the survival setting.

d. Don't overtreat! Overly vigorous treatment can do more harm than good.

e. Use cold applications for relief of pain and to minimize disability from burns and soft tissue strains or sprains.

f. Clean all wounds by gentle irrigation with large amounts of the cleanest water available.

g. Leave wounds open.

h. Splint fractures in a functional position.

i. After the bone has healed, begin an exercise program to restore function.

j. Remember that even improperly healed wounds or fractures may be improved by cosmetic or rehabilitative surgery and treatment upon rescue or repatriation.

Part Four

PERSONAL PROTECTION

Chapter 9

PROPER BODY TEMPERATURE

9-1. Introduction. In a survival situation the two key requirements for personal protection are maintenance of proper body temperature and prevention of injury. The means for providing personal protection are many and varied. They include the following general categories: clothing, shelter, equipment, and fire. These individual items are not necessary for survival in every situation; however, all four will be essential in some environments. In this part of the regulation, the conditions which affect the body temperature, the physical principles of heat transfer, and the methods of coping with these conditions will be covered.

9-2. Body Temperature. The body functions best when core temperatures range from 96°F to 102°F. Preventing too much heat loss or gain should be a primary concern for survivors. Factors causing changes in body core temperature (excluding illness) are the climatic conditions of temperature, wind, and moisture.

a. Temperature. As a general rule, exposure to extreme temperatures can result in substantial decreases in physical efficiency. In the worst case, incapacitation and death can result.

b. Wind. Wind increases the chill effect (Figure 9-1), causes dissipation of heat, and accelerates loss of body moisture.

c. Moisture. Precipitation, Ground Moisture, or Immersion. Water provides an extremely effective way to transfer heat to and from the body. When a person is hot, the whole body may be immersed in a stream or other body of water to be cooled. On the other hand, in the winter, a hot bath can be used to warm the body. When water is around the body, it tends to bring the "body" to the temperature of the liquid. An example is when a hand is burned and then placed in cold water to dissipate the heat. One way to lower body temperature is by applying water to clothing and exposing the clothed body to the wind. This action causes the heat to leave the body 25 times faster than when wearing dry clothing. This rapid heat transfer is the reason survivors must always guard against getting wet in cold environments. Consider the result of a body totally submerged in water at a temperature of 50°F and determine how long a person could survive (Figures 9-2 and 9-3).

9-3. Heat Transfer. There are five ways body heat can be transferred. They are radiation, conduction, convection, evaporation, and respiration.

a. Radiation. Radiation is the primary cause of heat loss. It is defined as the transfer of heat waves from the body to the environment and(or) from the environment back to the body. For example, at a temperature of 50°F, 50 percent of the body's total heat loss can occur through an exposed head and neck. As the temperature drops, the situation gets worse. At 5°F, the loss can be 75 percent under the same circumstances. Not only can heat be lost from the head, but also from the other extremities of the body. The hands and feet radiate heat at a phenomenal rate due to the large number of capillaries present at the surface of the skin. These three areas

WIND SPEED		COOLING POWER OF WIND EXPRESSED AS "EQUIVALENT CHILL TEMPERATURE"																				
KNOTS	MPH	TEMPERATURE (°F)																				
Calm	Calm	40	35	30	25	20	15	10	5	0	-5	-10	-15	-20	-25	-30	-35	-40	-45	-50	-55	-60
		EQUIVALENT CHILL TEMPERATURE																				
3 - 6	5	35	30	25	20	15	10	5	0	-5	-10	-15	-20	-25	-30	-35	-40	-45	-50	-55	-65	-70
7 - 10	10	30	20	15	10	5	0	-10	-15	-20	-25	-35	-40	-45	-50	-60	-65	-70	-75	-80	-90	-95
11 - 15	15	25	15	10	0	-5	-10	-20	-25	-30	-40	-45	-50	-60	-65	-70	-80	-85	-90	-100	-105	-110
16 - 19	20	20	10	5	0	-10	-15	-25	-30	-35	-45	-50	-60	-65	-75	-80	-85	-95	-100	-110	-115	-120
20 - 23	25	15	10	0	-5	-15	-20	-30	-35	-45	-50	-60	-65	-75	-80	-90	-95	-105	-110	-120	-125	-135
24 - 28	30	10	5	0	-10	-20	-25	-30	-40	-50	-55	-65	-70	-80	-85	-95	-100	-110	-115	-125	-130	-140
29 - 32	35	10	5	-5	-10	-20	-30	-35	-40	-50	-60	-65	-75	-80	-90	-100	-105	-115	-120	-130	-135	-145
33 - 36	40	10	0	-5	-15	-20	-30	-35	-45	-55	-60	-70	-75	-85	-95	-100	-110	-115	-125	-130	-140	-150
WINDS ABOVE 40 HAVE LITTLE ADDITIONAL EFFECT.		LITTLE DANGER					INCREASING DANGER (Flesh may freeze within 1 min)						GREAT DANGER (Flesh may freeze within 30 seconds)									
		DANGER OF FREEZING EXPOSED FLESH FOR PROPERLY CLOTHED PERSONS																				

INSTRUCTIONS

MEASURE LOCAL TEMPERATURE AND WIND SPEED IF POSSIBLE; IF NOT, ESTIMATE. ENTER TABLE AT CLOSEST 5°F INTERVAL ALONG THE TOP AND WITH APPROPRIATE WIND SPEED ALONG LEFT SIDE. INTERSECTION GIVES APPROXIMATE EQUIVALENT CHILL TEMPERATURE. THAT IS, THE TEMPERATURE THAT WOULD CAUSE THE SAME RATE OF COOLING UNDER CALM CONDITIONS.

NOTES

WIND

1. THIS TABLE WAS CONSTRUCTED USING MILES PER HOUR (MPH), HOWEVER, A SCALE GIVING THE EQUIVALENT RANGE IN KNOTS HAS BEEN INCLUDED ON THE CHART TO FACILITATE ITS USE WITH EITHER UNIT.
2. WIND MAY BE CALM BUT FREEZING DANGER GREAT IF PERSON IS EXPOSED IN MOVING VEHICLE, UNDER HELICOPTER ROTORS, IN PROPELLOR BLAST, ETC. IT IS THE RATE OF RELATIVE AIR MOVEMENTS THAT COUNTS AND THE COOLING EFFECT IS THE SAME WHETHER YOU ARE MOVING THROUGH THE AIR OR IT IS BLOWING PAST YOU.
3. EFFECT OF WIND WILL BE LESS IF PERSON HAS EVEN SLIGHT PROTECTION FOR EXPOSED PARTS. LIGHT GLOVES ON HANDS , PARKA HOOD SHIELDING FACE , ETC.

ACTIVITY DANGER IS LESS IF SUBJECT IS ACTIVE. A PERSON PRODUCES ABOUT 100 WATTS (341 BTUs) OF HEAT STANDING STILL BUT UP TO 1000 WATTS (3413 BTUs) IN VIGOROUS ACTIVITY LIKE CROSS-COUNTRY SKIING.

PROPER USE OF CLOTHING AND **ADEQUATE DIET** ARE BOTH IMPORTANT.

COMMON SENSE THERE IS NO SUBSTITUTE FOR IT. THE TABLE SERVES ONLY AS A GUIDE TO THE COOLING EFFECT OF THE WIND ON BARE FLESH WHEN THE PERSON IS FIRST EXPOSED. GENERAL BODY COOLING AND MANY OTHER FACTORS AFFECT THE RISK OF FREEZING INJURY.

Figure 9-1. Windchill Chart.

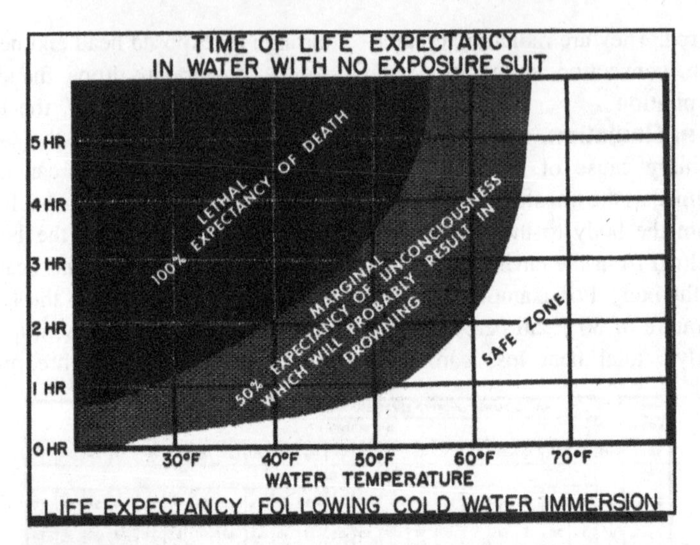

Figure 9-2. Life Expectancy Following Cold-Water Immersion.

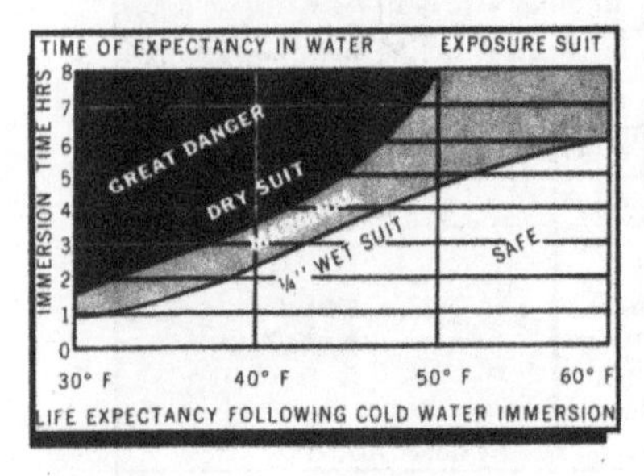

Figure 9-3. Life Expectancy Following Cold-Water Immersion (Exposure Suit.)

of the body must be given particular attention during all periods of exposure to temperature extremes.

b. Conduction:

(1) Conduction is defined as the movement of heat from one molecule to another molecule within a solid object. Extreme examples of how heat is lost and gained quickly are deep frostbite and third-degree burns, both gained from touching the same piece of metal at opposite extremes of cold and heat. Heat is also lost from the body in this manner by touching objects in the cold with bare hands, by sitting on a cold log, or by kneeling on snow to build a shelter. These are practices which survivors should avoid since they can lead to overchilling the body.

(2) Especially dangerous is the handling of liquid fuel at low temperatures. Unlike water which freezes at 32°F, fuel exposed to the outside temperatures will reach the same temperature as the air. The temperature of the fuel may be 10°F to 30°F below zero or colder. Spilling the fluid on exposed skin will cause instant frostbite, not only from the conduction of heat by the cold fluid, but by the further

cooling effects of rapid evaporation of the liquid as it hits the skin.

c. Convection. Heat movement by means of air or wind to or from an object or body is known as convection. The human body is always warming a thin layer of air next to the skin by radiation and conduction. The temperature of this layer of air is nearly equal to that of the skin. The body stays warm when this layer of warm air remains close to the body. However, when this warm layer of air is removed by convection, the body cools down. A major function of clothing is to keep the warm layer of air close to the body; however, by removing or disturbing this warm air layer, wind can reduce body temperature. Therefore, wind can provide beneficial cooling in dry, hot conditions, or be a hazard in cold, wet conditions.

d. Evaporation. Evaporation is a process by which liquid changes into vapor, and during this process, heat within the liquid escapes to the environment. An example of this process is how a "desert water bag" works on the front of a jeep while driving in the hot desert. The wind created by the jeep helps to accelerate evaporation and causes the water in the bag to be cooled. The body also uses this method to regulate core temperature when it perspires and air circulates around the body. The evaporation method works any time the body perspires regardless of the climate. For this reason, it is essential that people wear fabrics that "breathe" in cold climates. If water vapor cannot evaporate through the clothing, it will condense, freeze, and reduce the insulation value of the clothing and cause the body temperature to go down.

e. Respiration. The respiration of air in the lungs is also a way of transferring heat. It works on the combined processes of convection, evaporation, and radiation. When breathing, the air inhaled is rarely the same temperature as the lungs. Consequently, heat is either inhaled or expelled with each breath. A person's breath can be seen in the cold as heat is lost to the outside. Because this method is so efficient at transferring heat, warm, moist oxygen is used to treat hypothermia patients in a clinical environment. Understanding how heat is transferred and the methods by which that transfer can be controlled can help survivors keep the body's core temperature in the 96°F to 102°F range. (Figure 9-4.)

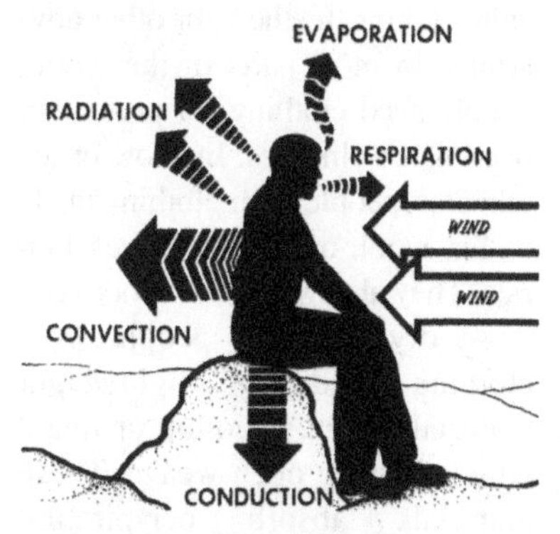

Figure 9-4. Heat Transfer.

Chapter 10

CLOTHING

10-1. Introduction. Every time people go outside they probably neglect to think about one of the most important survival-oriented assets—clothing. Clothing is often taken for granted; people tend to neglect those things which should be the most familiar to them. Clothing is an important asset to survivors and is the most immediate form of shelter. Clothing is important in staying alive, especially if food, water, shelter, and fire are limited or unobtainable. This is especially true in the first stages of an emergency situation because survivors must work to satisfy other needs. If survivors are not properly clothed, they may not survive long enough to build a fire or shelter, to find food, or to be rescued.

10-2. Protection:

People have worn clothing for protection since they first put on animal skins, feathers, or other coverings. In most parts of the world, people need clothing for protection from harsh climates. In snow or ice climates, people wear clothing made of fur, wool, or closely woven fabrics. They also wear warm footwear.

In dry climates, people wear clothing made of lightweight materials, such as cotton or linen, which have an open weave. These materials absorb perspiration and allow air to circulate around the body. People in dry climates sometimes wear white or light-colored clothes to reflect the sun's rays. They may also wear sandals, which are cooler and more comfortable than shoes. To protect the head and neck, people wear hats as sunshades.

Clothing also provides protection from physical injuries caused by vegetation, terrain features, and animal life which may cause bites, stings, and cuts.

10-3. Clothing Materials:

Clothing is made from a variety of materials such as nylon, wool, cotton, etc. The type of material used has a significant effect on protection. Potential survivors must be aware of both the environmental conditions and the effectiveness of these different materials in order to select the best type of clothing for a particular region.

Clothing materials include many natural and synthetic fibers. As material is woven together, a "dead air" space is created between the material fibers. When two or three layers of material are worn, a layer of air is trapped between each layer of material creating another layer of "dead air" or insulation. The ability of these different fibers to hold "dead air" is responsible for differing insulation values.

10-4. Natural Materials. They include fur, leather, and cloth made from plant and animal fibers.

a. Fur and leather are made into some of the warmest and most durable clothing. Fur is used mainly for coats and coat linings. Leather has to be treated to make it soft and flexible and to prevent it from rotting.

b. Wool is somewhat different because it contains natural lanolin oils. Although wool is somewhat absorbent, it retains most of its insulating qualities when wet.

c. Cotton is a common plant fiber widely used to manufacture clothing. It absorbs moisture quickly and, with heat radiated from the body, will allow the moisture to pass away from the body. It does not offer much insulation when wet. It's used as an inner layer against the skin and as an outer layer with insulation (for example, wool, Dacron pile, synthetic batting) sandwiched between. The cotton protects the insulation and, therefore, provides warmth.

10-5. Synthetic Materials. Clothing manufacturers are using more and more of these materials. Many synthetic materials are stronger, more shrink-resistant, and less expensive than natural materials. Most synthetic fibers are derived from petroleum in the form of long fibers which consist of different lengths, diameters, and strengths, and sometimes have hollow cores. These fibers, woven into materials such as nylon, Dacron, and polyester, make very strong long-lasting clothing, tarps, tents, etc. Some fibers are spun into a batting type material with air space between the fibers, providing excellent insulation used inside clothing.

a. Many fabrics are blends of natural and synthetic fibers. For example, fabrics could be a mixture of cotton and polyester or wool and nylon. Nylon covered with rubber is durable and waterproof but is also heavy. There are other coverings on nylon which are waterproof but somewhat lighter and less durable. However, most coated nylon has one drawback — it will not allow for the evaporation of perspiration. Therefore, individuals may have to change the design of the garment to permit adequate ventilation (for example, wearing the garment partially unzipped).

b. Synthetic fibers are generally lighter in weight than most natural materials and have much the same insulating qualities. They work well when partially wet and dry out easily; however, they generally do not compress as well as down.

10-6. Types of Insulation:

a. Natural:

(1) Down is the soft plumage found between the skin and the contour feathers of birds. Ducks and geese are good sources for down. If used as insulation in clothing, remember that down will

absorb moisture (either precipitation or perspiration) quite readily. Because of the light weight and compressibility of down, it has wide application in cold-weather clothing and equipment. It is one of the warmest natural materials available when kept clean and dry. It provides excellent protection in cold environments; however, if the down gets wet it tends to get lumpy and loses its insulating value.

(2) Cattail plants have a worldwide distribution, with the exception of the forested regions of the far north. The cattail is a marshland plant found along lakes, ponds, and the backwaters of rivers. The fuzz on the tops of the stalks forms dead-air spaces and makes a good down-like insulation when placed between two pieces of material.

(3) Leaves from deciduous trees (those that lose their leaves each autumn) also make good insulation. To create dead-air space, leaves should be placed between two layers of material.

(4) Grasses, mosses, and other natural materials can also be used as insulation when placed between two pieces of material.

b. Synthetic:

(1) Synthetic filaments such as polyesters and acrylics absorb very little water and dry quickly. Spun synthetic filament is lighter then an equal thickness of wool and, unlike down, does not collapse when wet. It is also an excellent replacement for down in clothing.

(2) The nylon material in a parachute insulates well if used in the layer system because of the dead-air space. Survivors must use caution when using the parachute in cold climates. Nylon may become "cold soaked"; that is, the nylon will take on the temperature of the surrounding air. People have been known to receive frostbite when placing cold nylon against bare skin.

10-7. Insulation Measurement:

a. The next area to be considered is how well these fibers insulate from the heat or cold. The most scientific way to consider the insulating value of these fibers is to use an established criterion. The commonly accepted measurement used is a comfort level of clothing, called a "CLo" factor.

b. The CLo factor is defined as the amount of insulation that maintains normal skin temperature when the outside ambient air temperature is 70 °F with a light breeze. However, the CLo factor alone is not sufficient to determine the amount of clothing required. Such variables as metabolic rate, wind conditions, and the physical makeup of the individual must be considered.

c. The body's rate of burning or metabolizing food to produce heat varies among individuals. Therefore, some may need more insulation than others even though food intake is equal, and consequently, the required CLo value must be increased. Physical activity also causes an increase in the metabolic rate and the rate of blood circulation through the body. When a person is physically active, less clothing or insulation is needed than when standing still or sitting. The effect of the wind, as shown on the windchill chart, must be considered (Figure 10-1). When the combination of temperature and wind drops the chill factor to minus 100°F or lower, the prescribed CLo for protecting the body may be inapplicable (over a long period of time) without relief from the wind. For example, when the temperature is minus 60 °F, the wind is blowing 60 to 70 miles per hour, and the resultant chill factor exceeds minus 150°F, clothing alone is inadequate to sustain life. Shelter is essential.

d. The physical build of a person also affects the amount of heat and cold that can be endured. For example, a very thin person will not be able to endure as low a temperature as one who has a layer of fat below the skin. Conversely, heavy people will not be able to endure extreme heat as effectively as thinner people.

e. In the Air Force clothing inventory, there are many items that fulfill the need for insulating the body. They are made of the different fibers previously mentioned, and when worn in layers, provide varying degrees of insulative CLo value. The following average zone temperature chart is a guide in determining the best combination of clothing to wear.

TEMPERATURE RANGE	CLo REQUIRED
86 to 68°F	1 - Lightweight
68 to 50°F	2 - Intermediate Weight
50 to 32°F	3 - Intermediate Weight
32 to 14°F	3.5 - Heavyweight
14 to –4°F	4.0 - Heavyweight
–4 to –40°F	4.0 - Heavyweight

The amount of CLo value per layer of fabric is determined by the loft (distance between the inner and outer surfaces) and the amount of dead air held within the fabric. Some examples of the CLo factors and some items of clothing are:

LAYERS:		
	1 - Aramid underwear (1 layer)	0.6 CLo
	2 - Aramid underwear (2 layers)	1.5 CLo
	3 - Quilted liners	1.9 CLo
	4 - Nomex coveralls	.6 CLo
	5 - Winter coveralls	1.2 CLo
	6 - Nomex jacket	1.9 CLo

This total amount of insulation should keep the average person warm at a low temperature. When comparing items 1 and 2 in the above example, it shows when doubling the layer of underwear, the CLo value more than doubles. This is true not only for item 1 but between all layers of any clothing system. Therefore, one gains added protection by using several very thin layers of insulation rather than two thick layers. The air held between these thin layers increases the insulation value.

The use of many thin layers also provides (through removal of desired number of layers) the ability to closely regulate the amount of heat retained inside the clothing. The ability to regulate body temperature helps to alleviate the problem of overheating and sweating, and preserves the effectiveness of the insulation.

The principle of using many thin layers of clothing can also be applied to the "sleeping system" (sleeping bag, liner, and bed). This system uses many layers of synthetic material, one inside the other, to form the amount of dead air needed to keep warm. To improve this system, a survivor should wear clean and dry clothing in layers (the layer system) in cold climates. While discussing the layer system, it is important to define the COLDER principle. This acronym is used to aid in remembering how to use and take care of clothing.

C - Keep clothing *C*lean.
O - Avoid *O*verheating.
L - Wear clothing *L*oose and in *L*ayers.
D - Keep clothing *D*ry.
E - *E*xamine clothing for defects or wear.
R - Keep clothing *R*epaired.

(1) Clean. Dirt and other materials inside fabrics will cause the insulation to be ineffective, abrade and cut the fibers which make up the fabric, and cause holes. Washing clothing in the field may be impractical; therefore, survivors should concentrate on using proper techniques to prevent soiling clothing.

(2) Overheating. Clothing best serves the purpose of preserving body heat when worn in layers as follows: absorbent material next to the body, insulating layers, and outer garments to protect against wind and rain. Because of the rapid change in temperature, wind, and physical exertion, garments should allow donning and removal quickly and easily. Ventilation is essential when working because enclosing the body in an airtight layer system results in perspiration which wets clothing, thus reducing its insulating qualities.

(3) Loose. Garments should be loose fitting to avoid reducing blood circulation and restricting body movement. Additionally, the garment should overhang the waist, wrists, ankles, and neck to reduce body heat loss.

(4) Dry. Keep clothing dry since a small amount of moisture in the insulation fibers will cause heat losses up to 25 times faster than dry clothing. Internally produced moisture is as damaging as is externally dampened clothing. The outer layer should protect the inner layers from moisture as well as from abrasion of fibers; for example, wool rubbing on logs or rocks, etc. The outer shell keeps dirt and other contaminants out of

the clothing. Clothing can be dried in many ways. Fires are often used; however, take care to avoid burning the items. The "bare hand" test is very effective. Place one hand near the fire in the approximate place the wet items will be and count to three slowly. If this can be done without feeling excessive heat, it should be safe to dry items there. Never leave any item unattended while it is drying. Leather boots, gloves, and mitten shells require extreme care to prevent shrinkage, stiffening, and cracking. The best way to dry boots is upright beside the fire (not upside down on sticks because the moisture does not escape the boot) or simply walk them dry in the milder climates. The sun and wind can be used to dry clothing with little supervision except for checking occasionally on the incoming weather and to make sure the article is secure. Freeze-drying is used in subzero temperatures with great success. Survivors let water freeze on or inside the item and then shake, bend, or beat it to cause the ice particles to fall free from the material. Tightly woven materials work better with this method than do open fibers.

(5) Examine. All clothing items should be inspected regularly for signs of damage or soil.

(6) Repair. Eskimos set an excellent example in the meticulous care they provide for their clothing. When damage is detected, immediately repair it.

f. The neck, head, hands, armpits, groin, and feet lose more heat than other parts of the body and require greater protection. Work with infrared film shows tremendous heat loss in those areas when not properly clothed. Survivors in a cold environment are in a real emergency situation without proper clothing. figure 10-1 shows some examples of how military clothing works to hold body heat.

g. Models wearing samples of aircrew attire appear as spectral figures in a thermogram, an image revealing differences in infrared heat radiated from their clothing and exposed skin. White is warmest; red, yellow, green, blue, and magenta form a declining temperature scale spanning about 15 degrees; while black represents all lower temperatures. Almost the entire scale is seen on the model in boxer shorts. Warm, white spots appear on the underarm and neck. Only the shorts block radiation from the groin. Temperatures cool along the arm to dark blue fingertips far from the heat-producing torso. The addition of the next layer of clothing (Aramid long underwear) prevents heat loss except where it is tight against the body. As more layers are added, it is easy to see the areas of greatest concern are the head, hands, and feet. These areas are difficult for crewmembers to properly insulate while flying an aircraft. Mittens are ineffective due to the degraded manual dexterity. Likewise, it is dif-

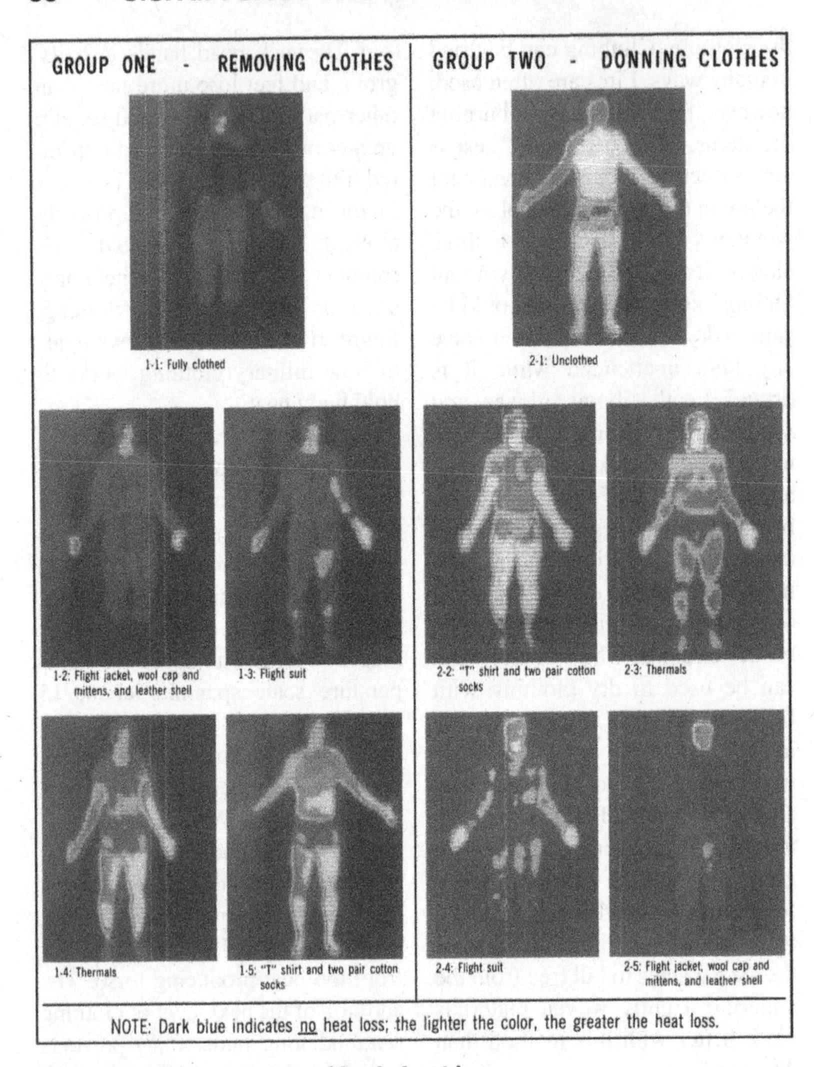

Figure 10-1. Thermogram of body heat loss.

ficult to feel the rudder pedal action while wearing bulky warm boots. These problems require inclusion of warm hats, mittens, and footgear (mukluk type) in survival kits during cold weather operation. Research has shown when a CLo value of 10 is used to insulate the head, hands, and feet and the rest of the body is only protected by one CLo, the average individual can be exposed to low temperatures (–10°F) comfortably for a reasonable period of time (30 to 40 minutes). When the amount of CLo value placed on the individual is reversed, the amount of time a survivor can spend in cold weather is greatly reduced due to the heat loss from their extremities. This same principle works in reverse in hot parts of the world if one submerges the head, hands, or feet in cold water, it lets the most vascular parts of the body lose heat quickly.

10-8. Clothing Wear in Snow and Ice Areas:

a. The survivor should:

(1) Avoid restricting the circulation. Clothing should not be worn so tight that it restricts the flow of blood that distributes the body heat and helps prevent frostbite. When wearing more than one pair of socks or gloves, ensure that each succeeding pair is large enough to fit comfortably over the other. Don't wear three or four pairs of socks in a shoe fitted for only one or two pairs. Release any restriction caused by twisted clothing or a tight parachute harness.

(2) Keep the head and ears covered. Survivors will lose as much as 50 percent of their total body heat from an unprotected head at 50°F.

(3) When exerting the body, prevent perspiration by opening clothing at the neck and wrists and loosening it at the waist. If the body is still warm, comfort can be obtained by taking off outer layers of clothing, one layer at a time. When work stops, the individual should put the clothing on again to prevent chilling.

(4) If boots are big enough, use dry grass, moss, or other material for added insulation around the feet. Footgear can be improvised by wrapping parachute cloth or other fabric lined with dry grass or moss for insulation.

b. Felt booties and mukluks with the proper socks and insoles are best for dry, cold weather. Rubber-bottomed boot shoepacs with leather tops are best for wet weather. Mukluks should not be worn in wet weather. The vapor-barrier rubber boots can be worn under both conditions and are best at extremely low temperatures. The air release valve should be closed at ground level. These valves are designed to release pressure when airborne. Air should not be blown into the valves as the moisture could decrease insulation.

c. Clothing should be kept as dry as possible. Snow must be brushed from clothing before entering a shelter or going near a fire. The survivors should beat the frost out of garments before warming them and dry them on a rack near a fire. Socks should be dried thoroughly.

d. One or two pairs of wool gloves and(or) mittens should be worn inside a waterproof shell (Figure 10-2). If survivors have to expose their hands, they should warm them inside their clothing.

e. To help prevent sun or snow blindness, a survivor should wear

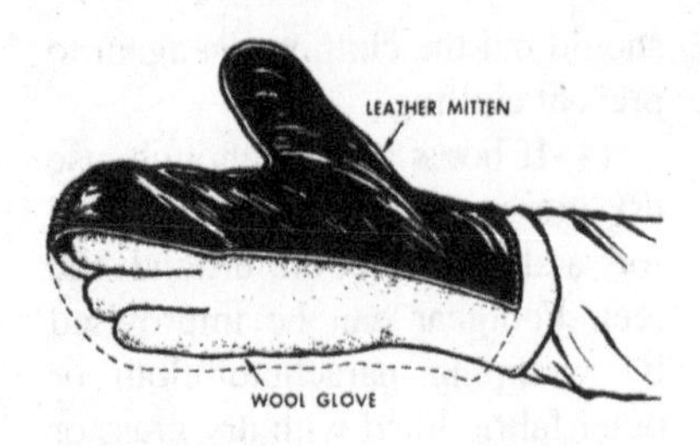

Figure 10-2. Layer system for hand.

sun or snow goggles or improvise a shield with a small horizontal slit opening (Figure 10-3).

f. In strong wind or extreme cold, as a last resort, a survivor should wrap up in parachute material, if available, and get into some type of shelter or behind a windbreak. Extreme care should be taken with hard materials, such as synthetics, as they may become cold soaked and require more time to warm.

g. At night, survivors should arrange dry spare clothing loosely around and under the shoulders and hips to help keep the body warm. Wet clothes should never be worn into the sleeping bag. The moisture destroys the insulation value of the bag.

h. If survivors fall into water, they should roll in dry snow to blot up moisture, brush off the snow, and roll again until most of the water is absorbed. They should not remove footwear until they are in a shelter or beside a fire.

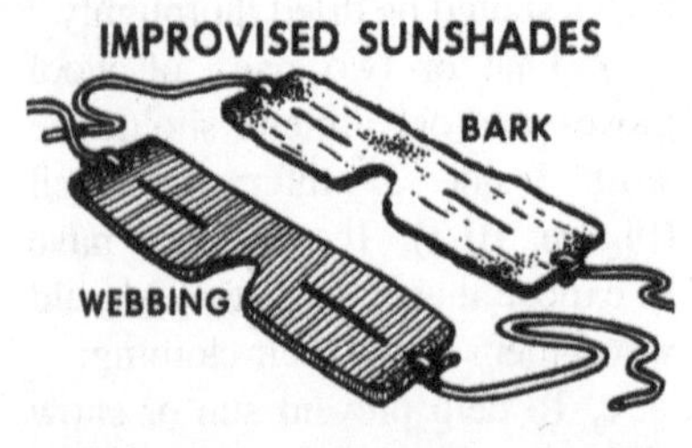

Figure 10-3. Improvised goggles.

i. All clothing made of wool offers good protection when used as an inner layer. When wool is used next to the face and neck, survivors should be cautioned that moisture from the breath will condense on the surface and cause the insulating value to decrease. The use of a wool scarf wrapped around the mouth and nose is an excellent way to prevent cold injury, but it needs to be de-iced on a regular basis to prevent freezing flesh adjacent to it. An extra shell is generally worn over the warming layers to protect them and to act as a windbreak.

j. Other headgear includes the pile cap and hood. These items are most effective when used with a covering for the face in extreme cold. The pile cap is extremely warm where it is insulated, but it offers little protection for the face and back of the neck.

k. The hood is designed to funnel the radiant heat rising from the rest of the body and to recycle it to keep the neck, head, and face warm (Figure 10-4). The individual's ability to tolerate cold should dictate the size of the front opening of the hood. The "tunnel" of a parka hood is usually lined with fur of some kind to act as a protecting device for the face. This same fur also helps to protect the hood from the moisture expelled during breathing. The

Figure 10-4. Proper wear of parka.

closed tunnel holds heat close to the face longer; the open one allows the heat to escape more freely. As the frost settles on the hair of the fur, it should be shaken from time to time to keep it free of ice buildup.

l. Sleeping systems (sleeping bag, liner, and bed) are the transition "clothing" used between normal daytime activities and sleep (Figure 10-5).

m. The insulating material in the sleeping bag may be synthetic or it may be down and feathers. (Feathers and down lining require extra protection from moisture.) However, the covering is nylon. Survivors must realize that sleeping bags are compressed when packed and must be fluffed before use to restore insulation value. Clean and dry socks, mittens, and other clothing can be used to provide additional insulation.

n. Footgear is critical in a survival situation because walking is the only means of mobility. Therefore, care of footgear is essential both before and during a survival situation. Recommendations for care are:

(1) Ensure footgear is properly "broken-in" before flying.

(2) "Treat" footgear to ensure water-repellency (follow manufacturer's recommendations).

(3) Keep leather boots as dry as possible.

o. Mukluks have been around for thousands of years and have proven their worth in extremely cold weather. The Air Force mukluks are made of cotton duck with rubber-cleated soles and heels (Figure 10-6). They have slide fasteners from instep to collar, laces at instep and collar, and are 18 inches high. They are used

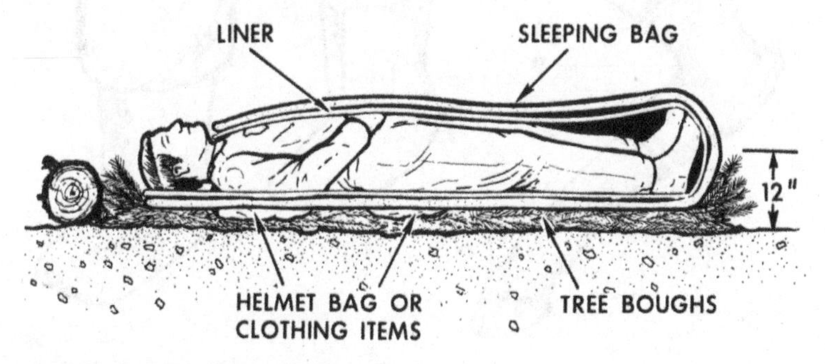

Figure 10-5. Sleeping system.

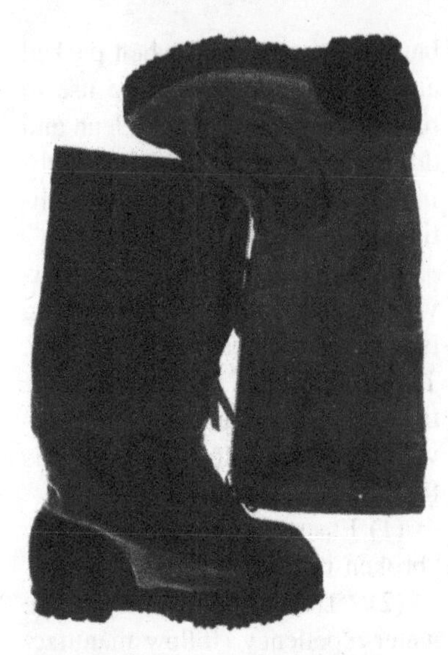

Figure 10-6. Issued mukluks.

by flying and ground personnel operating under dry, cold conditions in temperatures below +15°F. Survivors should change liners daily when possible.

10-9. Care of the Feet. Foot care is critical in a survival situation. Improvising foot gear may be essential to caring for feet.

a. Moose Hock Shoe. The hock skin of a moose or caribou will provide a suitable pair of shoes (Figure 10-7). Cut skin around leg at A and B. Separate from the leg and pull it over the hoof. Shape and sew up small end C. Slit skin from A to B; bore holes on each side of cut for lacing; turn inside out, and lace with rawhide, suspension line, or other suitable material.

b. Grass Insoles. Used extensively by northern natives to construct inner soles. Grass is a good insulator and will collect moisture from the feet. The survivor should use the following procedure to prepare grass for use as inner soles: Grasp a sheaf of tall grass, about one- half inch in diameter, with both hands. Rotate the hands in opposite directions. The grass will break up or "fluff'

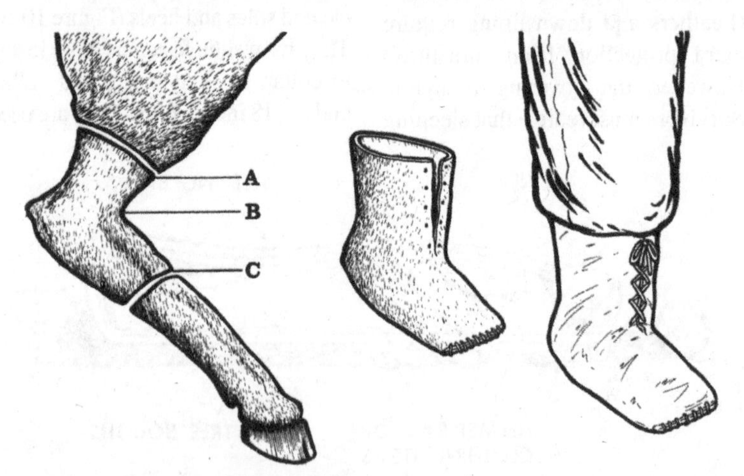

Figure 10-7. Moose Hock Shoes.

into a soft mass. Form this fluff into oblong shapes and spread it evenly throughout the shoes. The inner soles should be about an inch thick. Remove these inner soles at night and make new ones the following day.

c. Hudson Bay Duffel. A triangular piece of material used as a foot covering. To improvise this foot covering, a survivor can use the following procedures:

(1) Cut two to four layers of parachute cloth into a 30-inch square.

(2) Fold this square to form a triangle.

(3) Place the foot on this triangle with the toes pointing at one corner.

(4) Fold the front cover up over the toes.

(5) Fold the side corners, one at a time, over the instep. This completes the foot wrap. (Figure 10-8.)

d. Gaiters. Made from parachute cloth, webbing, or canvas. Gaiters help keep sand and snow out of shoes and protect the legs from bites and scratches (Figure 10-9).

e. Double Socks. Cushion padding, feathers, dry grass, or fur stuffed between layers of socks.

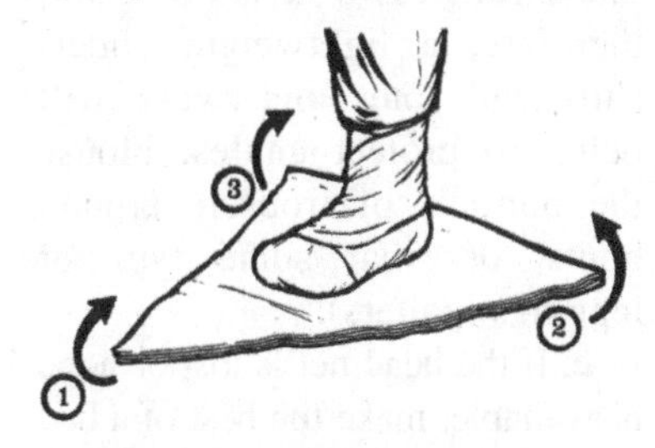

Figure 10-8. Hudson Bay Duffel.

Figure 10-9. Gaiters.

Wrap parachute or aircraft fabric around the feet and tie above the ankles. A combination of two or more types of improvised footwear may be more desirable and more efficient than any single type (Figure 10-10).

10.10. Clothing in the Summer Arctic:

a. In the summer arctic, there are clouds of mosquitoes and black flies so thick a person can scarcely see through them. Survivors can protect themselves by wearing proper clothing to ensure no bare skin is exposed. A good head net and gloves should be worn.

b. Head nets must stand out from the face so they won't touch the skin. Issued head nets are either black or green. If one needs to be improvised they can be sewn to the brim of the hat or can be attached with an elastic band that fits around the crown. Black is the best color, as it can be seen through more easily than green or white. A heavy tape encasing a drawstring should

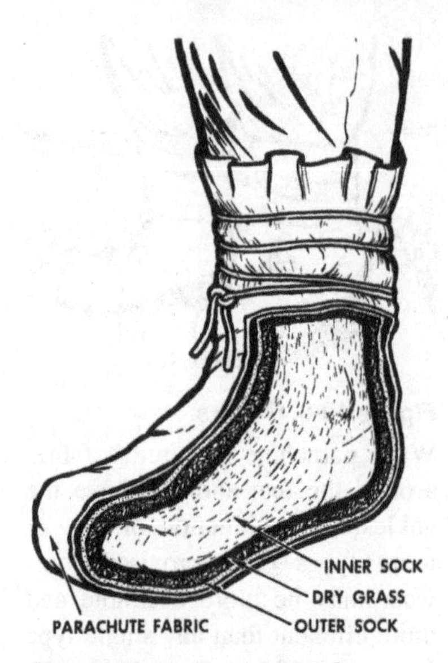

Figure 10-10. Double socks.

be attached to the bottom of the head net for tying snugly at the collar. Hoops of wire fastened on the inside will make the net stand out from the face and at the same time allow it to be packed flat. The larger they are, the better the ventilation. But very large nets will not be as effective in wooded country where they may become snagged on brush.

c. Gloves are hot, but are a necessity where flies are found in swamps. Kid gloves with a 6-inch gauntlet closing the gap at the wrist and ending with an elastic band halfway to the elbow are best. For fine work, kid gloves with the fingers cut off are good. Cotton/Nomex work gloves are better than no protection at all, but mosquitoes will bite through them. Treating

Figure 10-11. Insect protection.

the gloves with insect repellent will help. Smoky clothing may also help to keep insects away. (Figure 10-11.)

d. A survivor should remember that mosquitoes do not often bite through two layers of cloth; therefore, a lightweight undershirt and long underwear will help. To protect ankles, blouse the bottoms of trousers around boots, or wear some type of leggings (gaiters).

e. If the head net is lost or none is available, make the best of a bad situation by wearing sunglasses with improvised screened sides, plugging ears lightly with cotton, and tying a handkerchief around the neck. Treat clothing with insect repellent at night

10.11. Clothing at Sea. In cold oceans, survivors must try to stay dry and keep warm. If wet, they should use a wind screen to decrease the cooling effects of the wind. They should also remove, wring out, and replace outer garments or change into dry clothing. Hats, socks, and gloves should also be dried. If any survivors are dry, they should share extra clothes with those who are wet. Wet personnel should be given the most sheltered positions in the raft. Let them warm their hands and feet against those who are dry. Survivors should put on any extra clothing available. If no anti-exposure suits are provided, they can drape extra clothing around their shoulders and over their heads. Clothes should be loose and comfortable. Also, survivors should attempt to keep the floor of the raft dry. For insulation, covering the floor with any available material will help. Survivors should huddle together on the floor of the raft and spread extra tarpaulin, sail, or parachute material over the group. If in a 20- or 25-man raft, canopy sides can be lowered. Performing mild exercises to restore circulation may be helpful. Survivors should exercise fingers, toes, shoulders, and buttock muscles. Mild exercise will help keep the body warm, stave off muscle spasms, and possibly prevent medical problems. Survivors should warm hands under armpits and periodically raise feet slightly and hold them up for a minute or two. They should also move face muscles frequently to prevent frostbite. Shivering is the body's way of quickly generating heat and is considered normal. However, persistent shivering may lead to uncontrollable muscle spasms. They can be avoided by exercising muscles. If water is available, additional rations should be given to those suffering from exposure to cold. Survivors should eat small amounts frequently rather than one large meal.

10.12. Antiexposure Garments:

a. Assemblies. The antiexposure assemblies, both quick donning and constant wear, are designed for personnel participating in over-water flights where unprotected or prolonged exposure to the climatic conditions of cold air and(or) cold water (as a result of ditching or abandoning an aircraft) would be dangerous or could prove fatal. The suit provides protection from the wind and insulation against the chill of the ocean. The result of exposure in the water is illustrated in figures 10-2 and 10-3. Exposure time varies depending on the particular antiexposure assembly worn, the cold sensitiveness

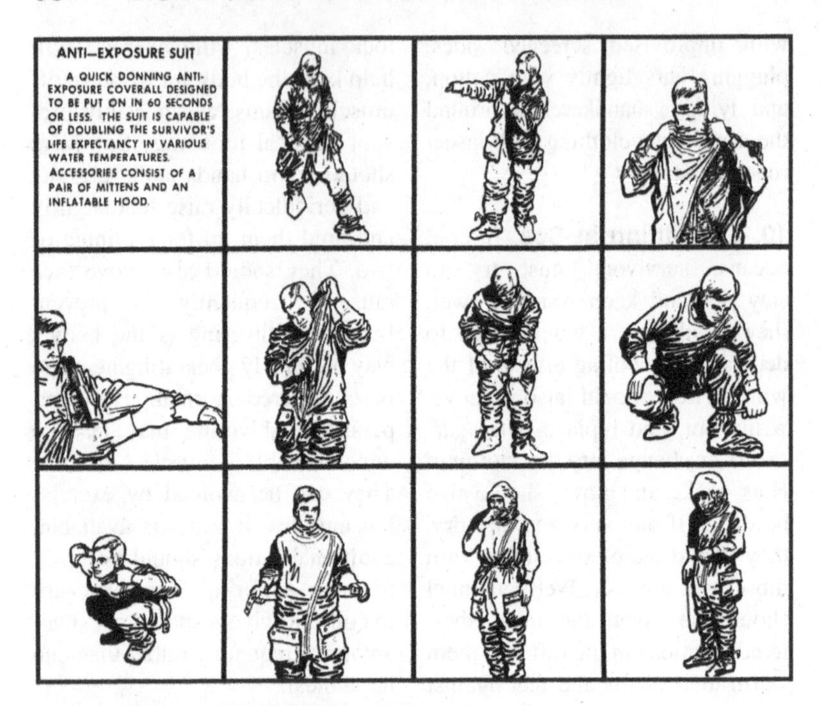

Figure 10-12. Donning Antiexposure Suit.

of the person, and survival procedures used.

b. Quick-Donning Antiexposure Flying Coverall.

Some antiexposure coveralls are designed for quick donning (approximately 1 minute) before emergency ditching. After ditching the aircraft, the coverall protects the wearer from exposure while swimming in cold water, and from exposure to wind, spray, and rain when adrift in a liferaft.

(1) The coverall is a one-size garment made from chloroprene-coated nylon cloth. It has two expandable- type patch pockets, an adjustable waist belt, and attached boots with adjustable ankle straps. One pair of insulated, adjustable wrist strap mittens, each with a strap attached to a pocket, is provided. A hood, also attached with a strap, is in the left pocket. A carrying case with instructions and a snap fastener closure is furnished for stowing in the aircraft.

(2) To use the coverall, personnel should wear it over regular flight clothing. It is large enough to wear over the usual flight gear. The gloves and hood are stowed in the pockets of the coverall and are normally worn after boarding the liferaft.

(3) The survivor should be extremely careful when donning the coverall to prevent damage by snagging, tearing, or puncturing it on projecting objects. After donning the coverall, the waist band and boot ankle straps should be adjusted to take up fullness. If possible, crewmembers should stoop while pulling the neck seal to expel air trapped in the suit. When jumping into the water, they should leap feet first with hands and arms close to sides or brought together above the head (Figure 10-12). Note there is a constant wear exposure suit designed to be worn continuously during overwater flights where the water temperature is 60° or below. The Command may waiver it to 51°.

10.13. Warm Oceans. Protection against the sun and securing drinking water are the most important problems. A survivor should keep the body covered as much as possible to avoid sunburn. A sunshade can be improvised out of any materials available or the canopy provided with the raft may be used. If the heat becomes too intense, survivors may dampen clothing with sea water to promote evaporation and cooling. The use of sunburn preventive cream or a lip balm is advisable. Remember, the body must be kept covered completely. Exposure to the sun increases thirst, wastes precious water, reduces the body's water content, and causes serious burns. Survivors should roll down their sleeves, pull up their socks, close their collars, wear a hat or improvised headgear, use a piece of cloth as a shield for the back of the neck, and wear sunglasses or improvise eye covers.

10.14. Tropical Climates:

a. In tropical areas, the body should be kept covered for prevention of insect bites, scratches, and sunburn.

b. When moving through vegetation, survivors should roll down their sleeves, wear gloves, and blouse the legs of their pants or tie them over their boot tops. Improvised puttees (gaiters) can be made from parachute material or any available fabric. This will protect legs from ticks and leeches.

c. Loosely worn clothing will keep survivors cooler, especially when subjected to the direct rays of the sun.

d. Survivors should wear a head net or tie material around the head for protection against insects. The most active time for insects is at dawn and dusk. An insect repellent should be used at these times.

e. In open country or in high grass, survivors should wear a neck cloth or improvised head covering for protection from sunburn and(or) dust. They should also move carefully through tall grass, as some sharp-edged grasses can cut clothing to shreds. Survivors should dry clothing before nightfall. If an extra change of clothing is available, effort should be made to keep it clean and dry.

10.15. Dry Climates:

a. In the dry climates of the world, clothing will be needed for protection against sunburn, heat, sand, and insects. Survivors should not discard any clothing. They should keep their head and body covered and blouse the legs of pants over the tops of footwear during the day. Survivors should not roll up sleeves, but keep them rolled down and loose at the cuff to stay cool.

Figure 10-13. Protective desert clothing.

b. Survivors should keep in mind that the people who live in the hot dry areas of the world usually wear heavy white flowing robes which protect almost every inch of their bodies. The only areas open to the sun are the face and the eyes. This produces an area of higher humidity between the body and the clothing, which helps keep them cooler and conserves their perspiration (Figure 10-13). The white clothing also reflects the sunlight.

c. Survivors should wear a cloth neckpiece to cover the back of the neck and protect it from the sun. AT- shirt makes an excellent neck drape, with the extra material used as padding under the cap. If hats are not available, survivors can make headpieces like those worn by the Arabs, as shown in figure 10.13. During dust storms, they should wear a covering for the mouth and nose; parachute cloth will work,

d. If shoes are lost or if they wear out, survivors can improvise footgear. One example of this is the "Russian Sock." Parachute material can be used to improvise these socks. The parachute material is cut into strips approximately 2 feet long and 4 inches wide. These strips are wrapped bandage fashion around the feet and ankles. Socks made in this fashion will provide comfort and protection for the feet.

Chapter 11
SHELTER

11-1. Introduction. Shelter is anything that protects a survivor from the environmental hazards. The information in this chapter describes how the environment influences shelter site selection and factors which survivors must consider before constructing an adequate shelter. The techniques and procedures for constructing shelters for various types of protection are also presented.

11-2. Shelter Considerations. The location and type of shelter built by survivors vary with each survival situation. There are many things to consider when picking a site. Survivors should consider the time and energy required to establish an adequate camp, weather conditions, life forms (human, plant, and animal), terrain, and time of day. Every effort should be made to use as little energy as possible and yet attain maximum protection from the environment.

a.Time. Late afternoon is not the best time to look for a site which will meet that day's shelter requirements. If survivors wait until the last minute, they may be forced to use poor materials in unfavorable conditions. They must constantly be thinking of ways to satisfy their needs for protection from environmental hazards.

b. Weather. Weather conditions are a key consideration when selecting a shelter site. Failure to consider the weather could have disastrous results. Some major weather factors which can influence the survivor's choice of shelter type and site selection are temperature, wind, and precipitation.

(1).Temperature. Temperatures can vary considerably within a given area. Situating a campsite in low areas such as a valley in cold regions can expose survivors to low night temperatures and windchill factors. Colder temperatures are found along valley floors which are sometimes referred to as "cold air sumps." It may be advantageous to situate campsites to take advantage of the sun. Survivors could place their shelters in open areas during the colder months for added warmth, and in shaded areas for protection from the sun during periods of hotter weather. In some areas a compromise may have to be made. For example, in many deserts the daytime temperatures can be very high while low temperatures at night can turn water to ice. Protection from both heat and cold are needed in these areas. Shelter type and location should be chosen to provide protection from the existing temperature conditions.

(2) Wind. Wind can be either an advantage or a disadvantage depending upon the temperature of the area and the velocity of the wind. During the summer or on warm days, survivors can take advantage of the cool breezes and protection the wind provides from insects by locating their camps on knolls or spits of land. Conversely, wind can become an annoyance or even a hazard as blowing sand, dust, or snow can cause skin and eye irritation and damage to clothing and equipment. On cold days or during winter months, survivors should seek shelter sites which are protected from the effects of windchill and drifting snow.

(3) Precipitation. The many forms of precipitation (rain, sleet, hail, or snow) can also present problems for survivors. Shelter sites should be out of major drainages and other low areas to provide protection from flash floods or mud slides resulting from heavy rains. Snow can also be a great danger if shelters are placed in potential avalanche areas.

c. Life Forms. All life forms (plant, human, and animal) must be considered when selecting the campsite and the type of shelter that will be used. The "human" factor may mean the enemy or other groups from whom survivors wish to remain undetected. Information regarding this aspect of shelters and shelter site selection is in Part Six of this regulation (Evasion). For a shelter to be adequate, certain factors must be considered, especially if extended survival is expected.

(1) Insect life can cause personal discomfort, disease, and injury. By locating shelters on knolls, ridges, or any other area that has a breeze or steady wind, survivors can reduce the number of flying insects in their area. Staying away from standing water sources will help to avoid mosquitoes, bees, wasps, and hornets. Ants can be a major problem; some species will vigorously defend their territories with painful stings or bites or particularly distressing pungent odors.

(2) Large and small animals can also be a problem, especially if the camp is situated near their trails or waterholes.

(3) Dead trees that are standing and trees with dead branches should be avoided. Wind may cause them to fall, causing injuries or death. Poisonous plants, such as poison oak or poison ivy, must also be avoided when locating a shelter.

d. Terrain. Terrain hazards may not be as apparent as weather and animal life hazards, but they can be many times more dangerous. Avalanche, rock, dry streambeds, or mud-slide areas should be avoided. These areas can be recognized by either a clear path or a path of secondary vegetation, such as 1- to 15-foot tall vegetation or other new growth which extends from the top to the bottom of a hill or mountain. Survivors should not choose shelter sites at the bottom of steep slopes which may be prone to slides. Likewise, there is a danger in camping at the bottom of steep scree or talus slopes. Additionally, rock overhang must be checked for safety before using it as a shelter.

11-3. Location:

a. Four prerequisites must be satisfied when selecting a shelter location.

(1) The first is being near water, food, fuel, and a signal or recovery site.

(2) The second is that the area be safe, providing natural protection from environmental hazards.

(3) The third is that sufficient materials be available to construct the shelter. In some cases, the "shelter" may already be present. Survivors seriously limit themselves if they assume shelters must be a fabricated framework having predetermined dimensions and a cover of parachute material or a signal paulin. More appropriately, survivors should consider using sheltered places already in existence in the immediate area. This does not rule out shelters with a fabricated framework and parachute or other manufactured material covering; it simply enlarges the scope of what can be used as a survival shelter.

(4) Finally, the area chosen must be both large enough and level enough for the survivor to lie down. Personal comfort is an important fundamental for survivors to consider. An adequate shelter provides physical and mental well-being for sound rest. Adequate rest is extremely vital if survivors are to make sound decisions. Their need for rest becomes more critical as time passes and rescue or return is delayed. Before actually constructing a shelter, survivors must determine the specific purpose of the shelter. The following factors influence the type of shelter to be fabricated.

(a) Rain or other precipitation.

(b) Cold.

(c) Heat.

(d) Insects.

(e) Available materials nearby (manufactured or natural).

(f) Length of expected stay.

(g) Enemy presence in the area—evasion "shelters" are covered in Part Six of the regulation (Evasion).

(h) Number and physical condition of survivors.

b. If possible, survivors should try to find a shelter which needs little work to be adequate. Using what is already there, so that complete construction of a shelter is not necessary, saves time and energy. For example, rock overhangs, caves, large crevices, fallen logs, root buttresses, or snow banks can all be modified to provide adequate shelter. Modifications may include adding snow blocks to finish off an existing tree well shelter, increasing the insulation of the shelter by using vegetation or parachute material, etc., or building a reflector fire in front of a rock overhang or cave. Survivors must consider the amount of energy required to build the shelter. It is not really wise to spend a great deal of time and energy in constructing a shelter if nature has provided a natural shelter nearby which will satisfy the survivor's needs. See figure 11-1 for examples of naturally occurring shelters.

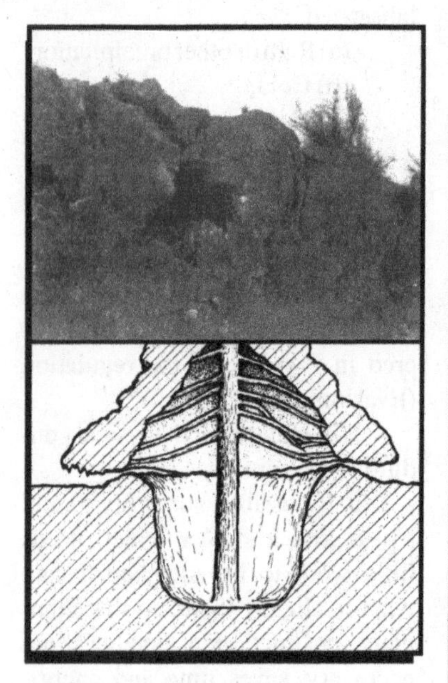

Figure 11-1. Natural Shelter.

c. The size limitations of a shelter are important only if there is either a lack of material on hand or if it is cold. Otherwise, the shelter should be large enough to be comfortable yet not so large as to cause an excessive amount of work. Any shelter, naturally occurring or otherwise, in which a fire is to be built must have a ventilation system which will provide fresh air and allow smoke and carbon monoxide to escape. Even if a fire does not produce visible smoke (such as heat tabs), the shelter must still be vented. See figure 11-2 for placement of ventilation holes in a snow cave. If a fire is to be placed outside the shelter, the opening of the shelter should be placed 90 degrees to the prevailing wind. This will

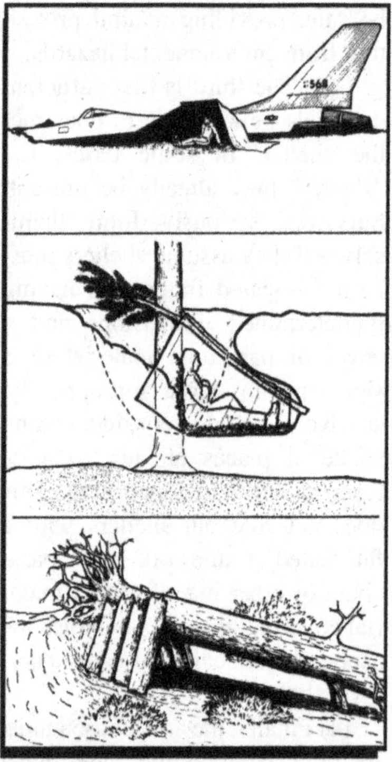

Figure 11-2. Immediate Action Shelters.

reduce the chances of sparks and smoke being blown into the shelter if the wind should reverse direction in the morning and evening. This frequently occurs in mountainous areas. The best fire to shelter distance is approximately 3 feet. One place where it would not be wise to build a fire is near the aircraft wreckage, especially if it is being used as a shelter. The possibility of igniting spilled lubricants or fuels is great. Survivors may decide instead to use materials from the aircraft to add to a shelter located a safe distance from the crash site.

11-4. Immediate Action Shelters. The first type of shelter that survivors may consider using, or the first type they may be forced to use, is an immediate action shelter. An immediate action shelter is one which can be erected quickly with minimum effort; for example, raft, aircraft parts, parachutes, paulin, and plastic bags. Natural formations can also shield survivors from the elements immediately, to include overhanging ledges, fallen logs, caves, and tree wells (Figure 11-2). It isn't necessary to be concerned with exact shelter dimensions. Survivors should remember that if shelter is needed, use an existing shelter if at all possible. They should improvise on natural shelters or construct new shelters only if necessary. Regardless of type, the shelter must provide whatever protection is needed and, with a little ingenuity, it should be possible for survivors to protect themselves and do so quickly. In many instances, the immediate action shelters may have to serve as permanent shelters for aircrew members. For example, many aircrew members fly without parachutes, large cutting implements (axes), and entrenching tools; therefore, multiperson liferafts may be the only immediate or long-term shelter available. In this situation, multiperson liferafts must be deployed in the quickest manner possible to ensure maximum advantages are attained from the following shelter principles:

a. Set up in areas which afford maximum protection from precipitation and wind and use the basic shelter principle in sections 11-2 and 11-3.

b. Anchor the raft for retention during high winds.

c. Use additional boughs, grasses, etc., for ground insulation.

11-5. Improvised Shelters. Shelters of this type should be easy to construct and(or) dismantle in a short period of time. However, these shelters usually require more time to construct than an immediate action shelter. For this reason, survivors should only consider this type of shelter when they aren't immediately concerned with getting out of the elements. Shelters of this type include the following:

a. The "A frame" design is adaptable to all environments as it can be easily modified; for example, tropical para-hammock, temperate area "A frame," arctic thermal "A frame," and fighter trench.

b. Simple shade shelter; these are useful in dry areas.

c. Various paratepees.

d. Snow shelters; includes tree-pit shelters.

e. All other variations of the above shelter types; sod shelters, etc.

11-6. Shelters for Warm Temperature Areas:

a. If survivors are to use parachute material, they should remember that "pitch and tightness" apply to shelters designed to shed rain or snow. Parachute material is porous and will not shed moisture unless it is stretched tightly at an angle of sufficient pitch which will encourage run-off instead of penetration. An angle of 40° to 60° is recommended for the "pitch" of the shelter. The material stretched over the framework should be wrinkle-free and tight. Survivors should not touch the material when water is running over it as this will break the surface tension at that point and allow water to drip into the shelter. Two layers of parachute material, 4 to 6 inches apart, will create a more effective water repellent covering. Even during hard rain, the outer layer only lets a mist penetrate if it is pulled tight. The inner layer will then channel off any moisture which may penetrate. This layering of parachute material also creates a dead-air space that covers the shelter. This is especially beneficial in cold areas when the shelter is enclosed. Adequate insulation can also be provided by boughs, aircraft parts, snow, etc. These will be discussed in more depth in the area of cold climate shelters. A double layering of parachute material helps to trap body heat, radiating heat from the earth's surface, and other heating sources.

b. The first step is deciding the type of shelter required. No matter which shelter is selected, the building or improvising process should be planned and orderly, following proven procedures and techniques. The second step is to select, collect, and prepare all materials needed before the actual construction; this includes framework, covering, bedding, or insulation, and implements used to secure the shelter ("deadmen," lines, stakes, etc.).

(1) For shelters that use a wooden framework, the poles or wood selected should have all the rough edges and stubs removed. Not only will this reduce the chances of the parachute fabric being ripped, but it will eliminate the chances of injury to survivors.

(2) On the outer side of a tree selected as natural shelter, some or all of the branches may be left in place as they will make a good support structure for the rest of the shelter parts.

(3) In addition to the parachute, there are many other materials

which can be used as framework coverings. Some of the following are both framework and covering all in one:

(a) Bark peeled off dead trees.

(b) Boughs cut off trees.

(c) Bamboo, palm, grasses, and other vegetation cut or woven into desired patterns.

(4) If parachute material is to be used alone or in combination with natural materials, it must be changed slightly. Survivors should remove all of the lines from the parachute and then cut it to size. This will eliminate bunching and wrinkling and reduce leakage.

c. The third step in the process of shelter construction is site preparation. This includes brushing away rocks and twigs from the sleeping area and cutting back overhanging vegetation.

d. The fourth step is to actually construct the shelter, beginning with the framework. The framework is very important. It must be strong enough to support the weight of the covering and precipitation buildup of snow. It must also be sturdy enough to resist strong wind gusts.

(1) Construct the framework in one of two ways. For natural shelters, branches may be securely placed against trees or other natural objects. For parachute shelters, poles may be lashed to trees or to other poles. The support poles or branches can then be layed and(or) attached depending on their function.

(2). The pitch of the shelter is determined by the framework. A 60-degree pitch is optimum for shedding precipitation and providing shelter room.

(3). The size of the shelter is controlled by the framework. The shelter should be large enough for survivors to sit up, with adequate room to lie down and to store all personal equipment.

(4) After the basic framework has been completed, survivors can apply and secure the framework covering. The care and techniques used to apply the covering will determine the effectiveness of the shelter in shedding precipitation.

(5) When using parachute material on shelters, survivors should remove all suspension line from the material. (Excess line can be used for lashing, sewing, etc.) Next, stretch the center seam tight; then work from the back of the shelter to the front, alternating sides and securing the material to stakes or framework by using buttons and lines. When stretching the material tight, survivors should pull the material 90 degrees to the wrinkles. If material is not stretched tight, any moisture will pool in the wrinkles and leak into the shelter.

(6) If natural materials are to be used for the covering, the shingle method should be used. Starting at the bottom and working toward the top of the shelter, the bottom

of each piece should overlap the top of the preceding piece. This will allow water to drain off. The material should be placed on the shelter in sufficient quantity so that survivors in the shelter cannot see through it.

11-7. Maintenance and Improvements. Once a shelter is constructed, it must be maintained. Additional modifications may make the shelter more effective and comfortable. Indian lacing (lacing the front of the shelter to the bipod) will tighten the shelter. A door may help block the wind and keep insects out. Other modifications may include a fire reflector, porch or work area, or another whole addition such as an opposing lean-to.

11-8. Construction of Specific Shelters:

a. A-Frame. The following is one way to build an A-frame shelter in a warm temperate environment using parachute material for the covering. There are as many variations of this shelter as there are builders. The procedures here will, if followed carefully, result in the completion of a safe shelter that will meet survivors' needs. For an example of this and other A-frame shelters, see figure 11-3.

b. Lean-To: See figure 11-4 for lean-to examples.

c. Paratepee, 9-Pole. The paratepee is an excellent shelter for protection from wind, rain, cold, and insects. Cooking, eating, sleeping, resting, signaling, and washing can all

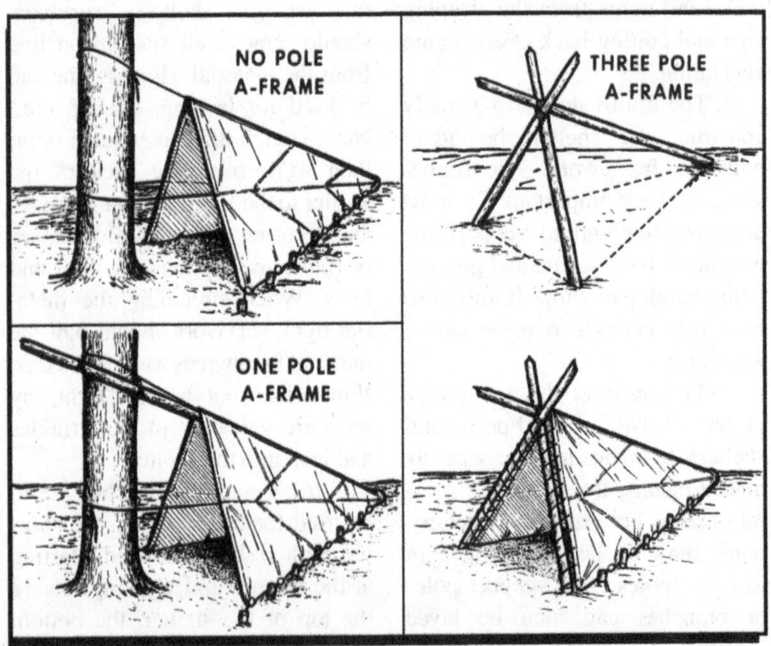

Figure 11-3. A-Frame Shelters.

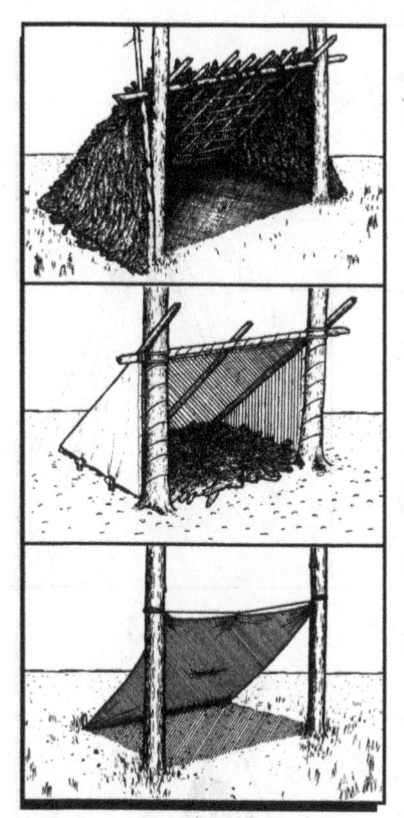

Figure 11-4. Lean-To Shelters

be done without going outdoors. The paratepee, whether 9-pole, 1-pole, or no-pole, is the only improvised shelter that provides adequate ventilation to build an inside fire. With a small fire inside, the shelter also serves as a signal at night. See figure 11-5.

d. Paratepee, 1-Pole: See figure 11-6.

e. Paratepee, No-Pole. For this shelter, the 14 gores of material are prepared the same way. A line is attached to the apex, thrown over a tree limb, etc., and tied off. The lower lateral band is then staked down starting opposite the door around a 12- to 14-foot circle. (Figure 11-7 for paratepee example.)

f. Sod Shelter. A framework covered with sod provides a shelter which is warm in cold weather and one that is easily made waterproof and insect-proof in the summer. The framework for a sod shelter must be strong, and it can be made of driftwood, poles, willow, etc. (Some natives use whale bones.) Sod, with a heavy growth of grass or weeds, should be used since the roots tend to hold the soil together. Cutting about 2 inches of soil along with the grass is sufficient. The size of the blocks are determined by the strength of the individual. A sod house is strong and fireproof.

11-9. Shelter for Tropical Areas. Basic considerations for shelter in tropical areas are as follows:

a. In tropical areas, especially moist tropical areas, the major environmental factors influencing both site selection and shelter types are:

(1) Moisture and dampness.

(2) Rain.

(3) Wet ground.

(4) Heat.

(5) Mud-slide areas.

(6) Dead standing trees and limbs.

(7) Insects.

b. Survivors should establish a campsite on a knoll or high spot

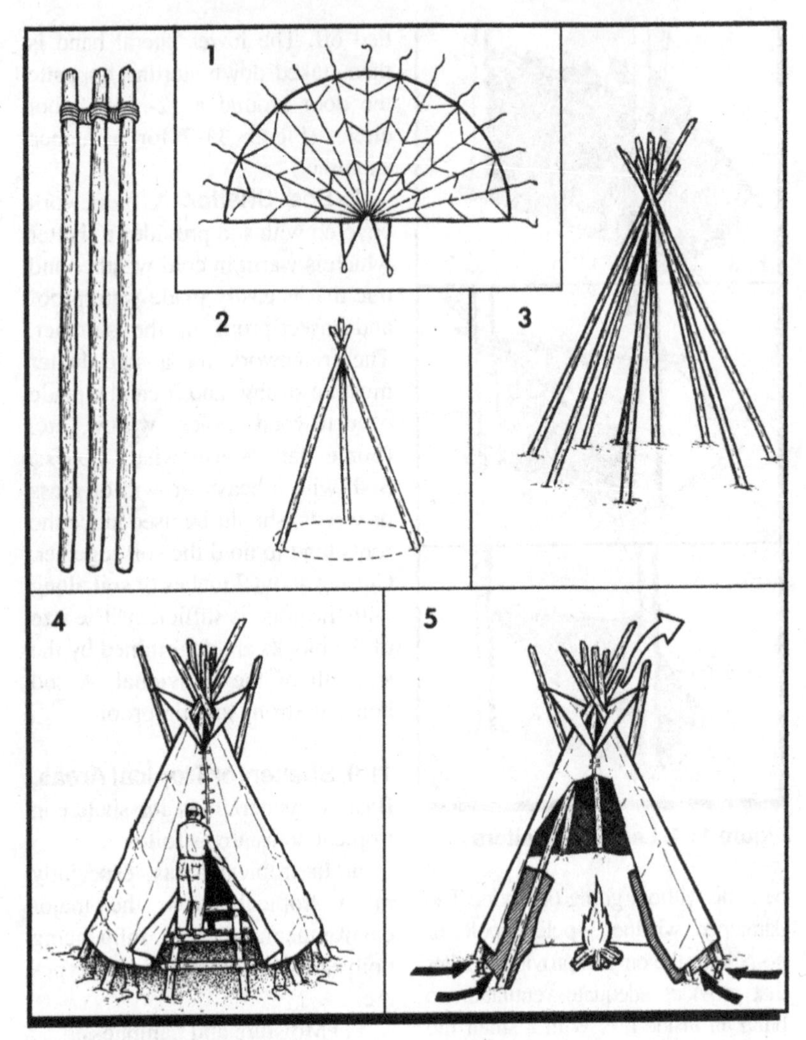

Figure 11-5. 9-Pole Tepee.

in an open area well back from any swamps or marshy areas. The ground in these areas is drier, and there may be a breeze which will result in fewer insects.

c. Underbrush and dead vegetation should be cleared from the shelter site. Crawling insects will not be able to approach survivors as easily due to lack of cover.

d. A thick bamboo clump or matted canopy of vines for cover reflects the smoke from the campfire and discourages insects. This

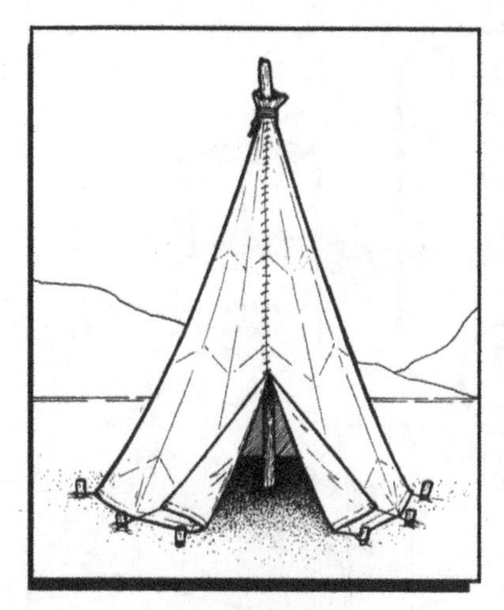

Figure 11-6.1-Pole Tepee.

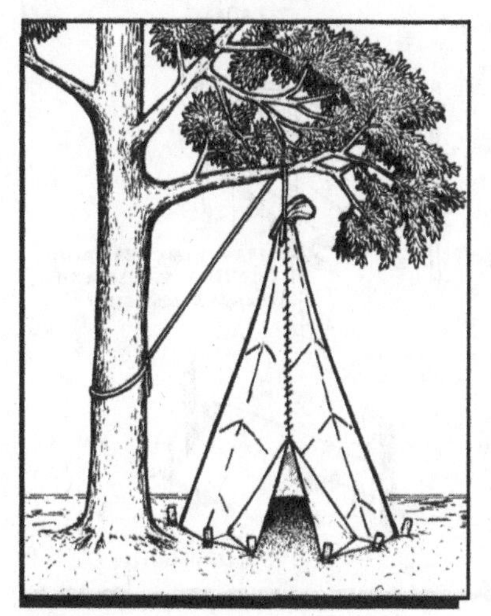

Figure 11-7. No-Pole Tepee.

cover will also keep the extremely heavy early morning dew off the bedding.

e. The easiest improvised shelter is made by draping a parachute, tarpaulin, or poncho over a rope or vine stretched between two trees. One end of the canopy should be kept higher than the other; insects are discouraged by few openings in shelters and smudge fires. A hammock made from parachute material will keep the survivor off the ground and discourage ants, spiders, leeches, scorpions, and other pests.

f. In the wet jungle, survivors need shelter from dampness. If they stay with the aircraft, it should be used for shelter. They should try to make it mosquito-proof by covering openings with netting or parachute cloth.

g. A good rain shelter can be made by constructing an A-type framework and shingling it with a good thickness of palm or other broad leaf plants, pieces of bark, and mats of grass (Figure 11-8).

Figure 11-8. Banana Leaf A Frame.

h. Nights are cold in some mountainous tropical areas. Survivors should try to stay out of the wind and build a fire. Reflecting the heat off a rock pile or other barrier is a good idea. Some natural materials that can be used in the shelters are green wood (dead wood may be too

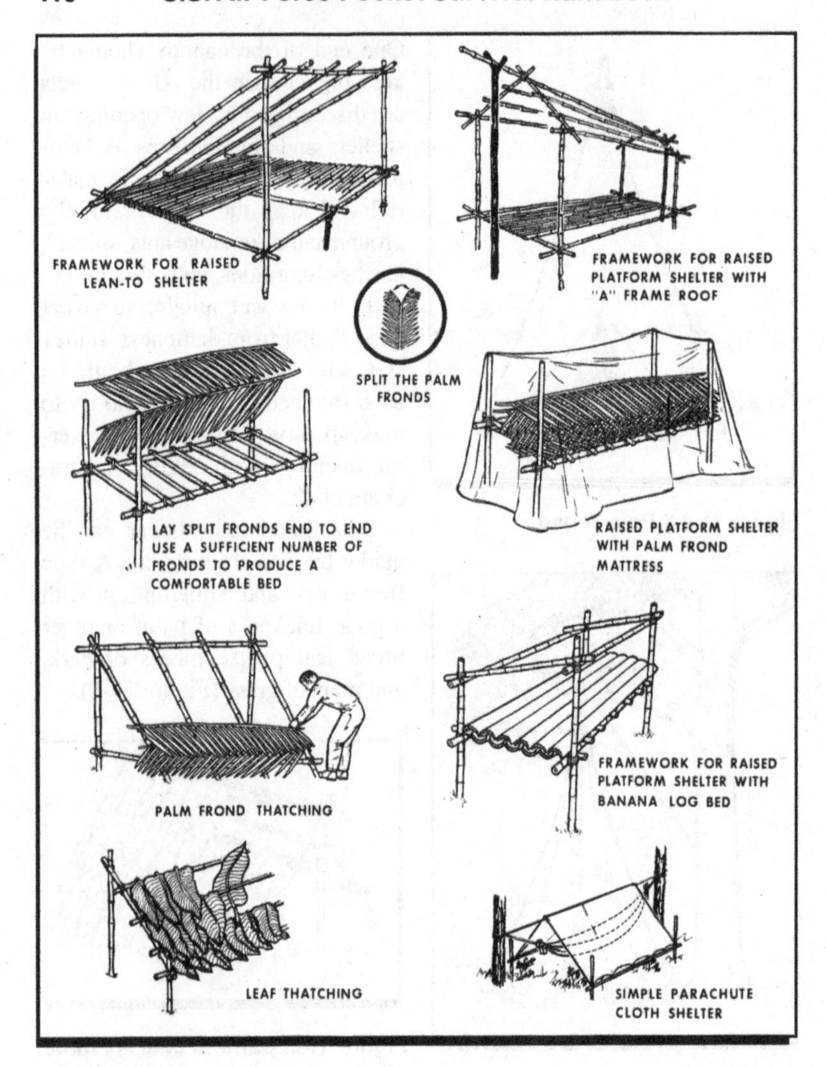

Figure 11.9. Raised Platform Shelter.

rotten), bamboo, and palm leaves. Vines can be used in place of suspension line for thatching roofs or floors, etc. Banana plant sections can be separated from the banana plant and fashioned to provide a mattress effect.

11-10. Specific Shelters for Tropical Environments:

a. Raised Platform Shelter (Figure 11-9). This shelter has many variations. One example is four trees or vertical poles in a rectangular pattern which is a little longer and a little wider than the survivor, keeping in mind the survivor will also need protection for equipment. Two long, sturdy poles are then square lashed between the trees or vertical poles, one on each side of the intended shelter. Cross pieces can then be secured across the two horizontal poles at 6- to 12-inch intervals. This forms the platform on which a natural mattress may be constructed. Parachute material can be used as an insect net and a roof can be built over the structure using A-frame building techniques. The roof should be waterproofed with thatching laid bottom to top in a thick shingle fashion. See figure 11-9 for examples of this and other platform shelters. These shelters can also be built using three trees in a triangular pattern. At the foot of the shelter, two poles are joined to one tree.

b. Variation of Platform Shelter. A variation of the platform-type shelter is the paraplatform. A quick and comfortable bed is made by simply wrapping material around the two "frame" poles. Another method is to roll poles in the material in the same manner as for an improvised stretcher (Figure 11-10).

c. Hammocks. Various parahammocks can also be made. They are more involved than a simple parachute wrapped framework and not quite as comfortable (Figure 11-11).

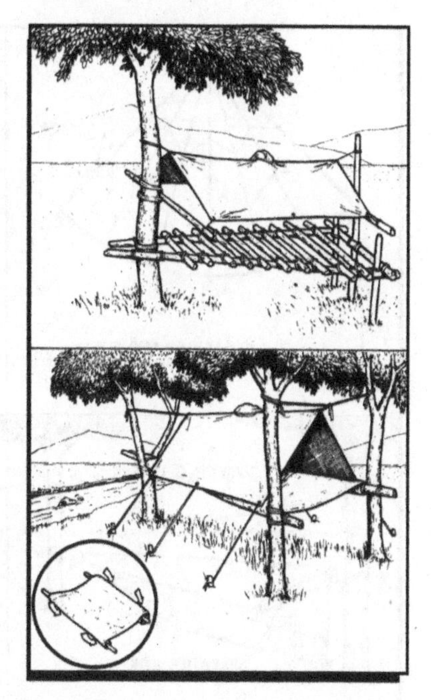

Figure 11-10. Raised Paraplatform Shelter.

d. Hobo Shelter. On tropical coasts and other coastal environments, if a more permanent shelter is desired as opposed to a simple shade shelter, survivors should build a "hobo" shelter. To build this shelter:

(1) Dig into the lee side of a sand dune to protect the shelter from the wind. Clear a level area large enough to lie down in and store equipment.

(2) After the area has been cleared, build a heavy driftwood framework to support the sand.

(3) Wall sides and top with strong material (boards, driftwood, etc.) that will support the sand; leave a door opening.

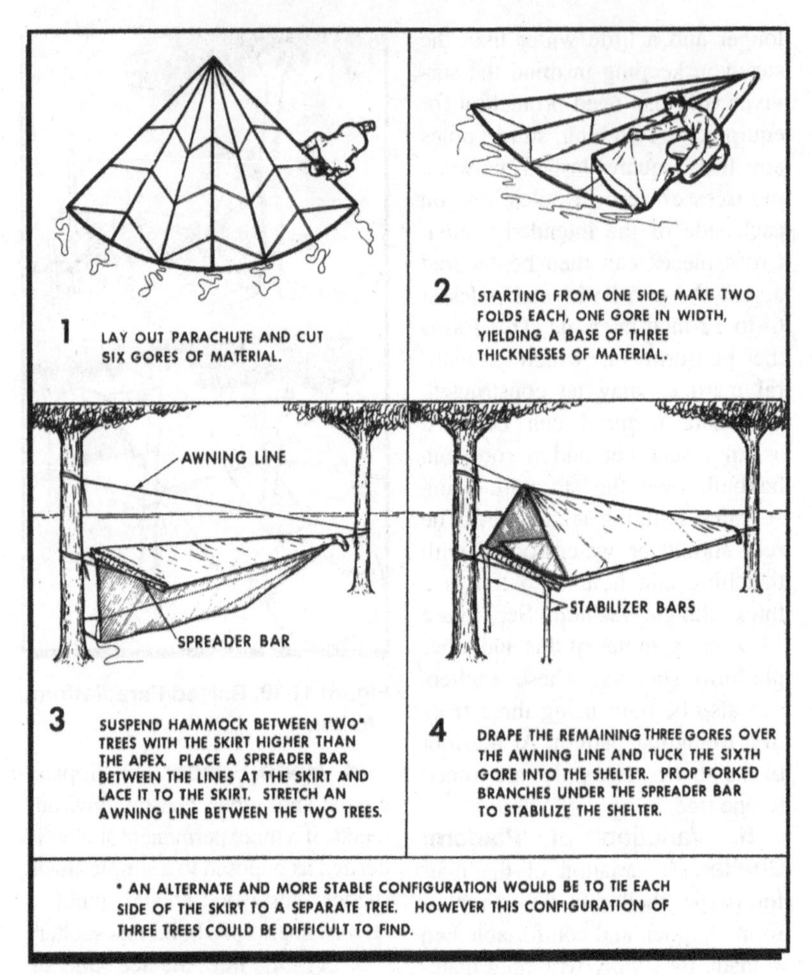

Figure 11-11. Parahammock.

(4) Slope the roof to equal the slope of the sand dune. Cover the entire shelter with parachute material to keep sand from sifting through small holes in the walls and roof.

(5) Cover with 6 to 12 inches of sand to provide protection from wind and moisture.

(6) Construct a door for the shelter (Figure 11-12).

11-11. Shelters for Dry Climates:

a. Natives of hot, dry areas make use of light-proof shelters with sides rolled up to take advantage of any breeze. Survivors should emulate these shade-type shelters if forced to survive in these areas. The extremes of heat *and* cold must be considered in hot areas, as most can become very cold during the

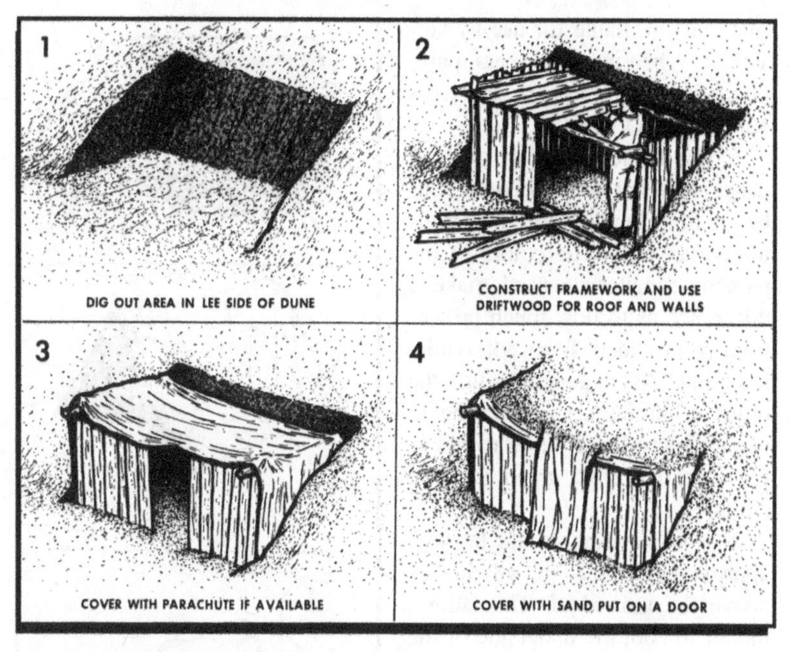

Figure 11-12. Hobo Shelter.

night. The major problem for survivors will be escaping the heat and sun rays.

b. Natural shelters in these areas are often limited to the shade of cliffs and the lee sides of hills, dunes, or rock formations. In some desert mountains, it is possible to find good rock shelters or cave-like protection under tumbled blocks of rocks which have fallen from cliffs. Use care to ensure that these blocks are in areas void of future rock falling activity and free from animal hazards.

c. Vegetation, if any exists, is usually stunted and armed with thorns. It may be possible to stay in the shade by moving around the vegetation as the sun moves. The hottest part of the day may offer few shadows because the sun is directly overhead. Parachute material draped over bushes or rocks will provide some shade.

d. Materials which can be used in the construction of desert shelters include:

(1) Sand, though difficult to work with when loose, may be made into pillars by using sandbags made from parachute or any available cloth.

(2) Rock can be used in shelter construction.

(3) Vegetation such as sage brush, creosote bushes, juniper trees, and desert gourd vines are valuable building materials.

(4) Parachute canopy and suspension lines. These are perhaps the most versatile building

materials available for use by survivors. When used in layers, parachute material protects survivors from the sun's rays.

a. The shelter should be made of dense material or have numerous layers to reduce or stop dangerous ultraviolet rays. The colors of the parachute materials used make a difference as to how much protection is provided from ultraviolet radiation. As a general rule, the order of preference should be to use as many layers as practical in the order of orange, green, tan, and white.

b. The material should be kept approximately 12 to 18 inches above the individual. This allows the air to cool the underside of the material.

c. Aircraft parts and liferafts can also be used for shade shelters. Survivors may use sections of the wing, tail, or fuselage to provide shade. However, the interior of the aircraft will quickly become superheated and should be avoided as a shelter. An inflatable raft can be tilted against a raft paddle or natural object such as a bush or rock to provide relief from the sun (Figure 11-13).

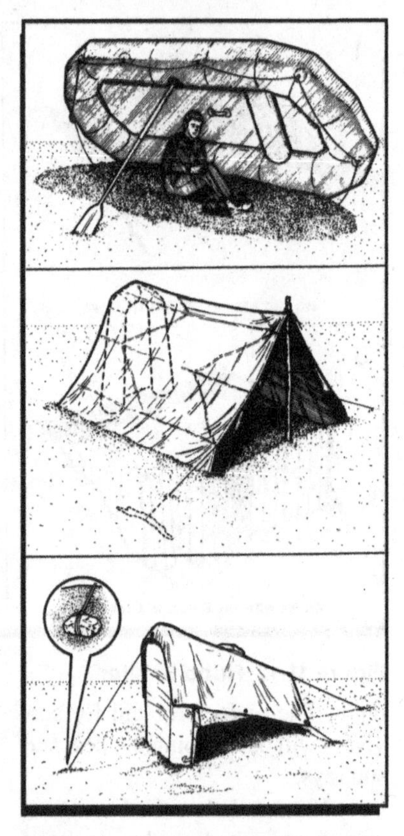

Figure 11-13. Improvised Natural Shade shelters.

11-12. Principles of Desert Shelters:

a. The roof of a desert shelter should be multilayered so the resulting airspace reduces the inside temperature of the shelter. The layers should be separated 12 to 18 inches apart (Figure 11-14).

b. Survivors should place the floor of the shelter about 18 inches above or below the desert surface to increase the cooling effect.

c. In warmer deserts, white parachute material should be used as an outer layer. Orange or sage green material should be used as an inner layer for protection from ultraviolet rays.

d. In cooler areas, multiple layers of parachute material should be used with sage green or orange

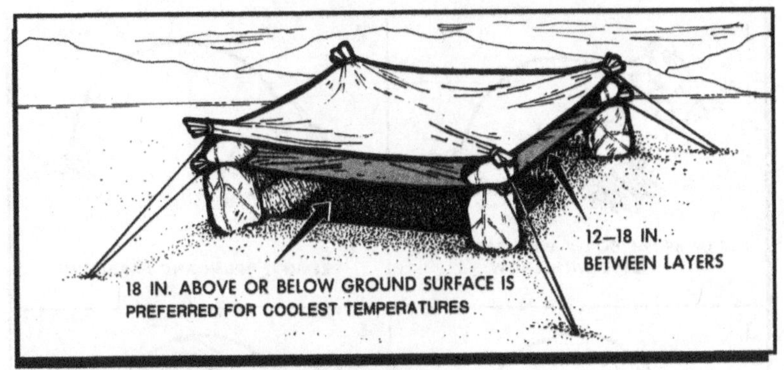

Figure 11-14. Parachute Shade Shelter.

material as the outer layer to absorb heat.

e. The sides of shelters should be movable in order to protect survivors during cold and(or) windy periods and to allow for ventilation during hot periods.

f. In a hot desert, shelters should be built away from large rocks which store heat during the day. Survivors may need to move to the rocky areas during the evening to take advantage of the warmth heated rocks radiate.

g. Survivors should:

(1) Build shelters on the windward sides of dunes for cooling breezes.

(2) Build shelters during early morning, late evening, or at night. However, potential survivors should recall that survivors who come down in a desert area during daylight hours must be immediately concerned with protection from the sun and loss of water. In this case, parachute canopy material can be draped over liferaft, vegetation, or a natural terrain feature for quick shelter.

11-13. Shelters for Snow and Ice Areas:

a. The differences in arctic and arctic-like environments create the need for different shelters. Basically, there are two types of environments which may require special shelter characteristics or building principles before survivors will have adequate shelter. They are:

(1) Barren lands which include some seacoasts, icecaps, sea ice areas, and areas above the tree line.

(2) Tree-line areas.

b. Barren lands offer a limited variety of materials for shelter construction. These are snow, small shrubs, and grasses. Ridges formed by drifting or wind-packed snow may be used for wind protection (survivors should build on the lee side). In some areas, such as sea

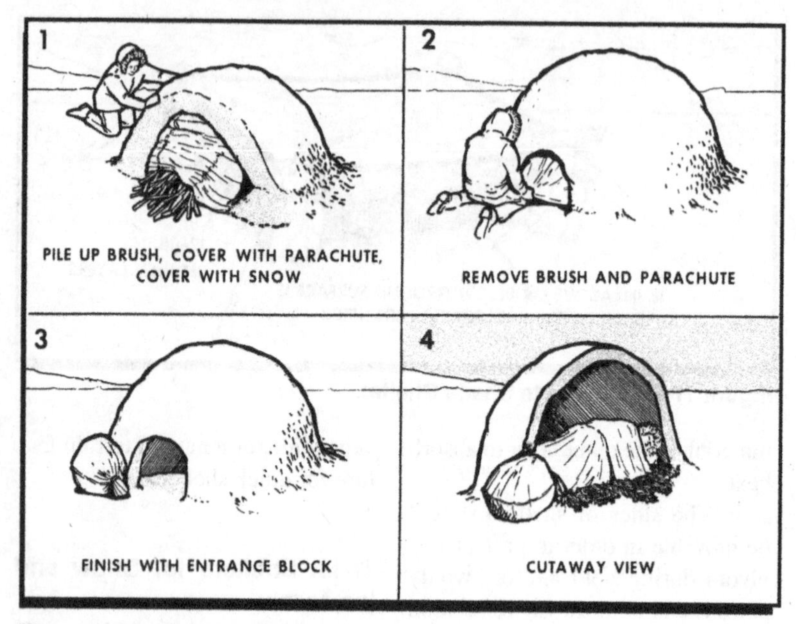

Figure 11-15. Modeled Dome Shelter.

ice, windy conditions usually exist and cause the ice to shift forming pressure ridges. These areas of unstable ice and snow should be avoided at all times. Shelters which are suitable for barren-type areas include:

(1) Molded dome (Figure 11-15).

(2) Snow cave (Figure 11-16).

(3) Fighter trench (Figure 11-17).

(4) Igloo (Figure 11-18).

(5) Para-snow house (Figure 11-19).

NOTE: Of these, the ones that are quick to construct and require minimum effort and energy are the molded dome, snow cave, and fighter trench. It is important to know which of these shelters is the easiest to build since reducing or eliminating the effect of the wind-chill factor is essential to remaining alive.

c. In tree-covered areas, sufficient natural shelter building materials are normally available. Caution is required. Shelters built near rivers and streams may get caught in the overflow.

d. Tree-line area shelter types include:

(1) Thermal A-Frame construction (Figure 11-20).

(2) Lean-to or wedge (Figure 11-21).

(3) Double lean-to (Figure 11-22).

(4) Fan (Figure 11-23).

(5) Willow frame (Figure 11-24).

(6) Tree well (Figure 11-25).

Figure 11-16. Snow Cave.

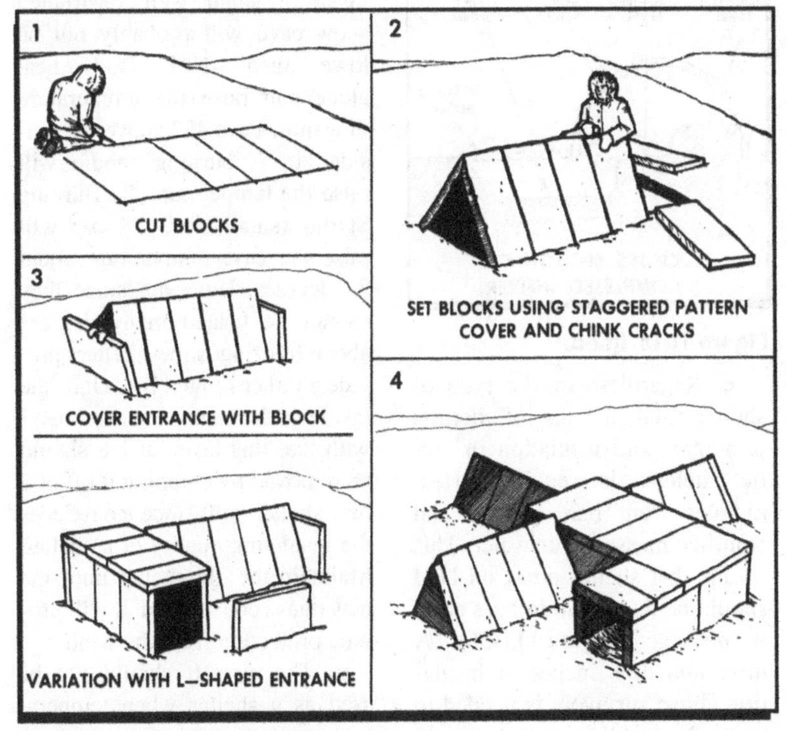

Figure 11-17. Fighter Trench.

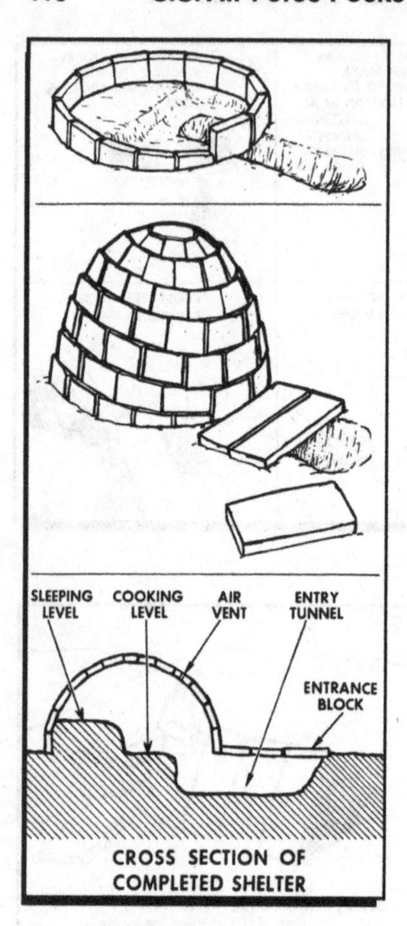

Figure 11-18. Igloo.

e. Regardless of the type of shelter used, the use of thermal principles and insulation in arctic shelters is required. Heat radiates from bare ground and from ice masses over water. This means that shelter areas on land should be dug down to bare earth if possible (Figure 11-26). A minimum of 8 inches of insulation above survivors is needed to retain heat. All openings except ventilation holes should be sealed to avoid heat loss. Leaving vent holes open is especially important if heat producing devices are used. Candles, sterno, or small oil lamps produce carbon monoxide. In addition to the ventilation hole through the roof, another may be required at the door to ensure adequate circulation of the air. (As a general rule, unless persons can see their breath, the snow shelter is too warm and should be cooled down to proclude melting and dripping.)

f. Regardless of how cold it may get outside, the temperature inside a small well-constructed snow cave will probably not be lower than –10°F. Body heat alone can raise the temperature of a snow cave 45° above the outside air. A burning candle will raise the temperature 4°. Burning Sterno (small size, 2⅝ oz) will raise the cave temperature about 28 degrees. However, since they cannot be heated many degrees above freezing, snow shelters provide a rather rugged life. Once the inside of the shelter "glazes" over with ice, this layer of ice should be removed by chipping it off or a new shelter built since ice reduces the insulating quality of a shelter. Maintain the old shelter until the new one is constructed. It will provide protection from the wind.

g. The aircraft should not be used as a shelter when temperatures are below freezing except in

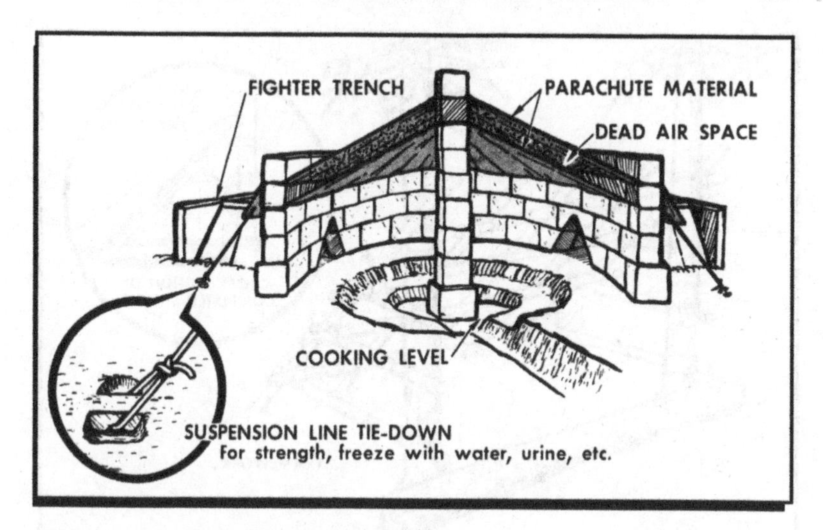

Figure 11-19. Para-Snow House.

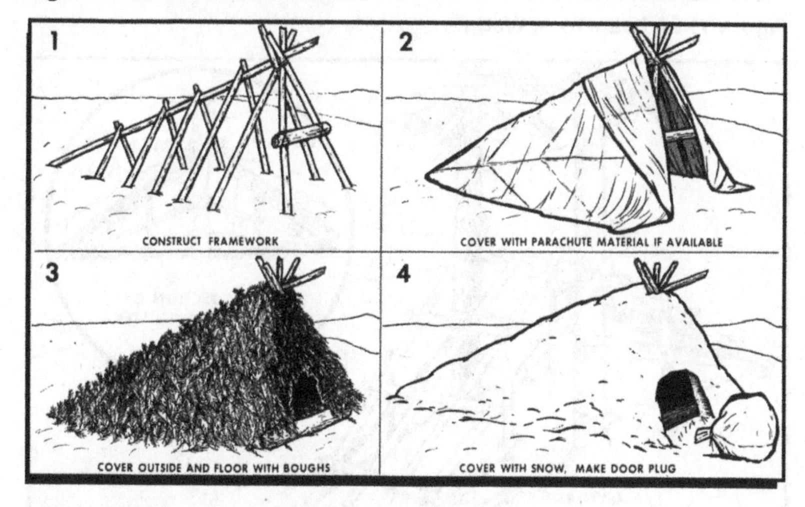

Figure 11-20. Thermal A-Frame.

high wind conditions. Even then a thermal shelter should be constructed as soon as the conditions improve. The aircraft will not provide adequate insulation, and the floor will usually become icy and hazardous.

11-14. General Construction Techniques:

a. All thermal shelters use a layering system consisting of the frame, parachute (if available), boughs or shrubs, and snow. The framework must be sturdy

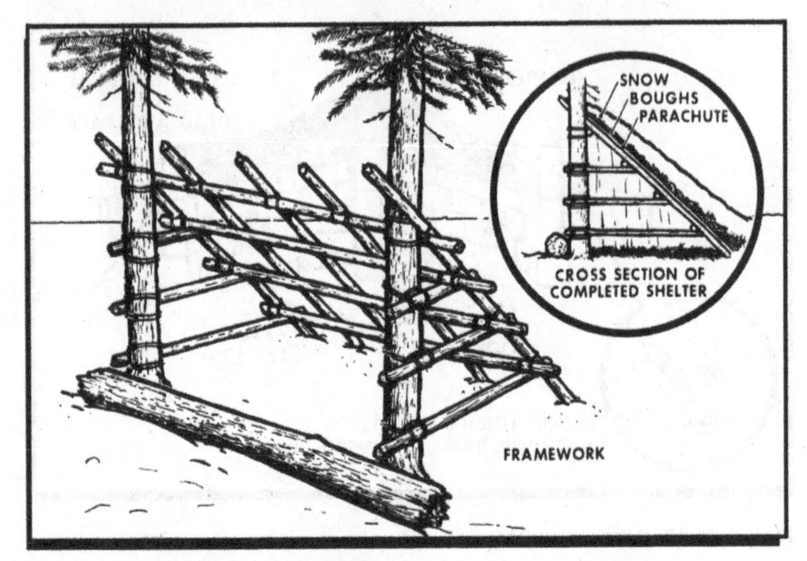

Figure 11-21. Lean-To or Wedge.

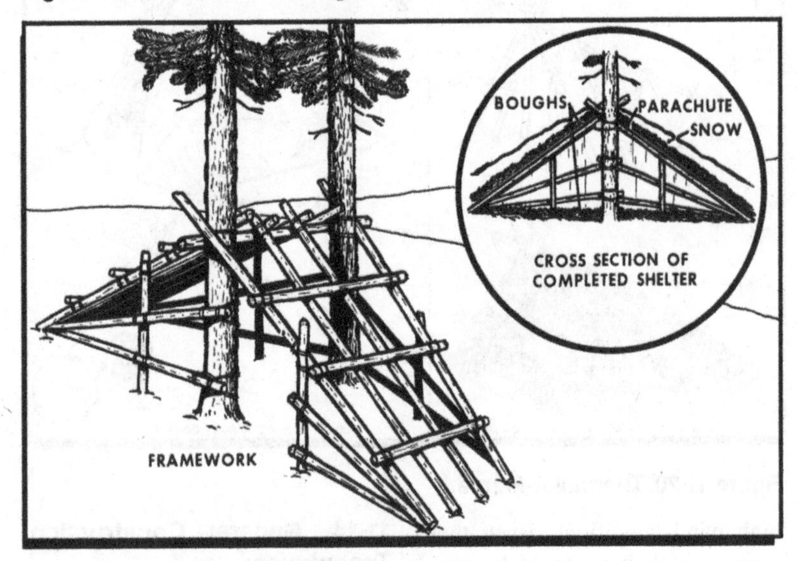

Figure 11-22. Double Lean-To.

enough to support the cover and insulation. A door block should be used to minimize heat loss. Insulation should be added on sleeping areas.

b. If a barren land-type shelter is being built with snow as the only material, a long knife or digging tool is a necessity. It normally takes 2 to 3 hours of hard work to dig a

Figure 11-23. Fan Shelter.

Figure 11-24. Willow Frame Shelter.

snow cave, and much longer for the novice to build an igloo.

c. Survivors should dress lightly while digging and working; they can easily become overheated and dampen their clothing with perspiration which will rapidly turn to ice.

d. If possible, all shelter types should have their openings 90 degrees to the prevailing wind. The entrance to the shelter should also be screened with snowblocks stacked in a L-shape.

e. Snow on the sea ice, suitable for cutting into blocks, will usually be found in the lee of pressure ridges or ice hummocks. The packed snow is often so shallow that the snowblocks have to be cut out horizontally.

f. No matter which shelter is used, survivors should take a digging tool into the shelter at night to cope with the great amount of snow which may block the door during the night.

11-15. Shelter Living:

a. Survivors should limit the number of shelter entrances to conserve heat. Fuel is generally scarce in the arctic. To conserve fuel, it is important to keep the

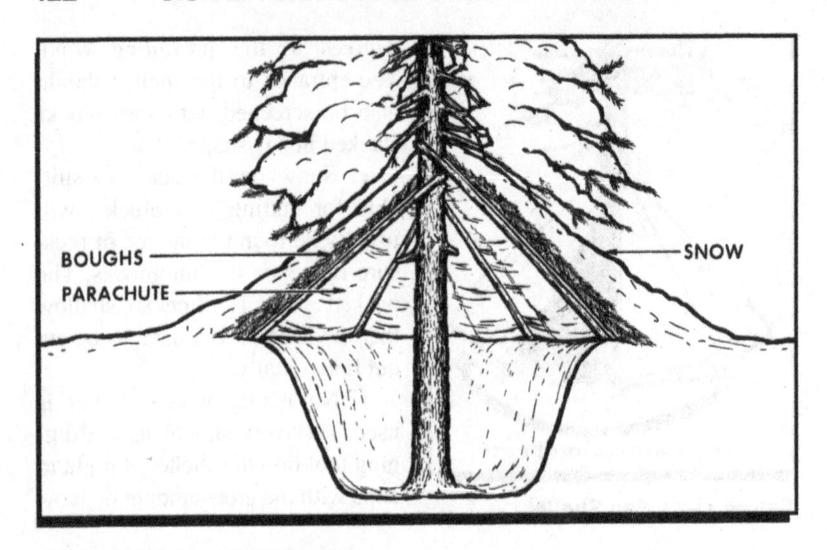

Figure 11-25. Tree Well Shelter.

Figure 11-26. Scraping Snow to Bare Earth.

shelter entrance sealed as much as possible (Figure 11-27). When it is necessary to go outside the shelter, activities such as gathering fuel, snow or ice for melting, etc., should be done. To expedite matters, a trash receptacle may be kept inside the door, and equipment may be stored in the entry way. Necessities which cannot be stored inside may be kept just outside the door. Any firearms (guns) the survivor may have must be stored outside the shelter to prevent condensation building which could cause them to malfunction.

b. A standard practice in snow shelter living is for people to relieve themselves indoors when possible. This practice conserves body heat. If the snowdrift is large enough to dig connecting snow caves, one

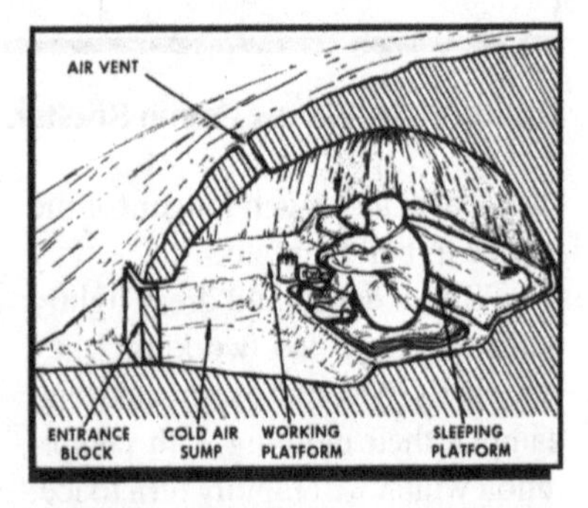

Figure 11-27. Snow Cave Shelter Living.

may be used as a toilet room. If not, tin cans may be used for urinals, and snowblocks for solid waste (fecal) matter.

c. Survivors should use thick insulation under themselves when sleeping or resting even if they have a sleeping bag. They can use a thick bough bed in shingle-fashion, seat cushions, parachute, or an inverted and inflated rubber raft.

d. Outer clothing makes good mattress material. A parka makes a good footbag. The shirt and inner trousers may be rolled up for a pillow. Socks and insoles can be separated and aired in the shelter. Drying may be completed in the sleeping bag by stowing around the hips. This drying method should only be used as a last resort.

e. Keeping the sleeping bag clean, dry, and fluffed will give maximum warmth. To dry the bag, it should be turned inside out, frost beaten out, and warmed before the fire—taking care that it doesn't burn.

f. To keep moisture (from breath) from wetting the sleeping bag, a moisture cloth should be improvised from a piece of clothing, a towel, or parachute fabric. It can then be lightly wrapped around the head in such a way that the breath is trapped inside the cloth. A piece of fabric dries easier than a sleeping bag. If cold is experienced during the night, survivors should exercise by fluttering their feet up and down or by beating the inside of the bag with their hands. Food or hot liquids can be helpful.

g. Snow remaining in clothing will melt in a warm shelter. When the clothing is again taken outside, the water formed will turn to ice and reduce the CLo value. Brush clothes before entering the shelter. Under living conditions where drying clothing is difficult, it is easier to keep clothing from getting wet than having to dry it out later.

h. If all the snow cannot be eliminated from outer clothing, survivors should remove the clothing and store it in the entry way or on the floor away from the source of heat so it remains cold. If ice should form in clothing, it may be beaten out with a stick.

i. In the cramped quarters of any small emergency shelter, pots of food or drink can be accidentally kicked over. The cooking area, even if it is only a Stemo stove, should be located out of the way, possibly in a snow alcove.

11-16. Summer Considerations for Arctic and Arctic-Like Areas:

a. Survivors need shelter against rain and insects. They should choose a campsite near water but on high, dry ground if possible. Survivors should also stay away from thick vegetation, as mosquitoes and flies will make life miserable. A good campsite is a ridge top, cold lake shore, or a spot which gets a breeze.

b. If survivors stay with the aircraft, it can be used for shelter during the summer. They should cover openings with netting or parachute cloth to keep insects out and cook outside to avoid carbon monoxide poisoning. Fires must be built a safe distance from the aircraft.

c. Many temperate area shelters are suitable for summer arctic conditions. The paratepee (of the 1- or nopole variety) is especially good. It will protect from precipitation and keep insects out.

11-17. Shelter for Open Seas. Personal protection from the elements is just as important on the seas as it is anywhere else. Some rafts come equipped with insulated floors, spray shields, and canopies to protect survivors from heat, cold, and water. If rafts are not so equipped or the equipment has been lost, survivors should try to improvise these items using parachute material, clothing, or other equipment.

Chapter 12

FIRECRAFT

12-1. Introduction:

The need for a fire should be placed high on the list of priorities. Fire is used for warmth, light, drying clothes, signaling, making tools, cooking, and water purification. When using fire for warmth, the body uses fewer calories for heat and consequently requires less food. Just having a fire to sit by is a morale booster. Smoke from a fire can be used to discourage insects.

Avoid building a very large fire. Small fires require less fuel, are easier to control, and their heat can be concentrated. Never leave a fire unattended unless it is banked or contained. Banking a fire is done by scraping cold ashes and dry earth onto the fire, leaving enough air coming through the dirt at the top to keep the fuel smoldering. This will keep the fire safe and allow it to be rekindled from the saved coals.

12-2. Elements of Fire:

a. The three essential elements for successful fire building are fuel, heat, and oxygen. These combined elements are referred to as the "fire triangle." By limiting fuel, only a small fire is produced. If the fire is not fed properly, there is too much or too little fire. Green fuel is difficult to ignite, and the fire must be burning well before it is used for fuel. Oxygen and heat must be accessible to ignite any fuel.

b. The survivor must take time and prepare well! Preparing all of the stages of fuel and all of the parts of the fire starting apparatus is the key. To be successful at firecraft, one needs to practice and be patient.

c. The fuels used in building a fire normally fall into three categories (Figure 12-1), relating to their size and flash point: tinder, kindling, and fuel.

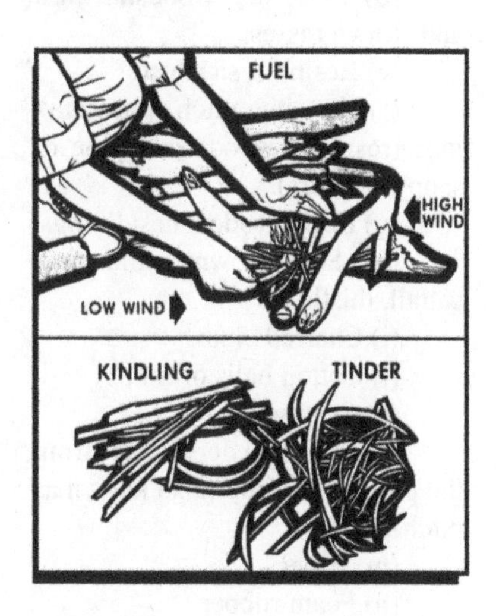

Figure 12-1. Stages of a Fire.

(1) Tinder is any type of small material having a low flash point. It is easily ignited with a minimum

of heat, even a spark. Tinder must be arranged to allow air (oxygen) between the hair-like, bone-dry fibers. The preparation of tinder for fire is one of the most important parts of firecraft. Dry tinder is so critical that pioneers used extreme care to have some in a waterproof "tinder box" at all times. It may be necessary to have two or three stages of tinder to get the flame to a useful size. Tinders include:

(a) The shredded bark from some trees and bushes.

(b) Cedar, birch bark, or palm fiber.

(c) Crushed fibers from dead plants.

(d) Fine, dry woodshavings, and straw/grasses.

(e) Resinous sawdust.

(f) Very fine pitch woodshavings (resinous wood from pine or sappy conifers).

(g) Bird or rodent nest linings.

(h) Seed down (milkweed, cattail, thistle).

(i) Charred cloth.

(j) Cotton balls or lint.

(k) Steel wool.

(l) Dry powdered sap from the pine tree family (also known as pitch).

(m) Paper.

(n) Foam rubber.

(2) Kindling is the next larger stage of fuel material. It should also have a high combustible point. It is added to, or arranged over, the tinder in such a way that it ignites when the flame from the tinder reaches it. Kindling is used to bring the burning temperature up to the point where larger and less combustible fuel material can be used. Kindling includes:

(a) Dead dry small twigs or plant fibers.

(b) Dead dry thinly shaved pieces of wood, bamboo, or cane (always split bamboo as sections can explode).

(c) Coniferous seed cones and needles.

(d) "Squaw wood" from the underside of coniferous trees; dead, small branches next to the ground sheltered by the upper live part of the tree.

(e) Pieces of wood removed from the insides of larger pieces.

(f) Some plastics such as the spoon from an inflight ration.

(g) Wood which has been soaked or doused with flammable materials; that is, wax, insect repellent, petroleum fuels, and oil.

(h) Strips of petrolatum gauze from a first-aid kit.

(i) Dry split wood burns readily because it is drier inside. Also the angular portions of the wood burn easier than the bark-covered round pieces because it exposes more surface to the flame. The splitting of all fuels will cause them to burn more readily.

(3) Fuel, unlike tinder and kindling, does not have to be kept completely dry as long as there is enough kindling to raise the fuel to a combustible temperature. It is

recommended that all fine materials be protected from moisture to prevent excessive smoke production. (Highly flammable liquids should not be poured on an existing fire. Even a smoldering fire can cause the liquids to explode and cause serious burns.) The type of fuel used will determine the amount of heat and light the fire will produce. Dry split hardwood trees (oak, hickory, monkey pod, ash) are less likely to produce excessive smoke and will usually provide more heat than soft woods. They may also be more difficult to break into usable sizes. Pine and other conifers are fast- burning and produce smoke unless a large flame is maintained. Rotten wood is of little value since it smolders and smokes. The weather plays an important role when selecting fuel. Standing or leaning wood is usually dry inside even if it is raining. In tropical areas, avoid selecting wood from trees that grow in swampy areas or those covered with mosses. Tropical soft woods are not usually a good fuel source. Trial and error is sometimes the best method to determine which fuel is best. After identifying the burning properties of available fuel, a selection can be made of the type needed. Recommended fuel sources are:

(a) Dry standing dead wood and dry dead branches (those that snap when broken). Dead wood is easy to split and break. It can be pounded on a rock or wedged between other objects and bent until it breaks.

(b) The insides of fallen trees and large branches may be dry even if the outside is wet. The heart wood is usually the last to rot.

(c) Green wood which can be made to burn is found almost anywhere, especially if finely split and mixed evenly with dry dead wood.

(d) In treeless areas, other natural fuels can be found. Dry grasses can be twisted into bunches. Dead cactus and other plants are available in deserts. Dry peat moss can be found along the surface of undercut streambanks. Dried animal dung, animal fats, and sometimes even coal can be found on the surface. Oil impregnated sand can also be used when available.

12-3. Fire Location. The location of a fire should be carefully selected. An old story is told of a mountain man who used his last match to light a fire built under a snow-covered tree. The heat from the fire melted the snow and it slid off the tree and put out the fire. For a survivor, this type of accident can be very demoralizing or even deadly. Locate and prepare the fire carefully.

12-4. Fire Site Preparation:

a. After a site is located, twigs, moss, grass, or duff should be cleaned away. Scrape at least a 3-foot diameter area down to bare soil for even a small fire. Larger fires require a larger area. If the fire must be built on snow, ice, or wet ground, survivors should build a platform of green logs or rocks. (Beware of wet or porous rocks, they may explode when heated.)

b. There is no need to dig a hole or make a circle of rocks in preparation for fire building. Rocks may be placed in a circle and filled with dirt, sand, or gravel to raise the fire above the moisture from wet ground. The purpose of these rocks is to hold the platform only.

c. To get the most warmth from the fire, it should be built against a rock or log reflector (Figure 12-2). This will direct the heat into the shelter. Cooking fires can be walled-in by logs or stones. This will provide a platform for cooking utensils and serve as a windbreak to help keep the heat confined.

d. After preparing the fire, all materials should be placed together and arranged by size (tinder, kindling, and fuel). As a rule of thumb, survivors should have three times the amount of tinder and kindling than is necessary for one fire. It is to their advantage to have too much than not enough. Having plenty of material on hand will prevent the possibility of the fire going out while additional material is gathered.

Figure 12-2. Fire Reflector.

12-5. Firemaking With Matches (or Lighter):

a. Survivors should arrange a small amount of kindling in a low pyramid, close enough together so flames can jump from one piece to another. A small opening should be left for lighting and air circulation.

b. Matches can be conserved by using a "shave stick," or by using a loosely tied fagot of thin, dry twigs. The match must be shielded from wind while igniting the shave stick. The stick can then be applied to the lower windward side of the kindling.

c. Small pieces of wood or other fuel can be laid gently on the kindling before lighting or can be added as the kindling begins to burn. The survivors can then place smaller pieces first, adding larger pieces of fuel as the fire begins to burn. They should avoid smothering the fire by crushing the kindling with heavy wood.

d. Survivors only have a limited number of matches or other instant fire-starting devices. In a long-term situation, they should use these devices sparingly or carry fire with them when possible. Many primitive cultures carry fire (fire bundles) by using dry punk or fiberous barks (cedar) encased in a bark. Others use torches. Natural fire bundles also work well for holding the fire (Figure 12-3).

e. The amount of oxygen must be just enough to keep the coals inside the dry punk burning slowly. This requires constant vigilance to control the rate of the burning process. The natural fire bundle is constructed in a cross section as shown in figure 12-3.

12-6. Heat Sources. A supply of matches, lighters, and other such devices will only last a limited time. Once the supply is depleted, they cannot be used again. If possible, before the need arises, survivors should become skilled at starting fires with more primitive means, such as friction, heat, or a sparking device. It is essential that they continually practice these procedures. The need to start a fire may arise at the most inopportune times. One of the greatest aids a survivor can have for rapid fire starting is the "tinder box" previously mentioned. Using friction, heat, and sparks are very reliable methods for those who use them on a regular basis. Therefore, survivors must practice these methods. Survivors must be aware of the problems associated with the use of primitive heat sources. If the humidity is high in the immediate area, a fire may be difficult to ignite even if all other conditions are favorable. For primitive methods to be successful, the materials must be BONE DRY. The primitive people who use these ignition methods take great care to keep their tinder, kindling, and other fuels dry, even to the point of wrapping many layers of waterproof materials around it. *PREPARATION, PRACTICE,* and

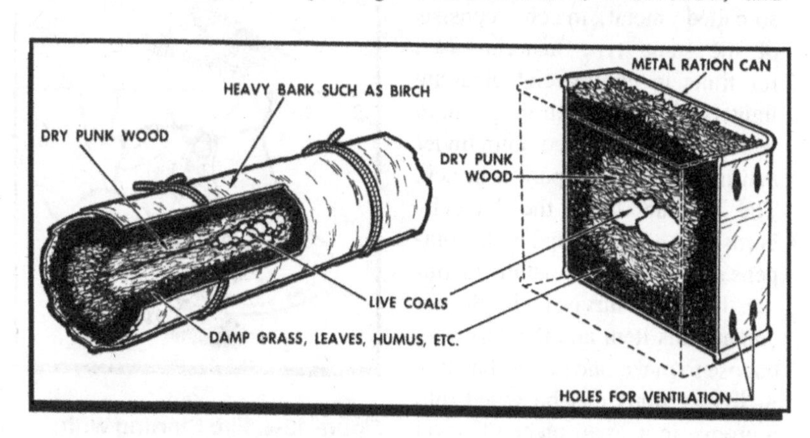

Figure 12.3 Fire Bundels

PATIENCE in the use of primitive fire-building techniques cannot be over emphasized. A key point in all primitive methods is to ensure that the tinder is not disturbed.

a. Flint and Steel:

(1) Flint and steel is one way to produce fire without matches.

(a) To use this method, survivors must hold a piece of flint in one hand above the tinder.

(b) Grasp the steel in the other hand and strike the flint with the edge of the steel in a downward glancing blow (Figure 12-4).

(2) True flint is not necessary to produce sparks. Iron pyrite and quartz will also give off sparks even if they are struck against each other. Check the area and select the best spark-producing stone as a backup for the available matches. The sparks must fall on the tinder and then be blown or fanned to produce a coal and subsequent flame.

(3) Synthetic flint, such as the so-called metal match, consists of the same type material used for flints in commercial cigarette lighters. Some contain magnesium which can be scraped into tinder and into which the spark is struck. The residue from the "match" burns hot and fast and will compensate for some moisture in tinder. If issued survival kits do not contain this item and the survivors choose to make one rather than buy it, lighter flints can be glued into a groove in a small piece of wood or plastic. The survivors can then practice striking a spark by scratching the flint with a knife blade. A 90-degree angle between the blade and flint works best. The device must be held close enough for the sparks to hit the tinder, but enough distance must be allowed to avoid accidentally extinguishing the fire. Cotton balls dipped in petroleum jelly make excellent tinder with

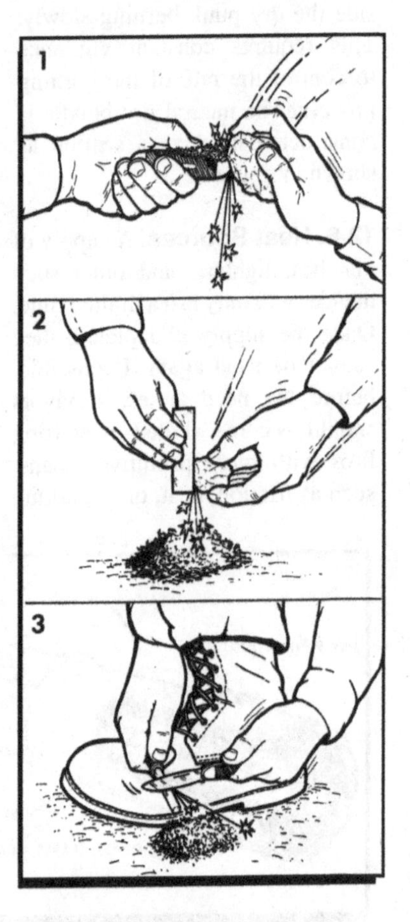

Figure 12-4. Fire Starting With Flint and Steel.

flint and steel. When the tinder ignites, additional tinder, kindling, and fuel can be added,

b. Batteries:

(1) Another method of producing fire is to use the battery of the aircraft, vehicle, storage batteries, etc. Using two insulated wires, connect one end of a wire to the positive post of the battery and the end of the other wire to the negative post. Touch the two remaining ends to the ends of a piece of noninsulated wire. This will cause a short in the electrical circuit and the noninsulated wire will begin to glow and get hot. Material coming into contact with this hot wire will ignite. Survivors should use caution when attempting to start a fire with a battery. They should ensure that sparks or flames are not produced near the battery because explosive hydrogen gas is produced and can result in serious injury (Figure 12-5).

Figure 12-5. Fire Starting With Batteries.

(2) If fine grade steel wool is available, a fire may be started by stretching it between the positive and negative posts until the wire itself makes a red coal.

c. Burning Glass. If survivors have sunlight and a burning glass, a fire can be started with very little physical effort (Figure 12-6). Concentrate the rays of the sun on tinder by using the lens of a lensatic compass, a camera lens, or the lens of a flashlight which magnifies; even a convex piece of bottle glass may work. Hold the lens so that the brightest and smallest spot of concentrated light falls on the tinder. Once a whisp of smoke is produced, the tinder should be fanned or blown upon until the smoking coal becomes a flame. Powdered charcoal in the tinder will decrease the ignition time. Add kindling carefully as in any other type of

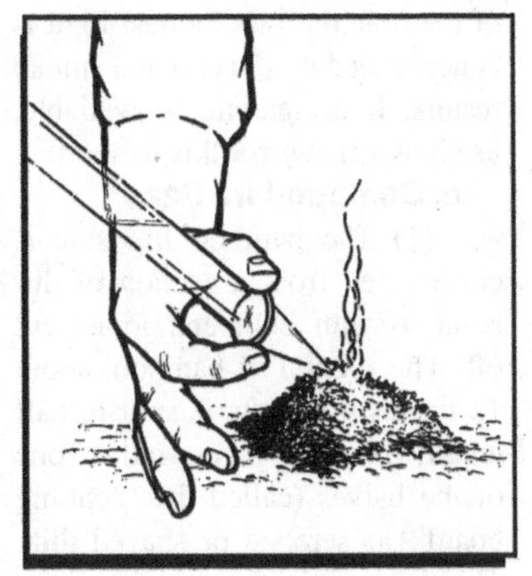

Figure 12-6. Fire Starting With Burning Glass.

fire. Practice will reduce the time it takes to light the tinder.

d. Flashlight Reflector. A flashlight reflector can also be used to start a fire (Figure 12-7).

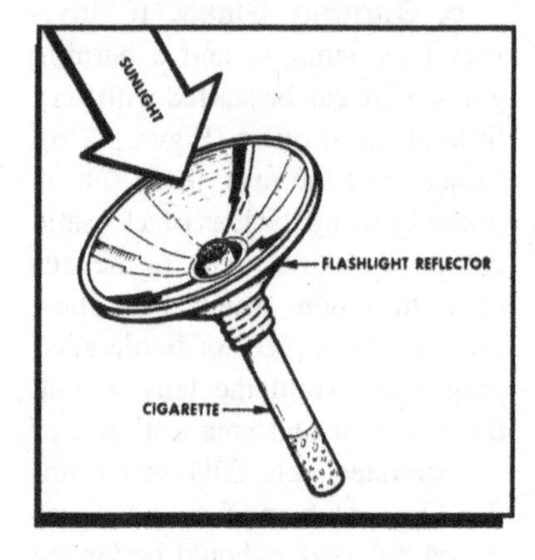

Figure 12-7. Fire Starting With Flashlight Reflector.

Place the tinder in the center of the reflector where the bulb is usually located. Push it up from the back of the hole until the hottest light is concentrated on the end and smoke results. If a cigarette is available, use it as a tinder for this method.

e. Bamboo Fire Saw:

(1) The bamboo fire saw is constructed from a section of dry bamboo with both end joints cut off. The section of bamboo, about 12 inches in length, is split in half lengthwise. The inner wall of one of the halves (called the "running board") is scraped or shaved thin. This is done in the middle of the running board. A notch to serve as a guide is cut in the outer sheath opposite the scraped area of the inner wall. This notch runs across the running board at a 90-degree angle (Figure 12-8).

(2) The other half of the bamboo joint is further split in half lengthwise, and one of the resultant quarters is used as a "baseboard." One edge of the baseboard is shaved down to make a tapered cutting edge. The baseboard is then firmly secured with the cutting edge up. This may be done by staking it to the ground in any manner which does not allow it to move (Figure 12-8).

(3) Tinder is made by scraping the outer sheath of the remaining quarter piece of the bamboo

Figure 12-8. Bamboo Fire Saw.

section. The scrapings (approximately a large handful) are then rubbed between the palms of the hands until all of the wood fibers are broken down and dust-like material no longer falls from the tinder. The ball of scrapings is then fluffed to allow maximum circulation of oxygen through the mass (Figure 12-8).

(4) The finely shredded and fluffed tinder is placed in the running board directly over the shaved area, opposite the outside notch. Thin strips of bamboo should be placed lengthwise in the running board to hold the tinder in place. These strips are held stationary by the hands when grasping the ends of the running board (Figure 12-8).

(5) A long, very thin sliver of bamboo (called the "pick") should be prepared for future use. One end of the running board is grasped in each hand, making sure the thin strips of bamboo are held securely in place. The running board is placed over the baseboard at a right angle, so that the cutting edge of the base board fits into the notch in the outer sheath of the running board. The running board is then slid back and forth as rapidly as possible over the cutting edge of the baseboard, with sufficient downward pressure to ensure enough friction to produce heat.

(6) As soon as "billows" of smoke rise from the tinder, the running board is picked up. The pick is used to push the glowing embers from the bottom of the running board into the mass of tinder. While the embers are being pushed into the tinder, they are gently blown upon until the tinder bursts into flames.

(7) As soon as the tinder bursts into flames, slowly add kindling in small pieces to avoid smothering the fire. Fuel is gradually added to produce the desired size fire. If the tinder is removed from the running board as soon as it flames, the running board can be reused by cutting a notch in the outer sheath next to the original notch and directly under the scraped area of the inner wall.

f. Bow and Drill:

(1) This is a friction method which has been used successfully for thousands of years. A spindle of yucca, elm, basswood, or any other straight grainwood (not softwood) should be made. The survivors should make sure that the wood is not too hard or it will create a glazed surface when friction is applied. The spindle should be 12 to 18 inches long and three-fourths inch in diameter. The sides should be octagonal, rather than round, to help create friction when spinning. Round one end and work the other end into a blunt point. The round end goes to the top upon which the socket is placed. The socket is made from a piece of hardwood large enough to hold comfortably in the palm of the hand with the curved part up and the flat side down to

hold the top of the spindle. Carve or drill a hole in this side and make it smooth so it will not cause undue friction and heat production. Grease or soap can be placed in this hole to prevent friction (Figure 12-9).

(2) The bow is made from a stiff branch about 3 feet long and about 1 inch in diameter. This piece should have sufficient flexibility to bend. It is similar to a bow used to shoot arrows. Tie a piece of suspension line or leather thong to both ends so that it has the same tension as that of a bow. There should be enough tension for the spindle to twist comfortably.

(3) The fireboard is made of the softwood and is about 12 inches long, three-fourths inch thick, and 3 to 6 inches wide. A small hollow should be carved in the fireboard. A V-shaped cut can then be made in from the edge of the board. This V-shape should extend into the center of the hollow where the spindle will make the hollow deeper. The object of this "V" cut is to create an angle which cuts off the edge of the spindle as it gets hot and turns to charcoal dust. This is the critical part of the fireboard and must be held steady during the operation of spinning the spindle.

(4) While kneeling on one knee, the other foot can be placed on the fireboard as shown in figure 12-9 and the tinder placed under the

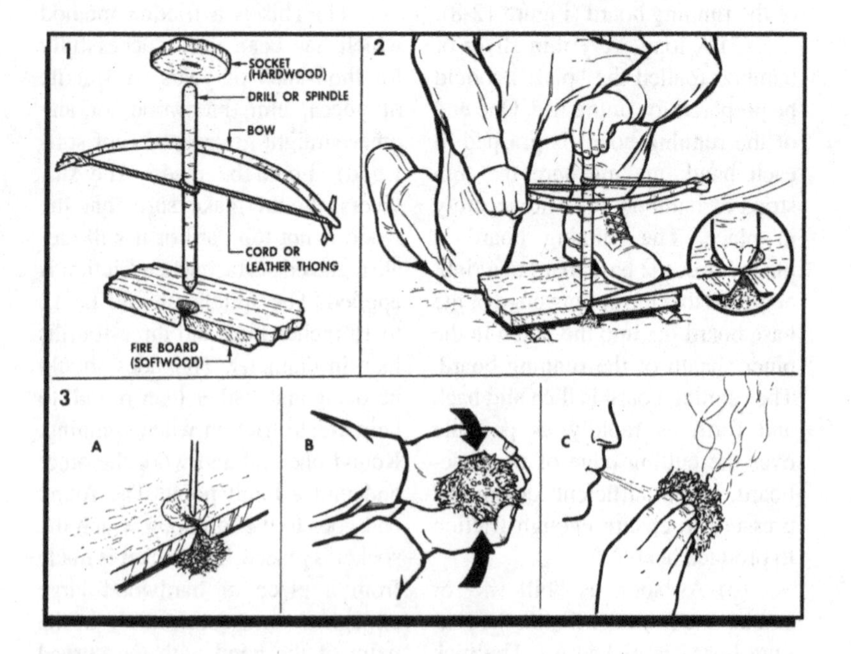

Figure 12-9. Bow and Drill.

fireboard just beneath the V- cut. Care should be taken to avoid crushing the tinder under the fireboard. Space can be obtained by using a small, three-fourths inch diameter stick to hold up the fireboard. This allows air into the tinder where the hot powder (spindle charcoal dust) is collected.

(5) The bow string should be twisted once around the spindle. The spindle can then be placed upright into the spindle hollow (socket). The survivor may press the socket down on the spindle and fireboard. The entire apparatus must be held steady with the hand on the socket braced against the leg or knee. The spindle should begin spinning with long, even slow strokes of the bow until heavy smoke is produced. The spinning should become faster until the smoke is very thick. At this point, hot powder, which can be blown into a glowing ember, has been successfully produced. The bow and spindle can then be removed from the fireboard and the tinder can be placed next to the glowing ember making sure not to extinguish it. The tinder must then be rolled gently around the burning ember, and blow into the embers, starting the tinder to burn. This part of the fire is most critical and should be done with care and planning.

(6) The burning tinder is then placed into the waiting fire "lay" containing more tinder and small kindling. At no time in this process should the survivor break concentration or change sequence. The successful use of these primitive methods of fire starting will require a great deal of patience. Success demands dedication and practice.

g. The Fire Thong. The fire thong, another friction method, is used in only those tropical regions where rattan is found. The system is simple and consists of a twisted rattan thong or other strong plant fiber, 4 to 6 feet long, less than 1 inch in diameter, and a 4-foot length of dry wood which is softer than rattan (dedicuous wood) (Figure 12-10). Rub with a steady but increasing rhythm.

h. The Plow. The plow is a method used by some primitives and basically follows the principles

Figure 12-10. Fire Thong.

Figure 12-11. Fire Plow.

of other friction methods. The wood used must not glaze with heat applied and must be able to produce powder with friction.

i. Ground Stake. Another variation can be constructed by driving a stake into the ground as shown in figure 12-11.

12.7. Firemaking With Special Equipment:

a. The night end of the day-night flare can be used as a fire starter. This means, however, that survivors must weigh the importance of a fire against the loss of a night flare.

b. Some emergency kits contain small fire starters, cans of special fuels, windproof matches, and other aids. Survivors should save the fire starters for use in extreme cold and damp (moist) weather conditions.

c. The white plastic spoon (packed in various in-flight rations) may be the type that burns readily. The handle should be pushed deep enough into the ground to support the spoon in an upright position. Light the tip of the spoon. It will burn for about 10 minutes (long enough to dry out and ignite small tinder and kindling).

d. If a candle is available, it should be ignited to start a fire and thus prevent using more than one match. As soon as the fire is burning, the candle can be extinguished and saved for future use.

e. Tinder can be made more combustible by adding a few drops of flammable fuel/material. An example of this would be mixing the powder from an ammunition cartridge with the tinder. After preparing tinder in this manner, it should be stored in a waterproof container for future use. Care must be used in handling this mixture because the flash at ignition could burn the skin and clothing.

f. For thousands of years, the Eskimos and other northern peoples have relied heavily upon oils from animals to heat their homes. A fat stove or "Koodlik" is used by the Eskimos to burn this fuel.

g. Survivors can improvise a stove from a ration can and burn any flammable oil-type liquid or animal fats available. Here again, survivors should keep in mind that if there is only a *limited* amount of animal fat, it should be eaten to produce heat inside the body.

12.8. Burning Aircraft Fuel. On barren lands in the arctic, aircraft fuel may be the only material survivors have available for fire.

a. A stove can be improvised to burn fuel, lubricating oil, or a mixture of both (Figure 12-12). The survivor should place 1 or 2 inches of sand or fine gravel in the bottom of a can or other container and add fuel. *Care should be used when lighting the fuel because it may explode.* Slots should be cut into the top of the can to let flame and smoke out, and holes punched just above the level of the sand to provide a draft. A mixture of fuel and oil will make the fire burn longer. If no can is available, a hole can be dug and filled with sand. Fuel is then poured on the sand and ignited. The survivor should not allow the fuel to collect in puddles.

b. Lubricating oil can be burned as fuel by using a wick arrangement. The wick can be made of string, rope, rag, sphagnum moss, or even a cigarette and should be placed on the edge of a receptacle filled with oil. Rags, paper, wood, or other fuel can be soaked in oil and thrown on the fire.

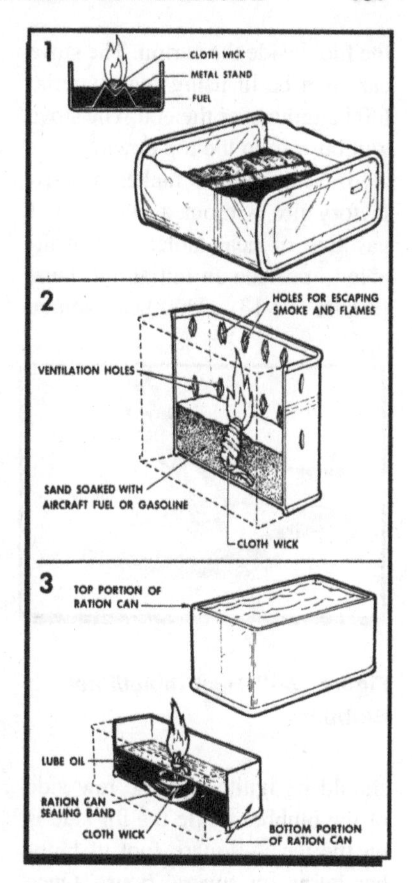

Figure 12-12. Fat and Oil Stoves.

c. A stove can be made of any empty waxed carton by cutting off one end and punching a hole in each side near the unopened end. Survivors can stand the carton on the closed end and loosely place the fuel inside the carton. The stove can then be lit using fuel material left hanging over the end. The stove will burn from the top down.

d. Seal blubber makes a satisfactory fire without a container if gasoline or heat tablets are available to provide an initial hot flame (Figure 12-13). The heat source should be ignited on the raw side of the blubber while the fur side is on the ice. A square foot of blubber burns for several hours. Once the blubber catches fire, the heat tablets can be recovered. Eskimos light a small piece of blubber and use it to kindle increasingly larger pieces. The smoke from a blubber fire is dirty, black, and heavy. The flame is very bright and can be seen for several miles. The smoke will penetrate clothing and blacken the skin.

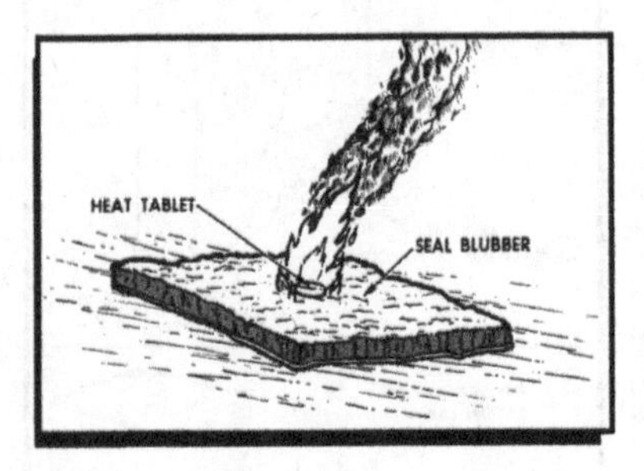

Figure 12-13. Heat Tablet/Seal Blubber

12.9. Useful Firecraft Hints:

a. Conserve matches by only using them on properly prepared fires. They should never be used to light cigarettes or for starting unnecessary fires.

b. Carry some dry tinder in a waterproof container. It should be exposed to the sun on dry days. Adding a little powdered charcoal will improve it. Cotton cloth is good tinder, especially if scorched or charred. It works well with a burning glass or flint and steel.

c. Remember that firemaking can be a difficult job in an arctic environment. The main problem is the availability of firemaking materials. Making a fire starts *WELL* before the match is lit. The fire must be protected from the wind. In wooded areas, standing timber and brush usually make a good windbreak but in open areas, some type of windbreak may have to be constructed. A row of snowblocks, the shelter of a ridge, or a pile of brush will work as a windbreak. It must be high enough to shield the fire from the wind. It may also act as a heat reflector if it is of solid material.

d. Remember, a platform will be required to prevent the fire from melting down through the deep snow and extinguishing it. A platform is also needed if the ground is moist or swampy. The platform can be made of green logs, metal, or any material that will not burn through very readily. Care must be taken when selecting an area for fire building. If the area has a large accumulation of humus material and(or) peat, a platform is needed to avoid igniting the material as it will tend to smolder long after the flames of the fire are extinguished. A smoldering peat fire is almost impossible to put out and may burn for years.

e. In forested areas, the debris on the ground and the lichen mat should be cleared away to mineral soil, if possible, to prevent the fire from spreading.

f. The ignition source used to ignite the fire must be quick and easily operated with hand protection such as mittens. Any number of devices will work well—matches, candle, lighter, fire starter, metal matches, etc.

12.10. Fire Lays. Most fires are built to meet specific needs or uses, either heat, light, or preparing food and water. The following configurations are the most commonly used for fires and serve one or more needs (Figure 12-14).

a. Tepee:

(1) The tepee fire can be used as a light source and has a concentrated heat point directly above the apex of the tepee which is ideal for boiling water. To build:

(a) Place a large handful of tinder on the ground in the middle of the fire site.

(b) Push a stick into the ground, slanting over the tinder.

(c) Then lean a circle of kindling sticks against the slanting stick, like a tepee, with an opening toward the windward side for draft.

(2) To light the fire:

(a) Crouch in front of the fire lay with the back to the wind.

(b) Feed the fire from the downwind side, first with thin pieces of fuel, then gradually with thicker pieces.

(c) Continue feeding until the fire has reached the desired size. The tepee fire has one big drawback. It tends to fall over easily. However, it serves as an excellent starter fire.

b. Log Cabin. As the name implies, this lay looks similar to a log cabin. Log cabin fires give off a great amount of light and heat primarily because of the amount of oxygen which enters the fire. The log cabin fire creates a quick and large bed of coals and can be used for cooking or as the basis for a signal fire. If one person or a group of people are going to use the coals for cooking, the log cabin can be modified into a long fire or a keyhole fire.

c. Long Fire. The long fire begins as a trench, the length of which is layed to take advantage of existing wind. The long fire can also be built above ground by using two parallel green logs to hold the coals together. These logs should be at least 6 inches in diameter and situated so the cooking utensils will rest upon the logs. Two 1-inch thick sticks can be placed under both logs, one at each end of the long fire. This is done to allow the coals to receive more air.

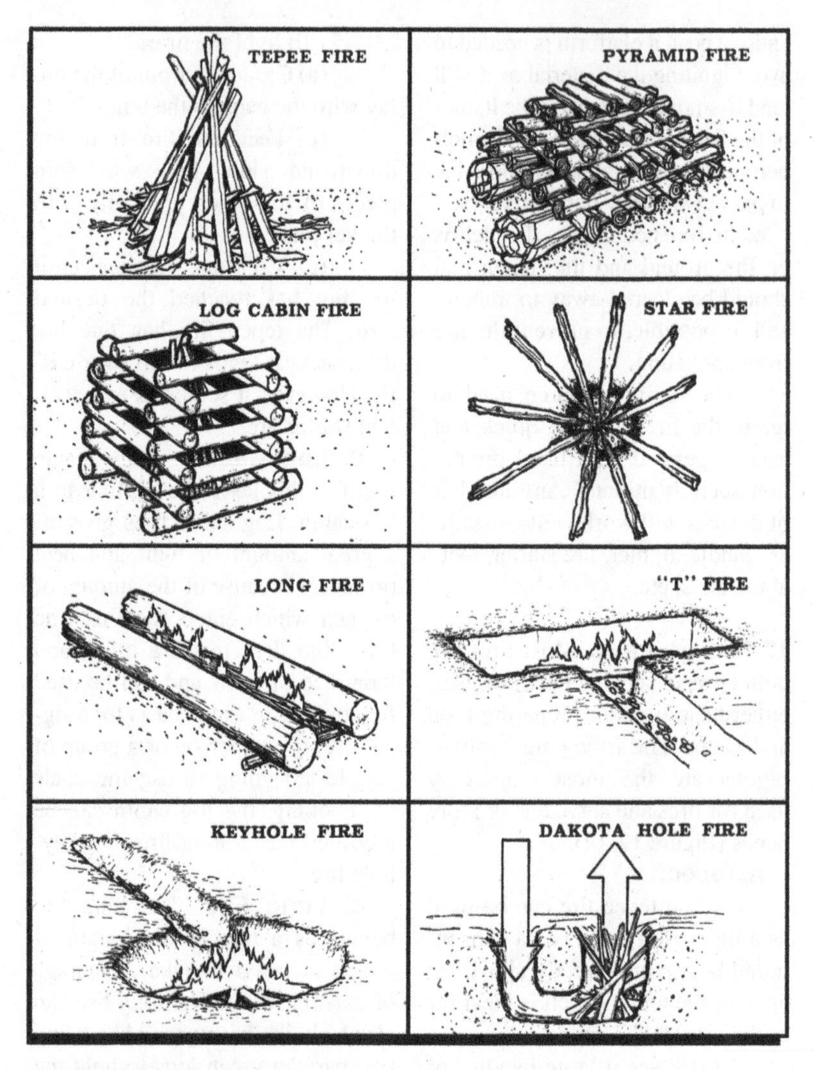

Figure 12-14. Fire Lays.

d. Keyhole Fire. To construct a keyhole fire, a hole is dug in the shape of an old style keyhole and does the same thing as the long fire.

e. Pyramid Fire. The pyramid fire looks similar to a log cabin fire except there are layers of fuel in place of a hollow framework. The advantage of a pyramid fire is that it burns for a long time resulting in a large bed of coals. This fire could possibly be used as an overnight fire when placed in front of a shelter opening.

f. Star Fire. This fire is used when conservation of fuel is necessary or a small fire is desired. It burns at the center of the "wheel" and must be constantly tended. Hardwood fuels work best with this type of fire.

g. "T" Fire. Used for large group cooking. The size of this lay may be adjusted to meet the group's cooking needs. In the top part of the "T," the fire is constructed and maintained as long as needed to provide hot coals for cooking in the bottom part of the "T" fire lay. The number of hot coals may be adjusted in the lower part of the "T" fire lay to regulate the cooking temperature.

h. "V" Fire. This fire lay is a modification of the long fire. The configuration allows a survivor to either block strong winds, or take advantage of light breezes. During high wind conditions, the vertex of the "V"—formed by the two outside logs—is placed in the direction from which the winds are coming, thereby sheltering the tinder (kindling) for ignition. Reversing the "lay" will funnel light breezes into the tinder (kindling) thereby facilitating ease of ignition (Figure12-1).

Chapter 13

EQUIPMENT

13.1. Introduction. Survivors in a survival situation have needs which must be met—food, water, clothing, shelters, etc. The survival kit contains equipment which can be used to satisfy these needs. Quite often, however, this equipment may not be available due to damage or loss. This chapter will address the care and use of issued equipment and improvising the needed equipment when not available. The uses of some issued items are covered in appropriate places throughout this regulation. The care and use of equipment (not covered elsewhere) will be addressed here.

2. Types of Kits:

a. All survival kits contain two types of equipment—mandatory and optional. The mandatory equipment for survival kits are:

(1) One-man liferaft (1 each).
(2) Compass (1 each).
(3) Smoke and illumination flares (2 each).
(4) Signal mirror (1 each).
(5) Hand-held launched flare (1 each).
(6) First-aid kit (1 each).
(7) Survival radio (1 each).

NOTE: These items of equipment may not be mandatory for raft kit.

b. Optional items are authorized by the major air commands; this authority is delegated to subordinate commanders. These optional items are directly related to climatic conditions and the type of terrain which is being flown over. There are over 40 optional items. Here are a few examples:

(1) Sleeping bag.
(2) Strobelight with lenses.
(3) Wire saw.
(4) Water container.
(5) Survival shovel.
(6) Matchbox container.

13.3. Issued Equipment. Survival equipment is designed to aid survivors throughout their survival episode. To maintain its effectiveness, the equipment must be well cared for.

a. Electronic Equipment:

(1) Electronic signaling devices are by far the survivors' most important signaling devices. Therefore, it is important for survivors to properly care for them to ensure their continued effectiveness. In cold temperatures, the electronic signaling devices must be kept warm to prevent the batteries from becoming cold soaked.

(2) In a cold environment, if survivors speak directly into the microphone, the moisture from their breath may condense and freeze on the microphone, creating communication problems.

(3) Caution must be used when using the survival radios in a cold environment. If the radio is placed against the side of the face to communicate, frostbite could result.

(4) In a wet environment, survivors should make every effort to keep their electronic signaling devices dry. In an open-sea environment, the only recourse may be to shake the water out of the microphone before transmitting.

(b) Firearms:

(1) A firearm is a precision tool. It will continue functioning only as long as it is cared for. Saltwater, perspiration, dew, and humidity can all corrode or rust a firearm until it is inoperable. If immersed in saltwater, the survivor should wash the parts in freshwater and then dry and oil them. As an expedient, one way to dry the firearm is to place it in boiling water and after removal, wipe off the excess moisture. The residual heat will evaporate most of the remaining moisture. Survivors should not use uncontrolled heat to dry the firearm as heat over 250°F can remove the temper from the springs in a short time and weaken the action.

(2) Any petroleum-based lubricants used in cold environments will stiffen or freeze causing the firearm to become inoperative. It would be better to thoroughly clean the firearm and remove all lubricant. Metal becomes brittle from cold and is, therefore, prone to breakage.

(3) A firearm was not intended for use as a club, hammer, or pry bar. To use it for any purpose other than for which it was designed, would only result in damage to the firearm.

(c) Cutting Tools:

(1) A file and sharpening stone are often packed in a survival kit. The file is normally used for axes, and the stone is normally used for knives.

(2) An old axiom states that a sharp cutting tool is a safe cutting tool. Control of a cutting tool is easier to maintain if it is sharp, and the possibility of accidental injury is reduced.

(3) One of the most valuable items in any survival situation is a knife, since it has a large number of uses. Unless the knife is kept sharp, however, it falls short of its potential.

(4) A knife should be sharpened only with a stone as repeated use of a file rapidly removes steel from the blade. In some cases, it may be necessary to use a file to remove plating from the blade before using the stone.

(5) One of two methods should be used to sharpen a knife. One method is to push the blade down the stone in a slicing motion. Then turn the blade over and draw the blade toward the body (Figure 13-1).

(6) The other method is to use a circular motion the entire length of the blade; turn the blade over and

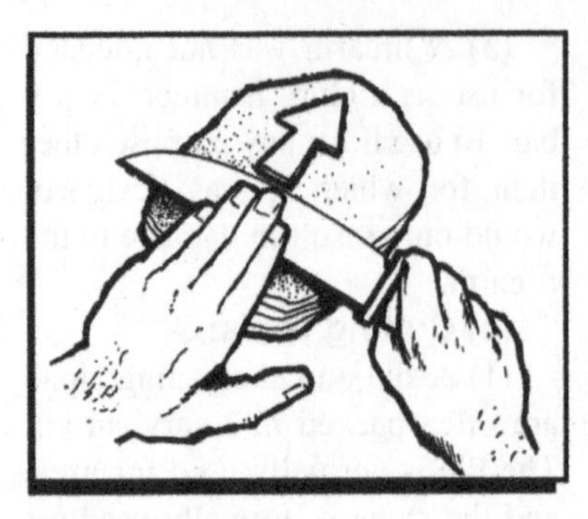

Figure 13-1. Knife Sharpening (Draw).

repeat the process. What is done to one side of the cutting edge should also be done to the other to maintain an even cutting edge (Figure 13-2).

(7) Most sharpening stones available to survivors will be whetstones. Water should be applied to these stones. The water will help to float away the metal removed by sharpening and make cleaning of the stone easier.

(8) If a commercial whetstone is not available, a natural whetstone can be used. Any standstone will sharpen tools, but a gray, clay-like sandstone gives better results. Quartzite should be avoided. Survivors can recognize quartzite instantly by scratching the knife blade with it—the quartz crystals will bite into steel. If no sandstone is available, granite or crystalline rock can be used. If granite is used, two pieces of the stone should be rubbed together to smooth the surface before use.

Figure 13-2. Knife Sharpening (Circular).

(9) As with a knife, a sharp axe will save time and energy and be much safer.

(10) A file should be used on an axe or hatchet. Survivors should file away from the cutting edge to prevent injury if the file should slip. The file should be worked from one end of the cutting edge to the other. The opposite side should be worked to the same degree. This will ensure that the cutting edge is even. After using a file, the stone may be used to hone the axe blade (Figure 13-3).

(11) When using an axe, don't try to cut through a tree with one blow. Rhythm and aim are more important than force. Too much power behind a swing interferes with aim. When the axe is swung properly, its weight provides all the power needed.

(12) Carving a new axe handle and mounting the axe head takes a great deal of time and effort. For this reason, a survivor should avoid actions which would require the handle to be changed. Using aim and paying attention to where the

axe falls will prevent misses which could result in a cracked or broken handle. Survivors should not use an axe as a pry bar and should avoid leaving the axe out in cold weather where the handle may become brittle.

Figure 13-3. Sharpening Axe.

(13) A broken handle is difficult to remove from the head of the axe. Usually the most convenient way is to burn it out (Figure 13-4). For a single-bit axe, bury the bit in the ground up to the handle, and build a fire over it. For a double-bit, a survivor should dig a small trench, lay the middle of the axe head over it, cover both "bits" with earth, and build the fire. The covering of earth keeps the flame from the cutting edge of the axe and saves its temper. A little water added to the earth will further ensure this protection.

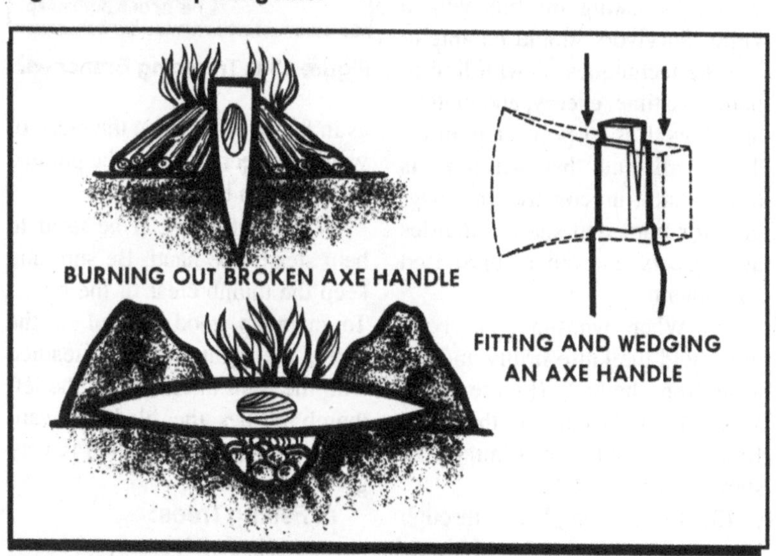

Figure 13-4. Removing Broken Axe Handle.

(14) When improvising a new handle, a survivor can save time and trouble by making a straight handle instead of a curved one like the original. Survivors should use a young, straight piece of hardwood without knots. The wood should be whittled roughly into shape and finished by shaving. A slot should be cut into the axe-head end of the handle. After it is fitted, a thin, dry wooden wedge can then be pounded into the slot. Survivors should use the axe awhile, pound the wedge in again, then trim it off flush with the axe. The handle must be smoothed to remove splinters. The new handle can be seasoned to prevent shrinkage by "scorching" it in the fire,

d. Whittling:

(1) Whittle means to cut, trim, or shape (a stick or piece of wood) by taking off bits with a knife. Survivors should be able to use the techniques of whittling to help save time, energy, and materials as well as to prevent injuries. They will find that whittling is a necessity in constructing triggers for traps and snares, shuttles and spacers, and other improvised equipment.

(2) When whittling, survivors must hold the knife firmly and cut away from the body (Figure 13-5). Wood should be cut with the grain. Branches should be trimmed as shown in figure 13-6.

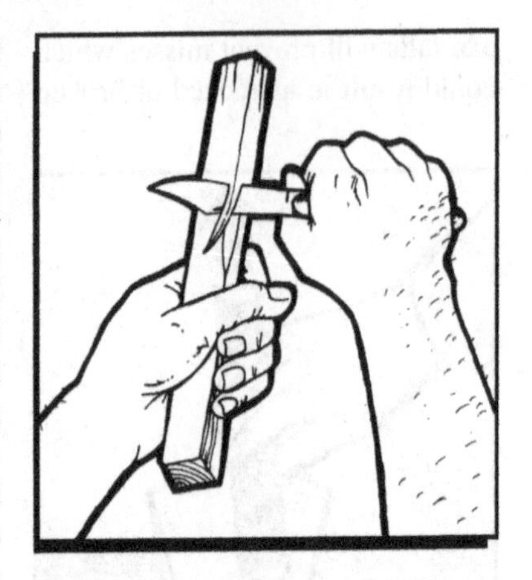

Figure 13-5. Whittling

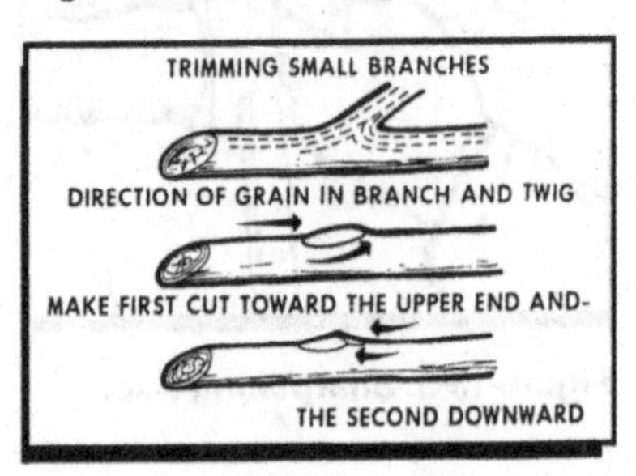

Figure 13-6. Trimming Branches.

(3) To cut completely through a piece of wood, a series of V-cuts should be made all the way around as in figure 13-7. Once the piece of wood has been severed, the pointed end can then be trimmed.

(4) The thumb can be used to help steady the hand. Be sure and keep the thumb clear of the blade. To maintain good control of the knife, the right hand is steadied with the right thumb while the left thumb pushes the blade forward (Figure 13-8). This method is very good for trimming.

e. Felling Trees:

(1) To fell a tree, the survivor must first determine the direc-

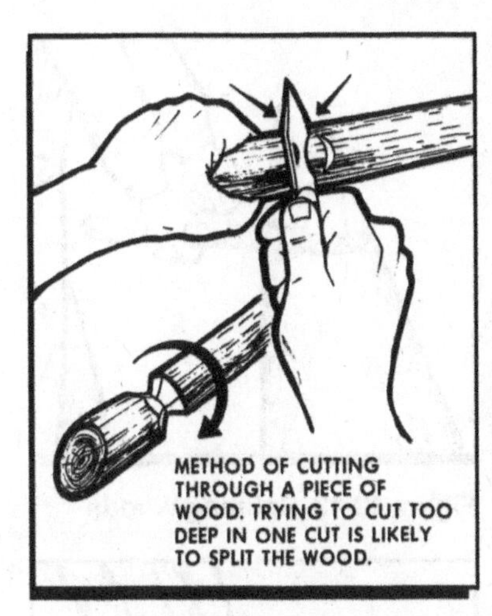

Figure 13-7. Cutting Through a Piece of Wood.

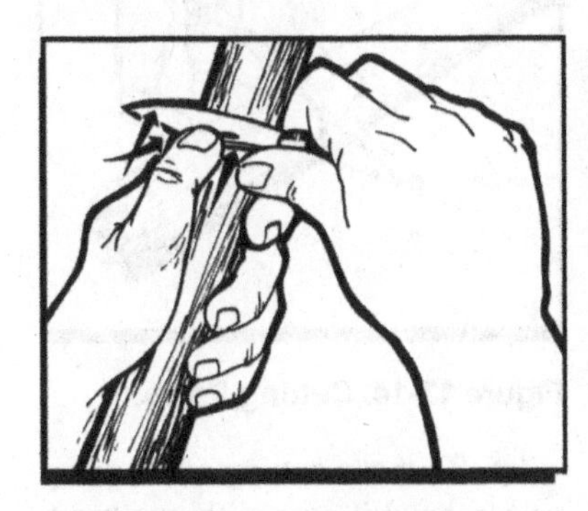

Figure 13-8. Fine Trimming.

Figure 13-9. Using the Axe as a Plumb Line.

Figure 13-10. Clearing Brush from Cutting Area.

tion in which the tree is to fall. It is best to fell the tree in the direction in which it is leaning. The lean of the tree can be found by using the axe as a plumb line (Figure 13-9). The survivor should then clear the area around the tree from underbrush and overhanging branches to prevent injury (Figure 13-10).

(2) The survivor should make two cuts. The first cut should be on the leaning side of the tree and close to the ground and the second cut on the opposite side and a little higher than the first cut (Figure 13-11).

(3) Falling trees often kick back and can cause serious injury (Figure 13-12), so survivors must ensure they have a clear escape route. When limbing a tree, start at the base of the tree and cut toward the top. This procedure will allow for easier limb removal and results in a smoother cut. For safety, the survivor should stand on one side

Figure 13-11. Felling Cuts.

Figure 13-12. Tree Kickbacks.

of the trunk with the limb on the other.

(4) To prevent damage to the axe head and possible physical injury, any splitting of wood should be done on a log as in figure 13-13. The log can also be used for cutting sticks and poles (Figure 13-14).

Figure 13-13. Splitting Woods

Figure 13-14. Cutting Poles.

(5) To make cutting of a sapling easier, bend it over with one hand, straining grain. A slanting blow close to the ground will cut the sapling (Figure 13-15).

13-4. Improvised Equipment:

a. If issued equipment is inoperative, insufficient, or nonexistent, survivors will have to rely upon their ingenuity to manufacture the needed equipment. Survivors must determine whether the need for the item outweighs the work involved

Figure 13-15. Cutting Saplings.

to manufacture it. They will also have to evaluate their capabilities. If they have injuries, will the injuries prevent them from manufacturing the item(s)?

b. Undue haste may not only waste materials, but also waste the survivors' time and energy. Before manufacturing equipment, they should have a plan in mind.

c. The survivors' equipment needs may be met in two different ways. They may alter an existing piece of equipment to serve more than one function, or they may also construct a new piece of equipment from available materials. Since the items survivors can improvise are limited only by their ingenuity, all improvised items cannot be covered in this regulation.

d. The methods of manufacturing the equipment referred to in this regulation are only ideas and do not have to be strictly adhered to. Many Air Force survivors have a parachute. This device can be used to improvise a variety of needed equipment items.

e. The parachute consists of (Figure 13-16):

(1) The pilot chute which deploys first and pulls the rest of the parachute out.

(2) The parachute canopy which consists of the apex (top) and the skirt or lower lateral band. The canopy material is divided by radial seams into 28 sections called gores. Each gore measures about 3 feet at the skirt and tapers to the apex.

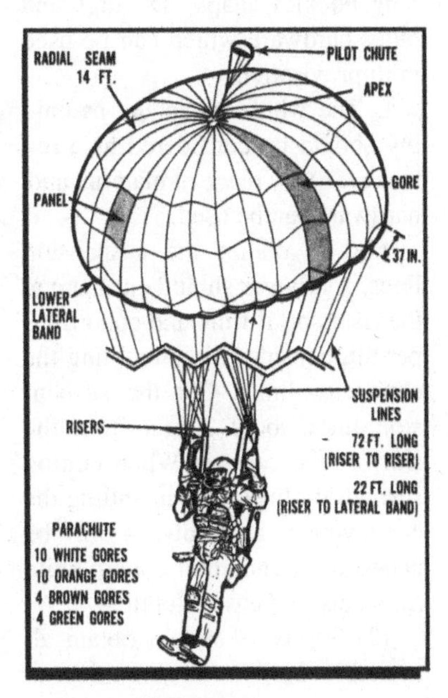

Figure 13-16. Parachute Diagram.

Each gore is further subdivided into four sections called panels. The canopy is normally divided into four colors. These colored areas are intended to aid the survivor in shelter construction, signaling, and camouflage.

(3) Fourteen suspension lines connect the canopy material to the harness assembly. Each piece of suspension line is 72 feet long from riser to riser and 22 feet long from riser to skirt and 14 feet from skirt to apex. The tensile strength of each piece of suspension line is 550 pounds. Each piece of suspension line contains seven to nine pieces of innercore with a tensile strength of 35 pounds. The harness assembly contains risers and webbing, buckles, snaps, "D" rings, and other hardware which can be used in improvisation.

f. The whole parachute assembly should be considered as a resource. Every piece of material and hardware can be used.

(1) To obtain the suspension lines, a survivor should cut them at the risers or, if time and conditions permit, consider disassembling the connector links. Cut the suspension lines about 2 feet from the skirt of the canopy. When cutting suspension lines or dismantling the canopy/pack assembly, it will be necessary to maintain a sharp knife for safety and ease of cutting.

(2) Survivors should obtain all available suspension line due to its many uses. Even the line within the radial seams of the canopy should be stripped for possible use. The suspension line should be cut above the radial seam stitches next to the skirt end of the canopy (two places). The cut should not go all of the way through the radial seam (Figure 13-17). At the apex of the canopy, and just below the radial seam stitching, a horizontal cut can be made and the suspension line extracted. The line can then be cut.

(3) For maximum use of the canopy, survivors must plan its disassembly. The quantity requirements for shelter, signaling, etc., should be thought out and planned for. Once these needs have been determined, the canopy may be cut up. The radial seam must be stretched tightly for ease of cutting. The radial seam can then be cut by holding the knife at an angle and following the center of the seam. With proper tension and the gentle pushing (or pulling) of a sharp blade there will be a controlled splitting of the canopy at the seam (Figure 13-17). It helps to secure the apex either to another individual or to an immobile object such as a tree.

(4) When stripping the harness assembly, the seams of the webbing should be split so the maximum usable webbing is obtained. The harness material and webbing should not be randomly cut as it will waste much needed materials.

g. One requirement in improvising is having available material. Parachute fabric, harness, suspen-

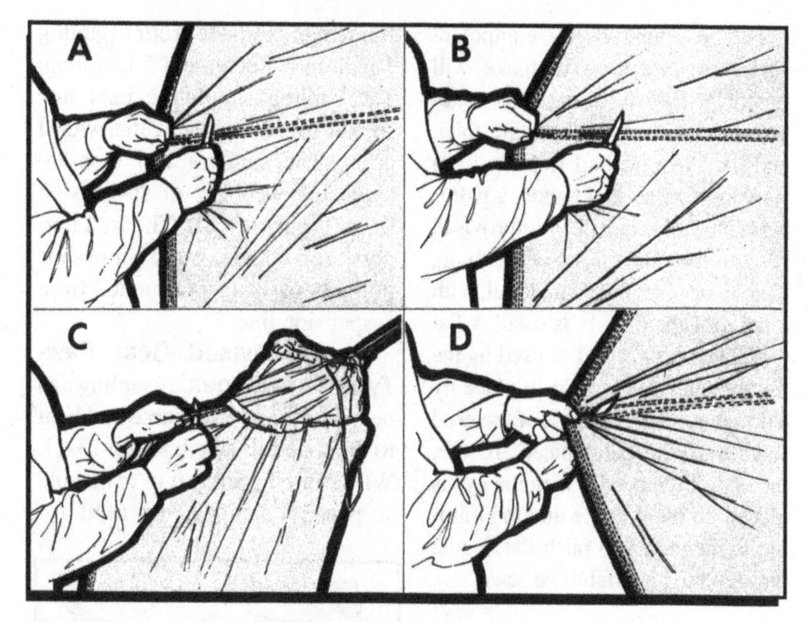

Figure 13-17. Cutting the Parachute.

sion lines, etc., can be used for clothing. Needles are helpful for making any type of emergency clothing. Wise survivors should always have extra sewing needles hidden somewhere on their person. A good needle or sewing awl can be made from the can-opening key from the ration tin (Figure 13-18) or, as the Eskimos do, from a sliver of bone. Thread is usually available in the form of innercore. It will be to the survivor's benefit to collect small objects which may "come in handy." Wire, nails, buttons, a piece of canvas, or animal skin should not be discarded. Any such object may be worth its weight in gold when placed in a hip pocket or a sewing kit. Any kind of animal skin can be used for making clothing such

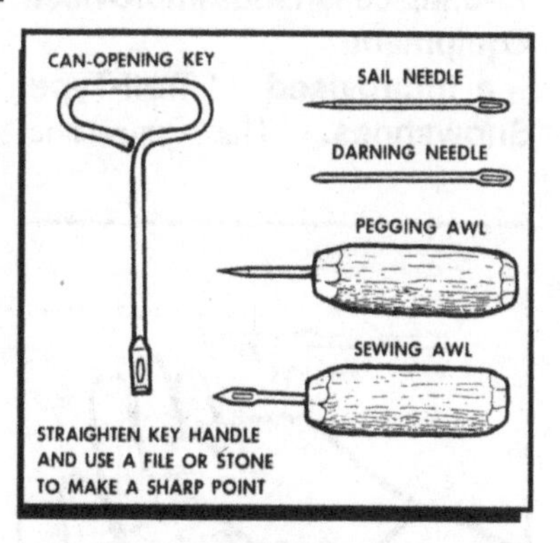

Figure 13-18. Needle and Sewing Awl.

as gloves or mittens or making a ground cover to keep the sleeping bag dry and clean. Small skins can be used for mending and for boot insoles. Mending and cleaning clothes when possible will pay dividends in health, comfort, and safety.

h. The improvised equipment survivors may need to make will probably involve sewing. The material to be sewn may be quite thick and hard to sew, and to keep from stabbing fingers and hands, a palm-type thimble can be improvised (Figure 13-19). A piece of webbing, leather, or other heavy material, with a hole for the thumb, is used. A flat rock, metal, or wood is used as the thimble and this is held in place by a doughnut-shaped piece of material sewn onto the palm piece. To use, the end of the needle with the eye is placed on the thimble and the thimble is then used to push the needle through the material to be sewn.

13-5. Miscellaneous Improvised Equipment:

a. Improvised Trail-Type Snowshoes. The snowshoe frame can be made from a sapling 1 inch in diameter and 5 feet long. The sapling should be bent and spread to 12 inches at the widest point. The survivor can then include the webbing of suspension lines (Figure 13-20). The foot harness, for attaching the snowshoe to the boot, is also fashioned from suspension line.

b. Improvised Bear Paw-Type Snowshoes. A sapling can be held over a heat source and bent to the shape shown in figure 13-21. Wire from the aircraft or parachute suspension line can be used for

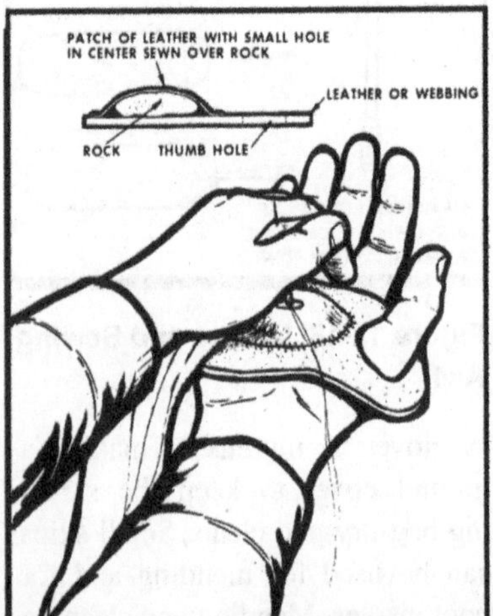

PATCH OF LEATHER WITH SMALL HOLE IN CENTER SEWN OVER ROCK
LEATHER OR WEBBING
ROCK
THUMB HOLE

Figure 13-19. Palm Thimble.

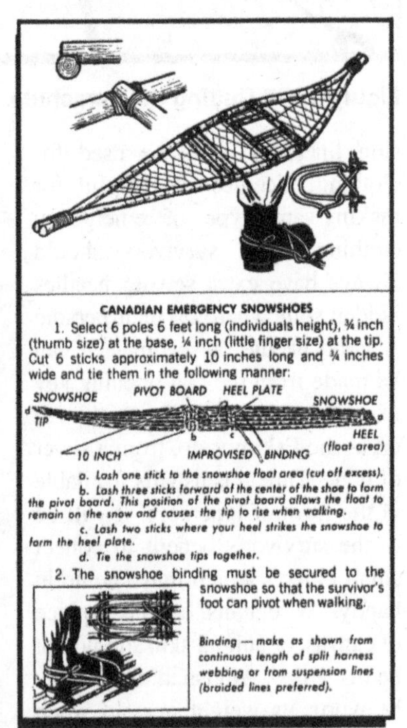

Figure 13-20. Improvised Trail Snowshoes.

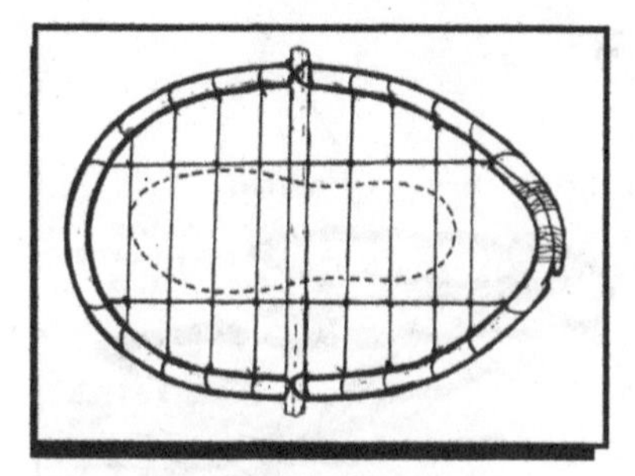

Figure 13-21. Improvised Bear Paws.

lashing and for making webbing. Snowshoes can also be quickly improvised by cutting a few pine boughs and lashing them together at the cut ends. The lashed boughs positioned with the cut ends forward can then be tied to the feet (Figure 13-22).

(1) Survivors should guard against frostbite and blistering while snowshoeing. Due to the design of the harness, the circulation of the toes is usually restricted, and the hazard of frostbite is greater. They should check the feet carefully, stop often, take off the harness, and massage the feet when they seem to be getting cold.

(2). Blistering between the toes or on the ball of the foot is sometimes unavoidable in a "tenderfoot" if much snowshoeing is done. To make blisters less likely, the survivor should keep socks and insoles dry and change them regularly.

c. Sleeping Bag. Immediate action should be to use the whole parachute until conditions allow for improvising. A sleeping bag can be improvised by using four gores of parachute material or an equivalent amount of other materials (Figure 13-23). The material should be folded in half lengthwise

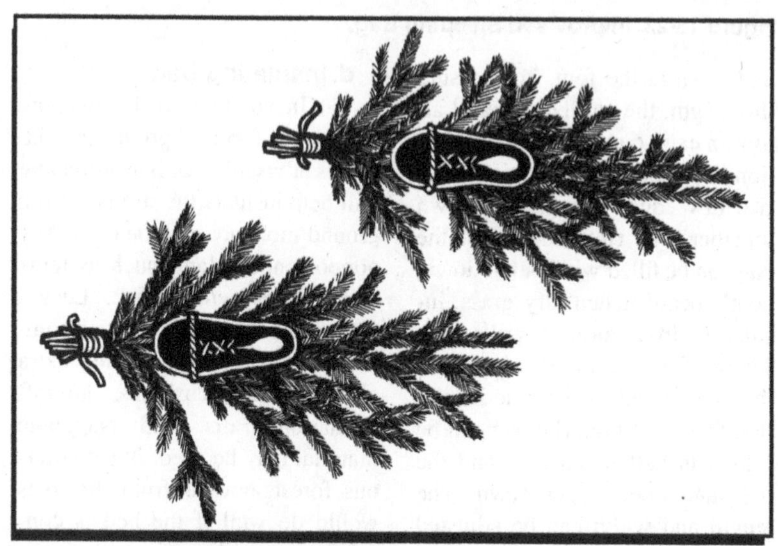

Figure 13-22. Bough Snowshoes.

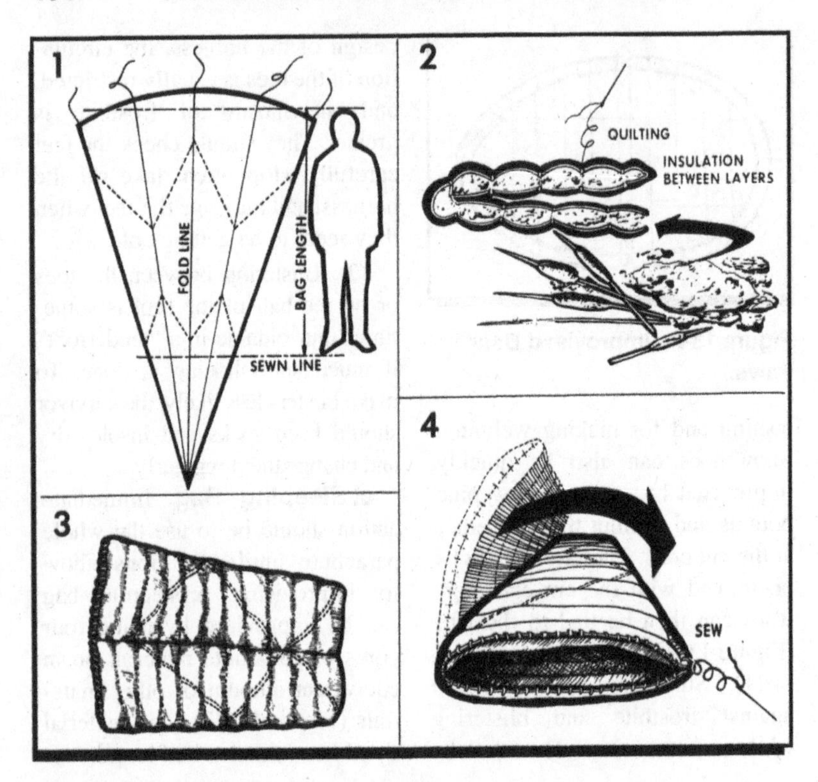

Figure 13-23. Improvised Sleeping Bag.

and sewn at the foot. To measure the length, the survivor should allow an extra 6 to 10 inches in addition to the individual's height. The two raw edges can then be sewn together. The two sections of the bag can be filled with cattail down, goat's beard lichen, dry grass, insulation from aircraft walls, etc. The stuffed sleeping bag should then be quilted to keep the insulation from shifting. The bag can be folded in half lengthwise and the foot and open edges sewn. The length and width can be adjusted for the individual.

d. Insulating Bed:

(1) In addition to the sleeping bag, some form of ground insulation is advisable. An insulation mat will help insulate the survivor from ground moisture and the cold. Any nonpoisonous plants such as ferns and grasses will suffice. Leaves from a deciduous tree make a comfortable bed. If available, extra clothing, seat cushions, aircraft insulation, rafts, and parachute material may be used. In a coniferous forest, boughs from the trees would do well if the bed is constructed properly.

(2) The survivor should start at the foot of the proposed bed and stick the cut ends in the ground at about a 45-degree angle and very close together. The completed bed should be slightly wider and longer than the body. If the ground is frozen, a layer of dead branches can be used on the ground with the green boughs placed in the dead branches, similar to sticking them in the ground.

(3) A bough bed should be a minimum of 12 inches thick before use. This will allow sufficient insulation between the survivor and the ground once the bed is compressed. The bough bed should be fluffed up and boughs added daily to maintain its comfort and insulation capabilities.

(4) Spruce boughs have many sharp needles and can cause some discomfort. Also the needles on various types of pines are generally located on the ends of the boughs, and it would take an abundance of pine boughs to provide comfort and insulation. Fir boughs on the other hand, have an abundance of needles all along the boughs and the needles are rounded. These boughs are excellent for beds, providing comfort and insulation (Figure 13-24).

e. Rawhide. Rawhide is a very useful material which can be made from any animal hide. Processing it is time consuming but the material obtained is strong and very durable. It can be used for making sheaths for cutting tools, lashing materials, ropes, etc.

(1) The first step in making rawhide is to remove all of the fat and muscle tissue from the hide. The large pieces can be cut off and the remainder scraped off with a dull knife or similar instrument.

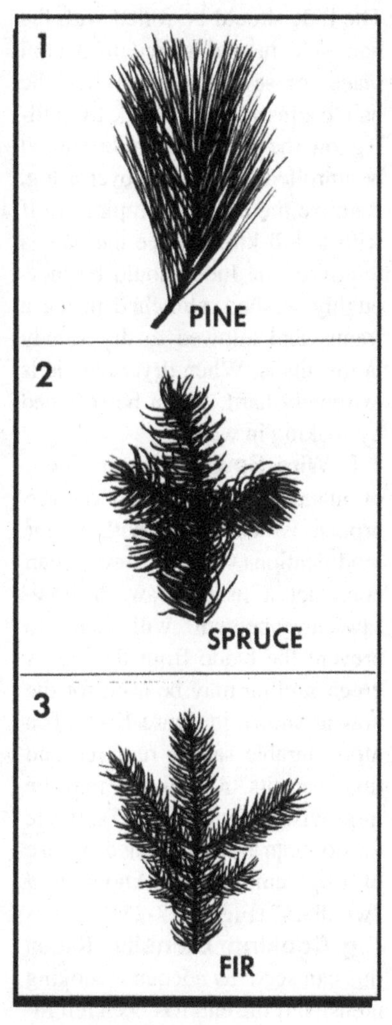

Figure 13-24. Boughs.

(2) The next step is to remove the hair. This can be done by applying a thick layer of wood ashes to the hair side. Ashes from a hardwood fire work best. Thoroughly sprinkle water all over the ashes. This causes lye to leach out of the ash. The lye will remove the hair. The hide should be rolled with the hair side in and stored in a cool place for several days. When the hair begins to slip (check by pulling on the hair), the hide should be unrolled and placed over a log. Remove the hair by scraping it off with a dull knife. Once the hair is removed, the hide should be thoroughly washed, stretched inside a frame, and allowed to dry slowly in the shade. When dry, rawhide is extremely hard. It can be softened by soaking in water.

f. Wire Saws. Wire or pieces of metal can be used to replace broken issued saws. With minor modifications, the survivor can construct a usable saw. A bow-saw arrangement will help to prevent the blade from flexing. A green sapling may be used for the bow as shown in figure 13-25. If a more durable saw is required and time permits, a bucksaw may be improvised (Figure 13-25). Blade tension can be maintained by use of a tightening device known as a "windlass" (Figure 13-25).

g. Cooking Utensils. Ration tins can serve as adequate cooking utensils. If the end has been left intact as in figure 13-26, use a green stick long enough to prevent burning the hand while cooking. If the side has been left intact, a forked stick may be used to add support to the container (Figure 13-26).

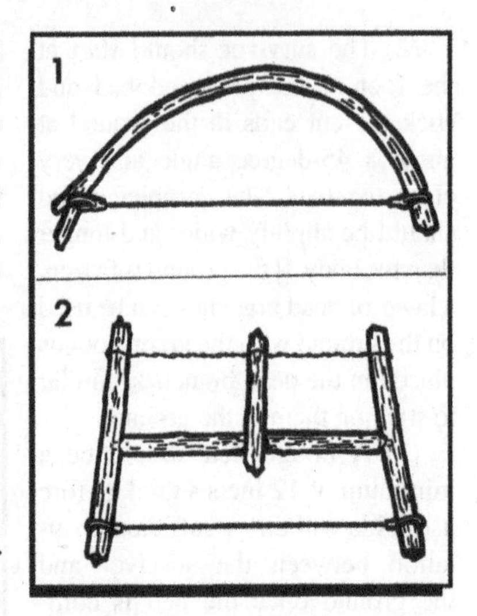

Figure 13-25. Bow Saw and Buck Saw.

13-6. Ropes and Knots:

a. Basic Knowledge of Tying a Knot. A basic knowledge of correct rope and knot procedures

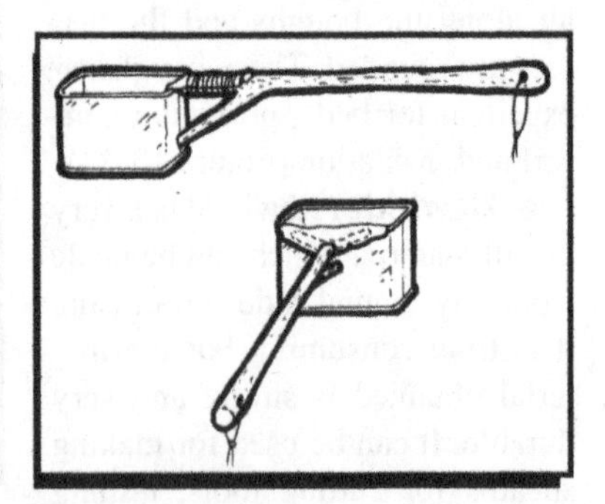

Figure 13-26. Cooking Utensils.

will aid the survivor to do many necessary actions. Such actions as improvising equipment, building shelters, assembling packs, and providing safety devices require the use of proven techniques. Tying a knot incorrectly could result in ineffective improvised equipment, injury, or death.

b. Rope Terminology: (See figure 13-27.)

(1) Bend. A bend (called a knot in this regulation) is used to fasten two ropes together or to fasten a rope to a ring or loop.

(2) Bight. A bight is a bend or U-shaped curve in a rope.

(3) Hitch. A hitch is used to tie a rope around a timber, pipe, or post so that it will hold temporarily but can be readily untied.

(4) Knot. A knot is an interlacement of the parts of bodies, as cordage, forming a lump or knot or any tie or fastening formed with a cord, rope, or line, including bends, hitches, and splices. It is often used as a stopper to prevent a rope from passing through an opening.

(5) Line. A line (sometimes called a rope) is a single thread, string, or cord.

(6) Loop. A loop is a fold or doubling of the rope through which another rope can be passed. A temporary loop is made by a knot or a hitch. A permanent loop is made by a splice or some other permanent means.

(7) Overhand Turn or Loop. An overhand loop is made when the running end passes over the standing part.

(8) Rope. A rope (often called a line) is made of strands of fiber twisted or braided together.

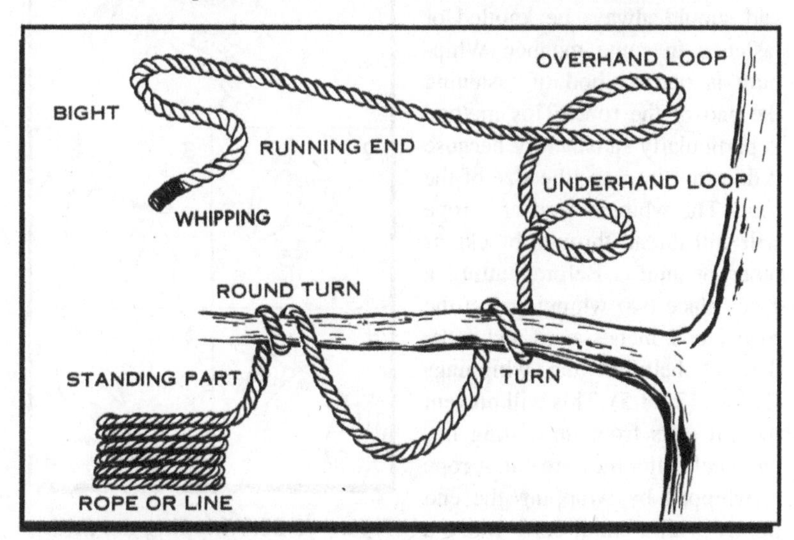

Figure 13-27. Elements of Ropes and Knots.

(9) Round Turn. A round turn is the same as a turn, with running end leaving the circle in the same general direction as the standing part.

(10) Running End. The running end is the free or working end of a rope.

(11) Standing End. The standing end is the balance of the rope, excluding the running end.

(12) Turn. A turn describes the placing of a rope around a specific object such as a post, rail, or ring with the running end continuing in the opposite direction from the standing end.

(13) Underhand Turn or Loop. An underhand turn or loop is made when the running end passes under the standing part.

c. Whipping the Ends of a Rope: The raw, cut end of a rope has a tendency to untwist and should always be knotted or fastened in some manner. Whipping is one method of fastening the end of the rope. This method is particularly satisfactory because it does not increase the size of the rope. The whipped end of a rope will still thread through blocks or other openings. Before cutting a rope, place two whippings on the rope 1 or 2 inches apart and make the cut between the whippings (Figure 13-28-5). This will prevent the cut ends from untwisting immediately after they are cut. A rope is whipped by wrapping the end tightly with a small cord. Make a bight near one end of the cord and lay both ends of the small cord along one side of the rope (Figure 13-28-1). The bight should project beyond the end of the rope about one-half inch. The running end (b) of the cord should be wrapped tightly around the rope and cord (Figure 13-28-2) starting at the end of the whipping which will be farthest from the end of the rope. The wrap should be in the same direction as the twist of the rope strands. Continue wrapping the cord around the rope, keeping it tight, to within about one-half inch of the end. At this point, slip the running end (b) through the bight of the cord (Figure 13-28-3). The standing part of the cord (a) can then be pulled until the bight of the cord is pulled under

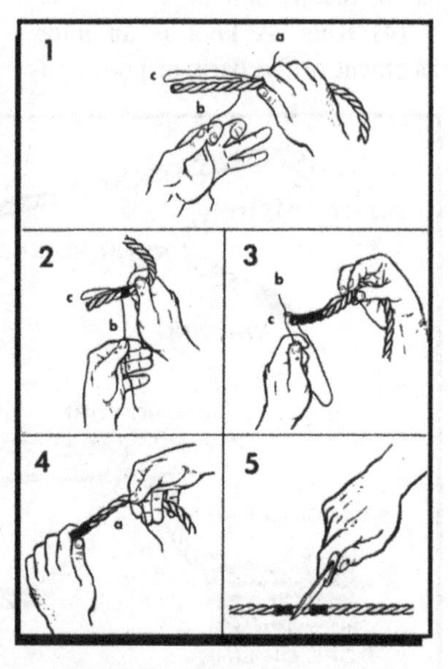

Figure 13-28. Whipping the End of a Rope.

the whipping and cord (b) is tightened (Figure 13-28-4). The ends of cord (a and b) should be cut at the edge of the whipping, leaving the rope end whipped,

d. Knots at End of the Rope:

(1) Overhand Knot. The overhand knot (Figure 13-29) is the most commonly used and the simplest of all knots. An overhand knot may be used to prevent the end of a rope from untwisting, to form a knot at the end of a rope, or as a part of another knot. To tie an overhand knot, make a loop near the end of the rope and pass the running end through the loop, pulling it tight.

Figure 13-29. Overhand Knot.

(2) Figure-Eight Knot. The figure-eight knot (Figure 13-30) is used to form a larger knot than would be formed by an overhand knot at the end of a rope. A figure-eight knot is used in the end of a rope to prevent the ends from slipping through a fastening or loop in another rope. To make the figure-eight knot, make a loop in the standing part, pass the running end aroundthe standing part back over one side of the loop, and down through the loop. The running end can then be pulled tight.

Figure 13-30. Figure-Eight Knot.

(3) Wall Knot. The wall knot (Figure 13-31) with a crown is used to prevent the end of a rope from untwisting when an enlargement is not objectionable. It also makes a desirable knot to prevent the end of the rope from slipping through small openings, as when rope handles are used on boxes. The crown or the wall knots may be used separately. To make the wall knot, untwist the strands for about five turns of the rope. A loop in strand "a" (Figure 13-32-1) should be used and strand "b" brought down (Figure 13-31-2) and around strand "a." Strand "c" (Figure 13-31-3) can then be brought around strand "b" and through the loop in strand "a." The knot can then be tightened (Figure 13-31-4) by grasping the rope in one hand and pulling each strand tight. The strands point up or away from the rope. To make a neat, round knot, the wall knot should be crowned.

(1) Crown on Wall Knot. To crown a wall knot, the end of

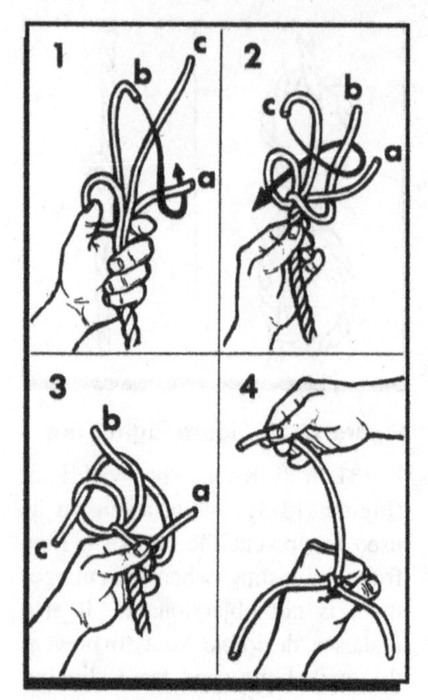

Figure 13-31. Wall Knot.

strand "a" (Figure 13-32-1) should be moved between strands "b" and "c." Next strand "c" is passed (Figure 13-32-2) between strand "b" and the loop in strand "a." Line "b" is then passed over line "a" and through the bight formed by line "c" (Figure 13-32-3). The knots can then be drawn tight and the loose strands cut. When the crown is finished, strands should point down or back along the rope.

e. Knots for Joining Two Ropes:

(1) Square Knot. The square knot (Figure 13-33) is used for tying two ropes of equal diameter together to prevent slippage. To tie the square knot, lay the running end of each rope together but pointing in opposite directions. The running end of one rope can be passed underthe standing part of the other rope. Bring the two running ends up away from the point where they cross and crossed again (Figure 13-33-1). Once each running end is parallel to its own standing part (Figure 13-33-2), the two ends can be pulled tight. If each running end does not come parallel to the standing part of its own rope, the knot is called a "granny knot" (Figure 13-34-1). Because it will

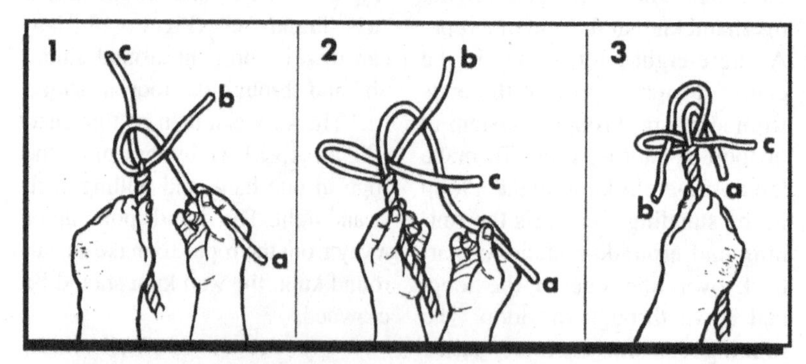

Figure 13-32. Crown on Wall Knot.

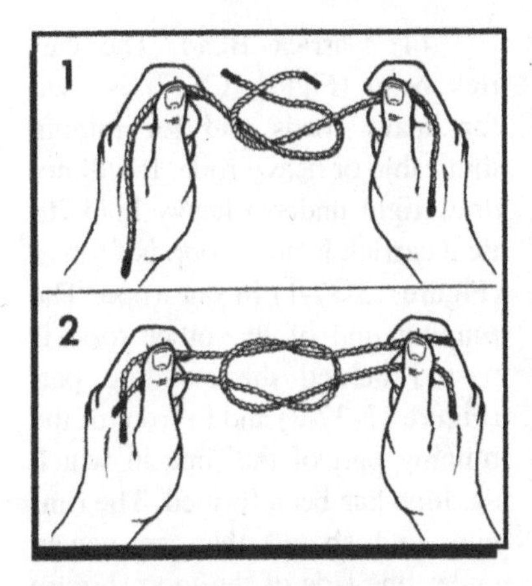

Figure 13-33. Square Knot.

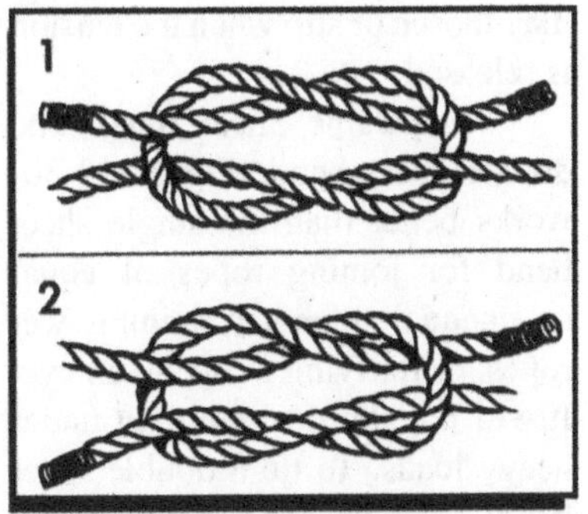

Figure 13-34. Granny and Thief Knots.

slip under strain, the granny knot should not be used. A square knot can also be tied by making a bight in the end of one rope and feeding the running end of the other rope through and around this bight. The running end of the second rope is routed from the standing side of the bight. If the procedure is reversed, the resulting knot will have a running end parallel to each standing part but the two running ends will not be opposite each other. This knot is called a "thief' knot (Figure 13-34-2). It will slip under strain and is difficult to untie. A true square knot will draw tighter understrain. A square knot can be untied easily by grasping the bends of the two bights and pulling the knot apart.

(2) Single Sheet Bend. The use of a single sheet bend (Figure 13-35), sometimes called a weaver's knot, is limited to tying together two dry ropes of unequal size. To tie the single sheet bend, the running end (a) (Figure 13-35-1) of the smaller rope should pass through a bight (b) in the larger rope. The running end should continue around both parts of the larger rope (Figure 13-35-2), and back under the smaller rope (Figure 13-35-3). The running end can then be pulled tight (Figure 13-35-4). This knot will draw tight under light loads but

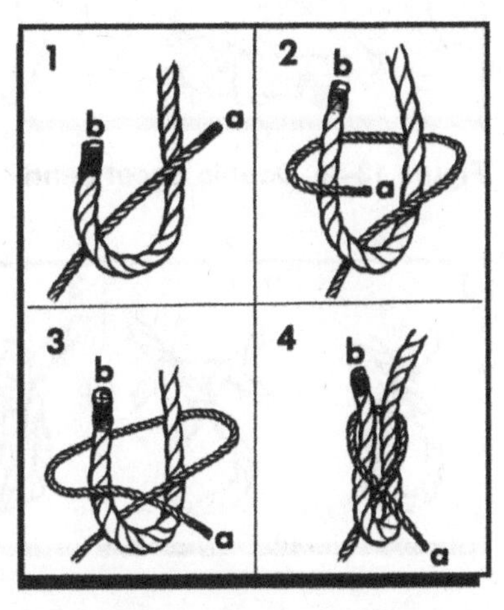

Figure 13-35. Single Sheet Knot.

may loosen or slip when the tension is released.

(3) Double Sheet Bend. The double sheet bend (Figure 13-36) works better than the single sheet bend for joining ropes of equal or unequal diameter, joining wet ropes, or for tying a rope to an eye. It will not slip or draw tight under heavy loads. To tie a double sheet bend, a single sheet bend is tied first. However, the running end is not pulled tight. One extra turn is taken around both sides of the bight in the larger rope with the running end for the smaller rope. Then tighten the knot.

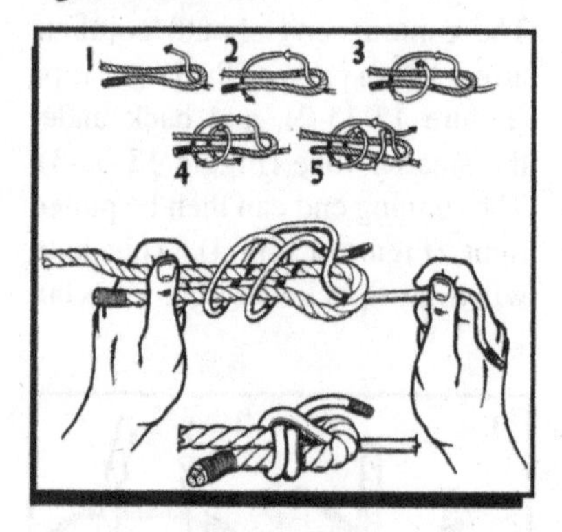

Figure 13-36. Double Sheet Bend

(4) Carrick Bend. The carrick bend (Figure 13-37) is used for heavy loads and for joining thin cable or heavy rope. It will not draw tight under a heavy load. To tie a carrick bend, a loop is formed (Figure 13-37-1) in one rope. The running end of the other rope is passed behind the standing part (Figure 13-37-2) and in front of the running part of the rope in which the loop has been formed. The running end should then be woven under one side of the loop (Figure 13-37-3), through the loop, over the standing part of its own rope (Figure 13-37-4), down through the loop, and under the remaining side of the loop (Figure 13-37-5).

f. Knots for Making Loops:

(1) Bowline. The bowline (Figure 13-38) is a useful knot for forming a loop in the end of a rope. It is also easy to untie. To tie the bowline, the running end (a) of the rope passes through the object to be affixed to the bowline and forms a loop (b) (Figure 13-38-1) in the standing part of the rope. The running end (a) is then passed through

Figure 13-37. Carrick Bend

the loop (Figure 13-38-2) from underneath and around the standing part (Figure 13-38-3) of the rope, and back through the loop from the top (Figure 13-38-4). The running end passes down through the loop parallel to the rope coming up through the loop. The knot is then pulled tight.

(2) Double Bowline. The double bowline (Figure 13-39) with a slip knot is a rigging used by tree surgeons who work alone in trees for extended periods. It can be made and operated by one person and is comfortable as a sling or boatswain's chair (Figure 13-40). A small board with notches as a seat adds to the personal comfort of the user. To tie a double bowline, the running end (a) (Figure 13-39) of a line should be bent back about 10 feet along the standing part. The bight (b) is formed as the new running end and a bowline tied as described and illustrated in figure 13-38. The new running end (b) (Figure 13-39) or loop is used to support the back and the remaining two loops (c) and (d) support the legs.

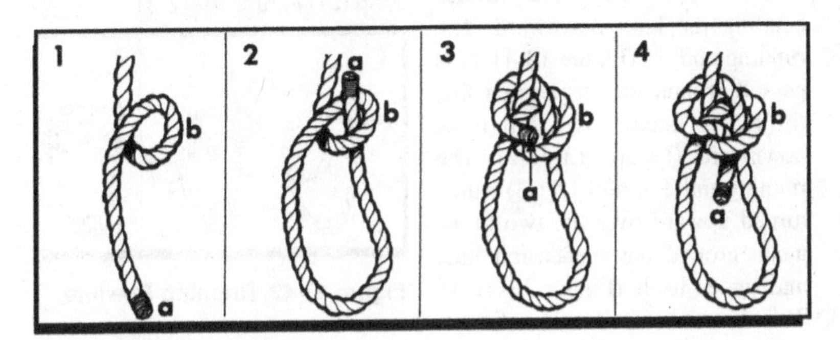

Figure 13-38. Bowline.

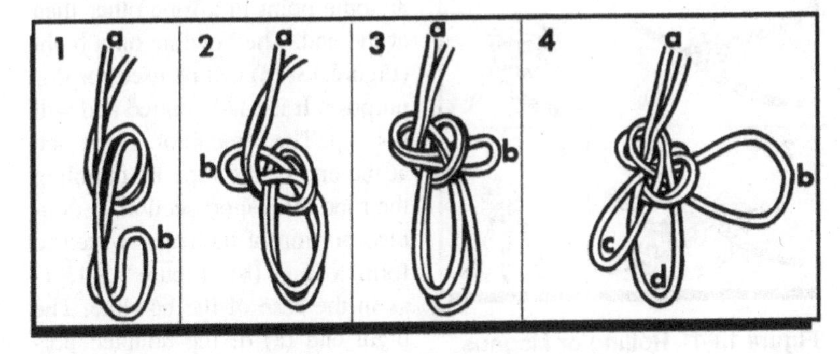

Figure 13-39. Double Bowline.

Figure 13-40. Boatswain's Chair.

(3) Rolling or Magnus Hitch (Figure 13-41). A rolling or Magnus hitch is a safety knot designed to make a running end fast to a suspension line with a nonslip grip yet it can be released by hand pressure bending the knot downward. The running end (a) (Figure 13-41-1) is passed around the suspension line (b) twice, making two full turns downward (Figure 13-41-2). The running end (Figure 13-41-3) is then turned upward over the two turns, again around the suspension line, and under itself (Figure 13-41-3). This knot is excellent for fastening a rope to itself, a larger rope, a cable, a timber, or a post.

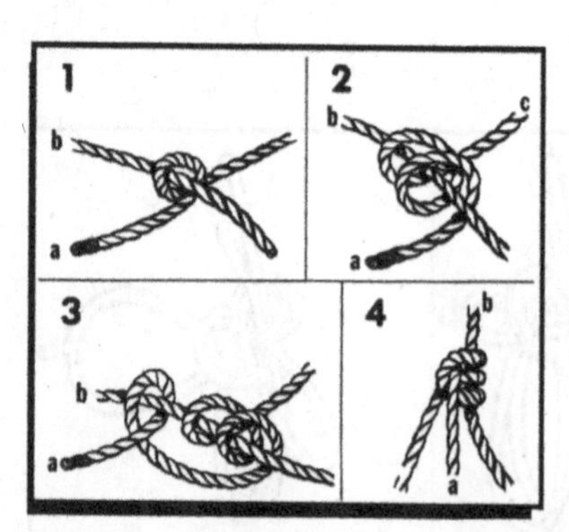

Figure 13-41. Rolling or Magnus Hitch.

(4) Running Bowline. The running bowline (Figure 13-42) is the basic air transport rigging knot. It provides a sling of the choker type at the end of a single line and is generally used in rigging. To tie a running bowline, make a bight (b) (Figure 13-42-1) with an overhand loop (c) made in the running end (a). The running end (a) is passed around the standing part, through the loop (c) (Figure 13-42-2), under, then back over the side of the bight, and back through the loop (c) (Figure 13-42-3).

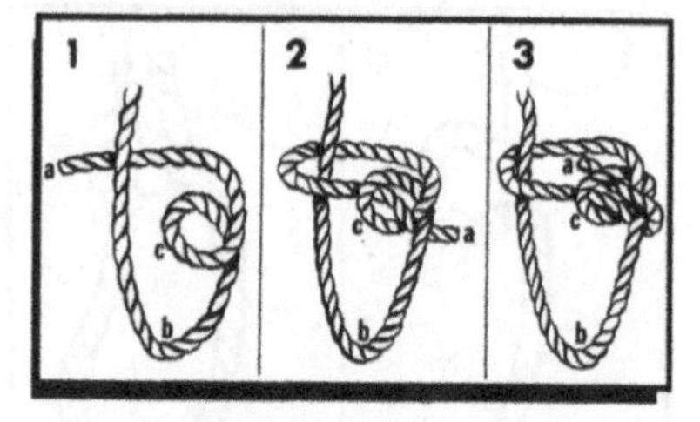

Figure 13-42. Running Bowline.

(5) Bowline on a Bight. It is sometimes desirable to form a loop at some point in a rope other than at the end. The bowline on a bight (Figure 13-43) can be used for this purpose. It is easily untied and will not slip. The same knot can be tied at the end of the rope by doubling the rope for a short section. A doubled portion of the rope is used to form a loop (b) (Figure 13-43-1) as in the case of the bowline. The bight end (a) of the doubled portion is passed up through the loop

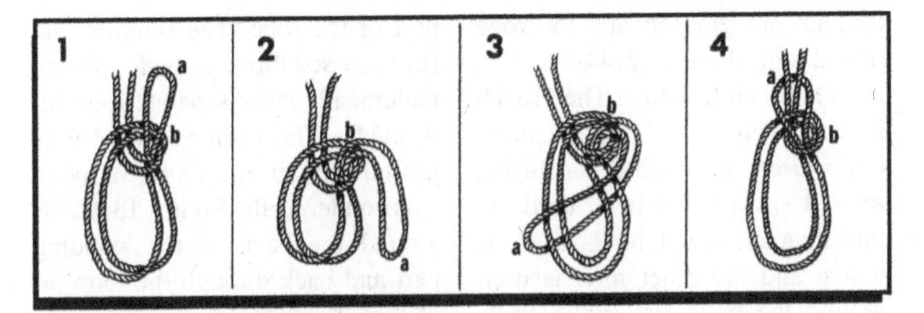

Figure 13-43. Bowline on a Bight.

(b), back down (Figure 13-43-2), up around the entire knot (Figure 13-43-3), and tightened (Figure 13-43-4).

(6) Spanish Bowline. A Spanish bowline (Figure 13-44) can be tied at any point in a rope, either at a place where the line is doubled or at an end which has been doubled back. The Spanish bowline is used in rescue work or to give a two-fold grip for lifting a pipe or other round object in a sling. To tie the Spanish bowline, a doubled portion of the rope is held in the left hand with the loop up and the center of the loop is turned back against the standing parts to form two loops (Figure 13-44-1) or "rabbit ears." The two rabbit ears (c) and (d) (Figure 13-44-2) are moved until they partly overlap each other. The top of the loop nearest the person is brought down toward the thumb of the left hand, being sure it is rolled over as it is brought down. The thumb is placed over this loop (Figure 13-44-5) to hold it in position. The top of the remaining loop is grasped and brought down, rolling it over and placing it under the thumb. There are now four small loops, (c, d, e, and f) in the rope. The lower left-hand loop (c) is turned one-half turn and inserted from front to back of the upper left-hand loop (e). The lower right-hand loop (d) is turned (Figure 13-44-4) and inserted through the upper right-hand loop (f). The two loops (c and d) which have been passed

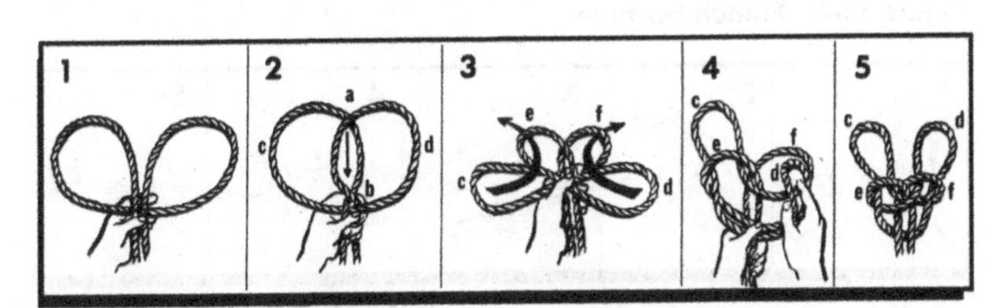

Figure 13-44. Spanish Bowline.

through are grasped and the rope pulled tight (Figure 13-44-5).

(7) French Bowline. The French bowline (Figure 13-45) is sometimes used as a sling for lifting injured people. When used in this manner, one loop is used as a seat and the other loop is used around the body under the arms. The weight of the injured person keeps the two loops tight so that the victim cannot fall out and for this reason, it is particularly useful as a sling for someone who is unconscious. The French bowline is started in the same way as the simple bowline. Make a loop (a) (Figure 13-45-1) in the standing part of the rope. The running end (b) is passed through the loop from underneath and a separate loop (c) is made. The running end (b) is passed through the loop (a), again from underneath (Figure 13-45-3), around the back of the standing part and back through the loop (a) so that it comes out parallel to the looped portion. The standing part of the rope is pulled to tighten the knot (Figure 13-45-4), leaving two loops (c and d).

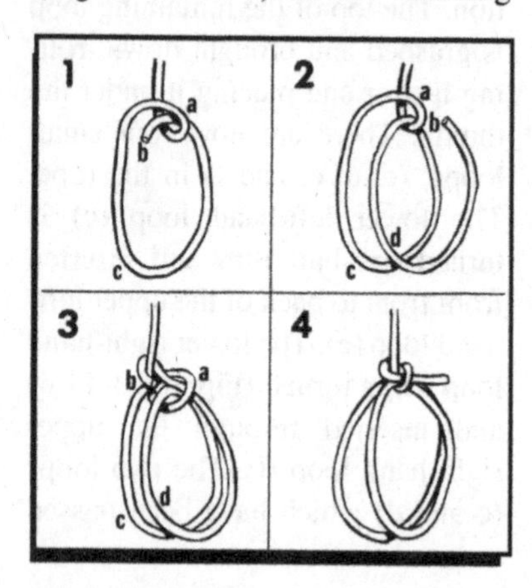

Figure 13-45. French Bowline.

(8) Harness Hitch. The harness hitch (Figure 13-46) is used to form a nonslipping loop in a rope. To make the harness hitch, form a bight (a) (Figure 13-46-1) in the running end of the rope. Hold this bight in the left hand and form a second bight (b) in the standing part of the rope. The right hand is used to pass bight (b) over bight (Figure 13-46-2). Holding all loops in place with the left hand, the right hand is inserted through bight (a) behind the upper part of bight (b) (Figure 13-46-3). The bottom (c) of the first loop is grasped and pulled up through the entire knot (Figure 13-46-4), pulling it tight.

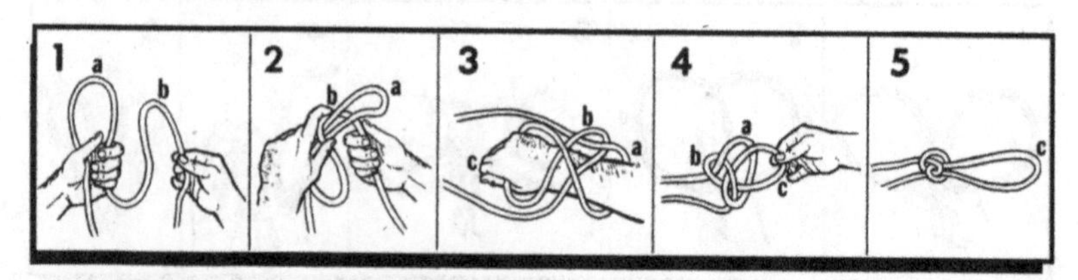

Figure 13-46. Harness Hitch.

g. Hitches:

(1) Half Hitch. The half hitch (Figure 13-47-1) is used to tie a rope to a timber or to another larger rope. It is not a very secure knot or hitch and is used for temporarily securing the free end of a rope. To tie a half hitch, the rope is passed around the timber, bringing the running end around the standing part, and back under itself.

(2) Timber Hitch. The timber hitch (Figure 13-47-2) is used for moving heavy timbers or poles. To make the timber hitch, a half hitch is made and similarly the running end is turned about itself at least another time. These turns must be taken around the running end itself or the knot will not tighten against the pull.

(3) Timber Hitch and Half Hitch. To get a tighter hold on heavy poles for lifting or dragging a timber hitch and half hitch are combined (Figure 13-47-3). The running end is passed around the timber and back under the standing part to form a half hitch. Further along the timber, a timber hitch is tied with the running end. The strain will come on the half hitch and the timber hitch will prevent the half hitch from slipping.

(4) Clove Hitch. A clove hitch (Figure 13-47-4) is used to fasten a rope to a timber, pipe, or post. It can be tied at any point in a rope. To tie a clove hitch in the center of the rope, two turns are made in the rope close together. They are twisted so that the two loops lay back-to-back. These two loops are slipped over the timber or pipe to form the knot. To tie the clove hitch at the end of a rope, the rope is passed around the timber in two turns so that the first turn crosses the standing part and the running end comes up under itself on the second turn.

(5) Two Half Hitches. A quick method for tying a rope to a timber or pole is the use of two half hitches. The running end of the rope is passed around the pole or timber, and a turn is taken around the standing part and under the running end. This is one

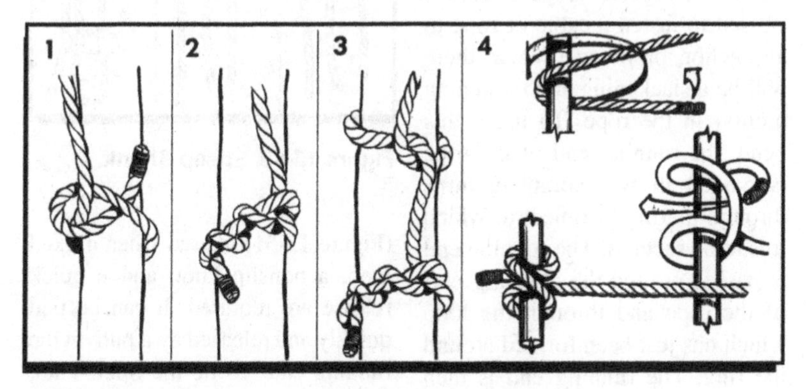

Figure 13-47. Half Hitch, Timber Hitch, and Clove Hitch.

half hitch. The running end is passed around the standing part of the rope and back under itself again.

(6) Round Turn and Two Half Hitches. Another hitch used for fastening a rope to a pole, timber, or spar is the round turn and two half hitches (Figure 13-48). The running end of the rope is passed around the pole or spar in two complete turns, and the running end is brought around the standing part and back under itself to make a half hitch. A second half hitch is made. For greater security, the running end of the rope should be secured to the standing part.

Figure 13-48. Round Turn and Two Half Hitches.

(7) Fisherman's Bend. The fisherman's bend (Figure 13-49) is used to fasten a cable or rope to an anchor, or for use where there will be a slackening and tightening motion in the rope. To make this bend, the running end of the rope is passed in two complete turns through thering or object to which it is to be secured. The running end is passed around the standing part of the rope and through the loop which has just been formed around the ring. The running end is then passed around the standing part in a

Figure 13-49. Fisherman's Bend.

half hitch. The running end should be secured to the standing part.

(8) Sheepshank. A sheepshank (Figure 13-50) is a method of shortening a rope, but it may also be used to take the load off a weak spot in the rope. To make the sheepshank (which is never made at the end of a rope), two bights are made in the rope so that three parts of the rope are parallel. A half hitch is made in the standing part over the end of the bight at each end.

(9) Speir Knot. A Speir knot

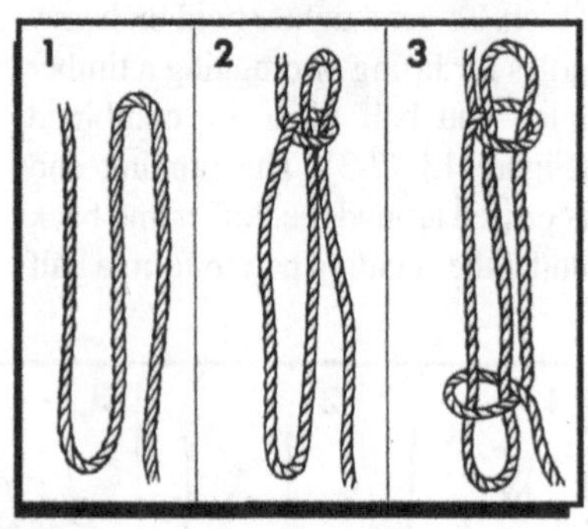

Figure 13-50. Sheep Shank.

(Figure 13-51) is used when a fixed loop, a nonslip knot, and a quick release are required. It can be tied quickly and released by a pull on the running end. To tie the Speir knot, the running end (a) is passed through a ring (Figure 13-51-1) or around a pipe or post and brought back on the left side of the standing part (b). Both hands are placed, palms up, under both parts of the rope with the left hand higher than the right hand; grasping the standing part (b) with the left hand and the running end (a) with the right hand. The left hand is moved to the left and the right hand to the right (Figure 13-51-3) to form two bights (c and d). The left hand is twisted a half turn toward the body so that bight (c) is twisted into a loop (Figure 13-51-3). Pass bight (d) over the rope and down through the loop (c). The Speir knot is tightened by pulling on the bight (d) and the standing part (b) (Figure 13-51-4).

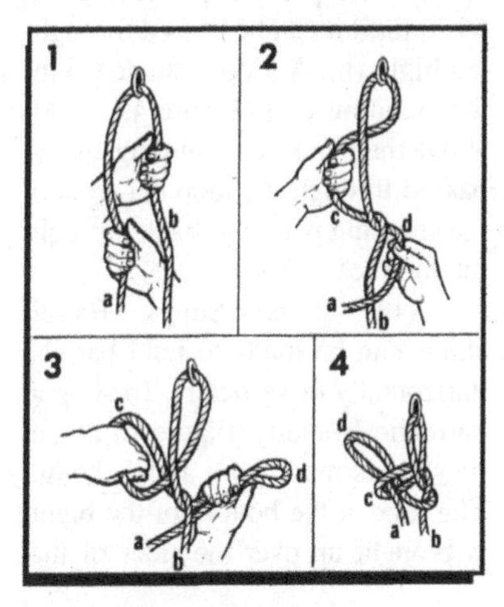

Figure 13-51. Speir Knot.

(10) Rolling Hitch (Pipe or Pole). The rolling hitch (pipe or pole) (Figure 13-52) is used to secure a rope to a pipe or pole so that the rope will not slip. The standing part (a) of the rope is placed along the pipe or pole (Figure 13-52-1) extending in the direction opposite to the direction the pipe or pole will be moved. Two turns (b) are taken with the running end around the standing part (a) and the pole (Figure 13-52-3). The standing part (a) of the rope is reversed so that it is leading off in the direction in which the pole will be moved (Figure 13-52-3) and two turns taken (c) (Figure 13-52-4) with the running end (d). On the second turn around, the running end (d) is passed under the first turn (c) to secure it. To make this knot secure, a half hitch (e) (Figure 13-52-6) is tied with the standing part of the rope 1 or 2 feet above the rolling hitch.

(11) Blackwall Hitch. The blackwall hitch (Figure 13-53) is

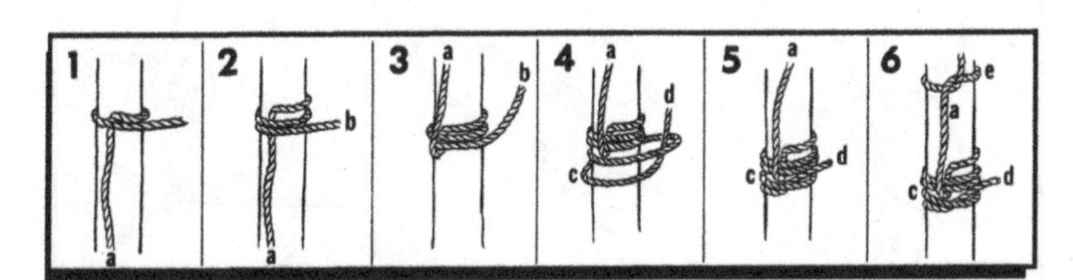

Figure 13-52. Rolling Hitch.

used for fastening a rope to a hook. To make the blackwall hitch, a bight of the rope is placed behind the hook. The running end (a) and standing part (b) are crossed through the hook so that the running end comes out at the opposite side of the hook and under the standing part.

(12) Catspaw. A catspaw can be made at the end of a rope (Figure 13-54) for fastening the rope to a hook. Grasp the running end (a) of the rope in the left hand and make two bights (c and d) in the standing part (b). Hold these two bights in place with the left hand and take two turns about the junction of the two bights with the standing part of the rope. Slip the two loops (c and d) so formed over the hook.

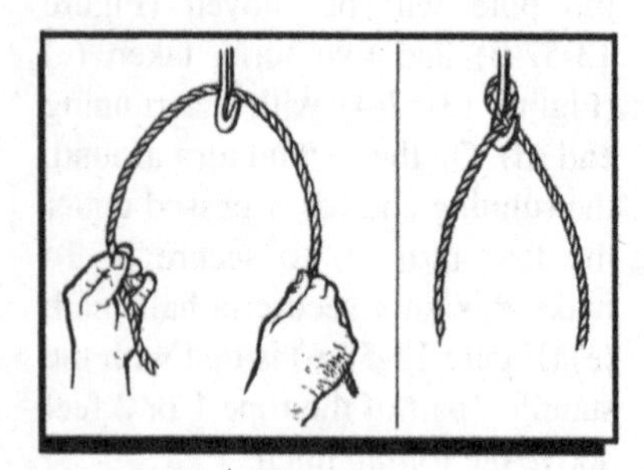

Figure 13-53. Blackwall Hitch.

(13) Scaffold Hitch. The scaffold hitch (Figure 13-55) is used to support the end of a scaffold plank with a single rope. To make the scaffold hitch, the running end of the rope is layed across the top and around the plank, then up and over the standing part (Figure 13-55-1). A doubled portion of the running end is brought back under the plank (Figure 13-55-2) to form a bight (b) at the opposite side of the plank. The running end is taken back across the top of the plank (Figure 55-3) until it can be passed through the bight (b). A loop is made (c) in the standing part (Figure 13-55-4) above the plank. The running end is passed through the loop (c) around the standing part, and back through the loop (c).

(14) Barrel Slings. Barrel slings can be made to hold barrels horizontally or vertically. To sling a barrel horizontally (Figure 13-56), a bowline is made with a long bight. The rope at the bottom of the bight is brought up over the sides of the

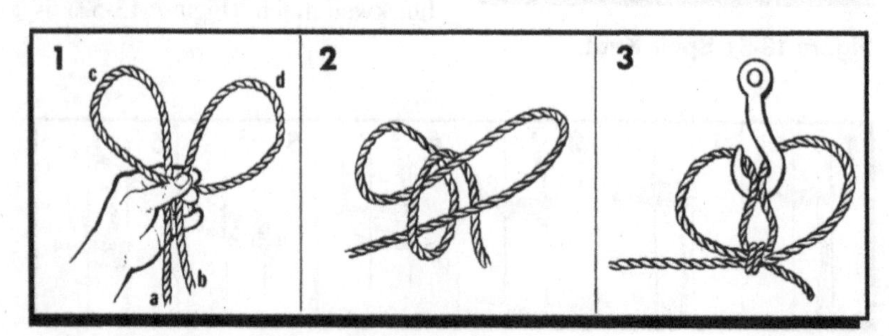

Figure 13-54. Catspaw.

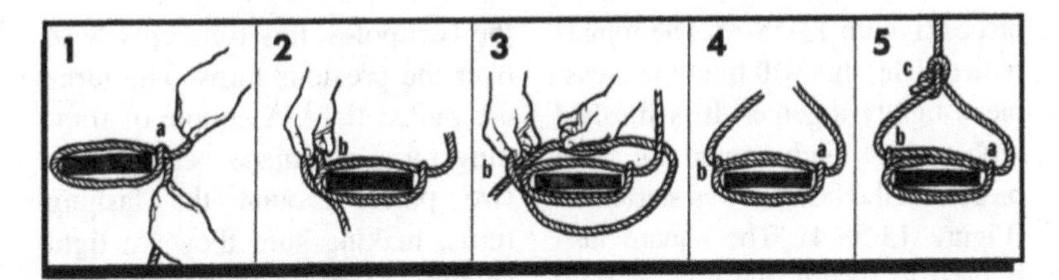

Figure 13-55. Scaffold Hitch.

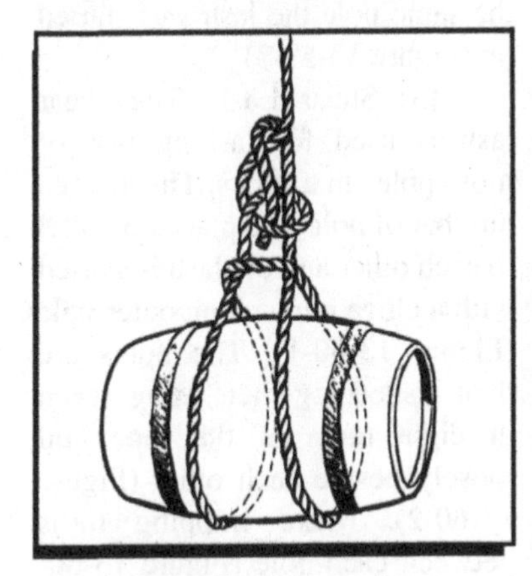

Figure 13-56. Bareel Slung Horizontally.

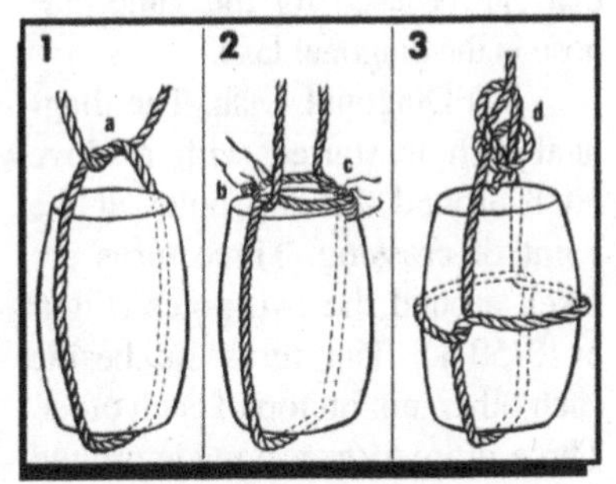

Figure 13-57. Bareel Slung Vertically.

bight. The two "ears" are thus moved foward over the end of the barrel. To sling a barrel vertically (Figure 13-57) the rope is passed under the barrel and up to the top. An overhand knot is made (a) on top (Figure 13-57-1). With a slight tension on the rope, the two parts (Figure 13-57-2) of the overhand knot are grasped, separated and pulled down to the center of the barrel (b and c). The rope is pulled snug and a bowline tied (d) over the top of the barrel (Figure13-57-3).

h. Lashing. There are numerous items which require lashings for construction; for example, shelters, equipment racks, and smoke generators. Three types of lash is used to secure one pole at right angles to another pole. Another lash that can be used for the same purpose is the diagonal lash.

(1) Square Lash. Square lashing is started with a clove hitch around the log, immediately under the place where the crosspiece is to be located (Figure 13-58-1). In laying the turns, the rope goes on the outside of the previous turn around the crosspiece, and on the inside of the previous turn around the log. The rope should be kept tight (Figure 13-58-2). Three or four turns are necessary. Two or three "frapping" turns are made between the cross-

pieces (Figure 13-58-3). The rope is pulled tight; this will bind the crosspiece tightly together. It is finished with a clove hitch around the same piece that the lashing was started on (Figure 13-58-4). The square lash is used to secure one pole at right angles to another pole. Another lash that can be used for the same purpose is the diagonal lash.

(2) Diagonal Lash. The diagonal lash is started with a clove hitch around the two poles at the point of crossing. Three turns are taken around the two poles (Figur e 13-59-1). The turns lie beside each other, not on top of each other. Three more turns are made around

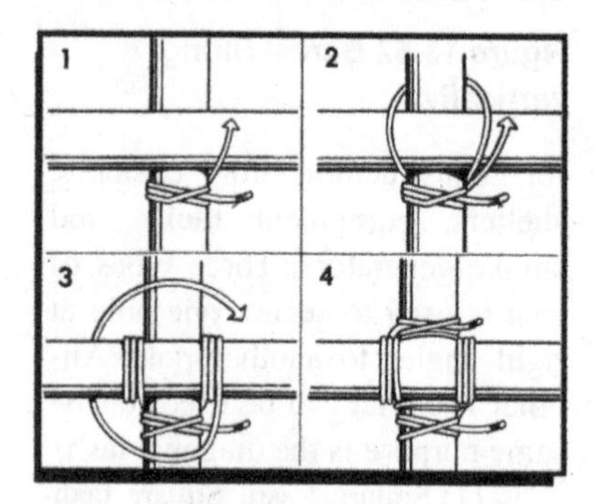

Figure 13-58. Square Lash.

the two poles, this time crosswise over the previous turns. The turns are pulled tight. A couple of frapping turns are made between the two poles, around the lashing turns, making sure they are tight (Figure 13-59-2). The lashing is finished with a clove hitch around the same pole the lash was started on (Figure 13-59-3).

(3) Shear Lash. The shear lash is used for lashing two or more poles in a series. The desired number of poles are placed parallel to each other and the lash is started with a clove hitch on an outer pole (Figure 13-60-1). The poles are then lashed together, using seven or eight turns of the rope laid loosely beside each other (Figure 13-60-2). Make frapping turns between each pole (Figure 13-60-3). The lashing is finished with a clove hitch on the pole opposite that on which the lash was started (Figure 13-60-4).

i. Making Ropes and Cords. Almost any natural fibrous material can be spun into good serviceable rope or cord, and many materi-

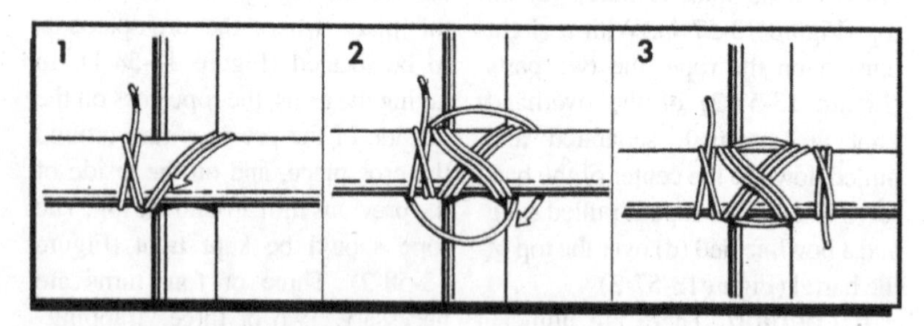

Figure 13-59. Diagonal Lash.

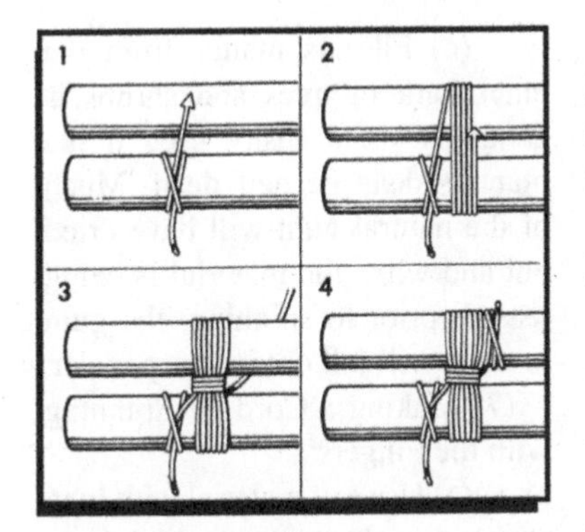

Figure 13-60. Shear Lash.

als which have a length of 12 to 24 inches or more can be braided. Ropes up to 3 and 4 inches in diameter can be "laid" by four people, and tensile strength for bush-made rope of 1-inch diameter range from 100 pounds to as high as 3,000 pounds.

(1) Tensile Strength. Using a three-lay rope of 1-inch diameter as standard, the following table of tensile strengths may serve to illustrate general strengths of various materials. For safety's sake, the lowest figure should always be regarded as the tensile strength.

Green Grass....... 100 lbs to 250 lbs
Bark Fiber ...... 500 lbs to 1,500 lbs
Palm Fiber...... 650 lbs to 2,000 lbs
Sedges .... 2,000 lbs to 2,500 lbs
Monkey Rope
(Lianas) .... 560 lbs to 700 lbs
Lawyer Vine
(Calamus).....½-inch diam, 1,200 lbs

NOTE: Doubling the diameter quadruples the tensile strength half the diameter reduces the tensile strength to one-fourth.

(2) Principles of Ropemaking Materials. To discover whether a material is suitable for rope making, it must have four qualities:

(a) It must be reasonably long in the fiber.

(b) It must have "strength."

(c) It must be pliable.

(d) It must have "grip" so the fibers will "bite" onto one another.

(3) Determining Suitability of Material. There are simple tests to determine if a material is suitable:

(a) First, pull on a length of the material to test for strength.

(b) Second, twist it between the fingers and "roll" the fibers together; if it will withstand this and not "snap" apart, an overhand knot is tied and gently tightened. If the material does not cut upon itself, but allows the knot to be pulled taut, it is suitable for ropemaking if the material will "bite" together and is not smooth or slippery.

(4) Where to Find Suitable Material. These qualities can be found in various types of plants, in ground vines, in most of the longer grasses, in some of the water reeds and rushes, in the inner barks of many trees and shrubs, and in the long hair or wool of many animals.

(5) Obtaining Fibers for Making Ropes. Some green freshly gathered materials may be "stiff" or unyielding. When this is the case, it should be passed through hot flames for a few moments. The heat treatment should cause the sap to burst through some of the cell structure,

and the material thus becomes pliable. Fibers for rope making may be obtained from many sources such as:

(a) Surface roots of many shrubs and trees have strong fibrous bark.

(b) Dead inner bark of fallen branches of some species of trees and in the new growth of many trees such as willows.

(c) The fibrous material of many water and swamp growing plants and rushes.

(d) Many species of grass and weeds.

(e) Some seaweeds.

(f) Fibrous material from leaves, stalks, and trunks of many palms.

(g) Many fibrous-leaved plants such as the aloes.

(6) Gathering and Preparing Materials. There may be a high content of vegetable gum in some plants. This can often be removed by soaking the plants in water, by boiling, or by drying the material and "teasing" it into thin strips.

(a) Some of the materials have to be used green if any strength is required. The materials that should be green include the sedges, water rushes, grasses, and lianas.

(b) Palm fiber is harvested in tropical or subtropical regions. It is found at the junction of the leaf and the palm trunk, or it will be found lying on the ground beneath many palms. Palm fiber is a "natural" for making ropes and cords.

(c) Fibrous matter from the inner bark of trees and shrubs is generally more easily used if the plant is dead or half dead. Much of the natural gum will have dried out and when the material is being teased, prior to spinning, the gum or resin will fall out in fine powder.

(7) Making a Cord by Spinning with the Fingers:

(a) Use any material with long strong threads or fibers which have been previously tested for strength and pliability. The fibers are gathered into loosely held strands of even thickness. Each of these strands is twisted clockwise. The twist will hold the fibers together. The strands should be formed one-eighth inch diameter. As a general rule, there should be about 15 to 20 fibers to a strand. Two, three, or four of these strands are later twisted together, and this twisting together or "laying" is done with a counterclockwise twist, while at the same time, the separate strands which have not yet been laid up are twisted clockwise. Each strand must be of equal twist and thickness.

(b) Figure 13-61 shows the general direction of twist and the method whereby the fibers are bonded into strands. In a similar manner, the twisted strands are put together into lays, and the lays into ropes.

(c) The person who twists the strands together is called the "layer" and must see that the twisting is even, the strands are uniform,

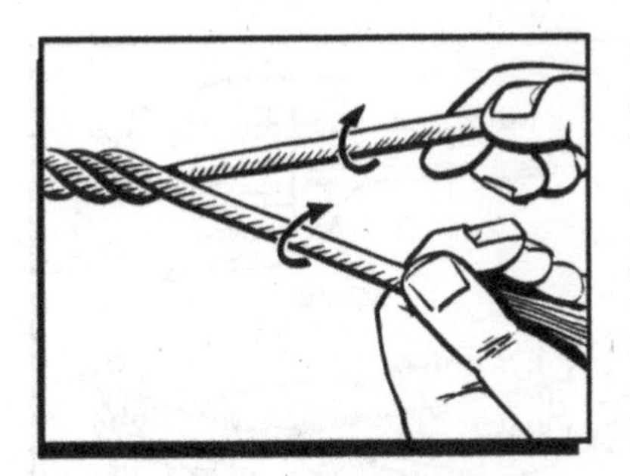

Figure 13-61. Twisting Fibers.

and the tension on each strand is equal. In "laying," care must be taken to ensure each of the strands is evenly "laid up;" that is, one strand does not twist around the other one.

(d) When spinning fine cords for fishing lines, snares, etc., considerable care must be taken to keep the strands uniform and the lay even. Fine thin cords of no more than 1/32-inch thickness can be spun with the fingers and are capable of taking a breaking strain of 20 to 30 pounds or more.

(e) Normally two or more people are required to spin and lay up the strands for cord. However, many native people spin cord unaided. They twist the material by running the flat of the hand along the thigh, with the fibrous material between hand and thigh; and with the free hand, they feed in fiber for the next "spin." Using this technique, one person can make long lengths of single strands. This method of making cord or rope with the fingers is slow if any considerable length of cord is required.

(f) An easier and simpler way to rapidly make lengths of rope from 50 to 100 yards or more in length is to make a rope machine and set up multiple spinners in the form of cranks. Figure 13-62 shows the details of rope spinning.

(g) To use a rope machine, each feeder holds the material under one arm and with one free hand feeds it into the strand which is being spun by the crank. The other hand lightly holds the fibers together till they are spun. As the lightly spun strands are increased in length, they must be supported on crossbars. They should not be allowed to lie on the ground. Spin strands from 20 to 100 yards before laying up. The material should not be spun in too thickly. Thick strands do not help strength in any way, rather, they tend to make a weaker rope.

(8) Setting Up a Rope Machine:

(a) When spinning ropes of 10 yards or longer, it is necessary to set crossbars every 2 or 3 yards to carry the strands as they are spun. If crossbars are not set up, the strands or rope will sag to the ground, and some of the fibers will tangle up with grass, twigs, or dirt on the ground. Also, the twisting of the free end may either be stopped or interrupted and the strand will be unevenly twisted.

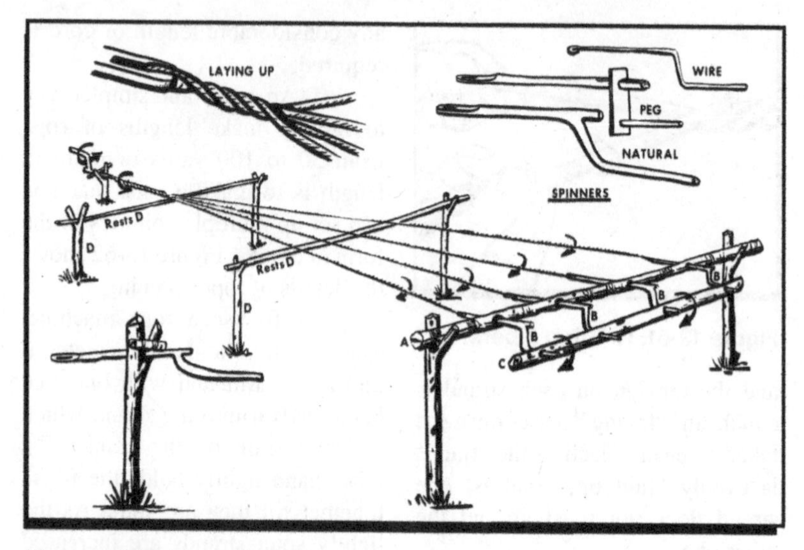

Figure 13-62. Rope Machine.

(b) The easiest way to set up crossbars for the rope machine is to drive pairs of stakes into the ground about 6 feet apart and at intervals of about 6 to 10 feet. The crossbars must be smooth and free from twigs and loose portions of bark that might twist in with the spinning strands.

(c) The crossbar (a) is supported by two uprights and pierced to take the cranks (b). These cranks can be made out of natural sticks, morticed slab, and pegs, or if available, bent wire. The connecting rod (c) enables one person to turn all cranks clockwise simultaneously. Crossbars supporting the strands as they are spun are shown (d). A similar crank handle to the previous ones (b) is supported on a forked stick at the end of the rope machine. This handle is turned in reverse (counterclockwise) to the cranks (c) to twist the connected strands together. These are "laid up" by one or more of the feeders.

(d) The first strand should be turned clockwise, then the laying up of the strands will be done counterclockwise and the next laying will again be clockwise. Proof that the rope is well made is that the individual fibers lay lengthways along the rope.

(e) In the process of laying up the strands, the actual twisting together or laying will take some of the original "twist" out of the strand which has not yet been laid. Therefore, it is necessary to keep twisting the strands while laying together.

(f) When making a rope too long to be spun and laid in one piece, a section is laid up and coiled on the ground at the end of the rope walk farthest from the cranks. Strands for a second length are spun, and these strands are married or spliced into the strands of the first section and then the laying up of the second section continues the rope.

(g) The actual "marrying" of the strands is done only in the last lay, which makes the rope when completed. The ends where the strands are married should be staggered in different places. By this means, rope can be made and extended in sections to a great length.

(h) After a complete length of rope is laid up, it should be passed through the fire to burn off the loose ends and fibers. This will make the rope smooth and more professional looking.

(9) Laying the Strands:

(a) The strands lie on the crossbar as they are spun. When the strands have been spun to the required length, which should not be more than about a hundred feet, they are joined together by being held at the far end. They are then ready for laying together. The turner, who is facing the cranks, twists the ends together counterclockwise, at the same time keeping full weight on the rope which is being layed up. The layer advances placing the strands side by side as they turn.

(b) It is important to learn to feed the material evenly, and lay up slowly, thereby getting a smooth even rope (Figure 13-63).

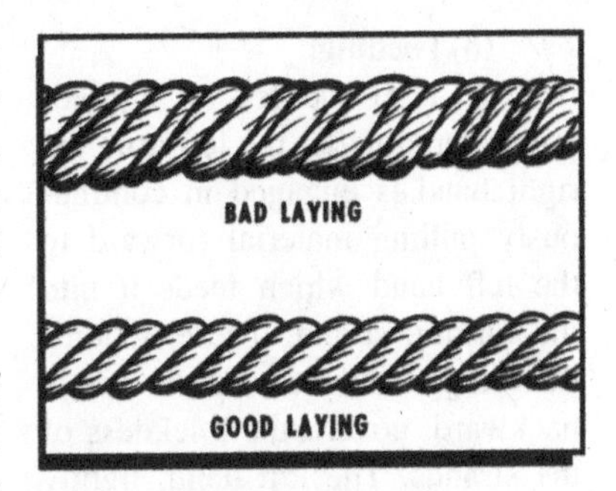

Figure 13-63. Rope Laying

Do not try to rush the ropemaking. Speed in ropemaking only comes with practice. At first it will take a team of three or four up to 2 hours or more to make a 50-yard length of rope of three lays, each of three strands; that is, nine strands for a rope with a finished diameter of about 1 inch. With practice, the same three or four people will make the same rope in 15 to 20 minutes. These times do not include time for gathering material. In feeding the free ends of the strands, twist in the loose material fed in by the feeder. As the feeders move backward, they must keep a slight tension on the strands.

(10) Making Rope with a Single Spinner:

(a) Using a Single Crank. Two people can make a rope, using a single crank. A portion of the material is fastened to the eye of the crank, as with the multiple crank. Supporting crossbars, as used in a ropewalk, are required when a length of more than 20 or 30 feet is being spun.

(b) Feeding:

-1. If the feeder is holding material under the left arm, the right hand is engaged in continuously pulling material forward to the left hand which feeds it into the turning strand. These actions, done together as the feeder walks backward, govern the thickness of the strands. The left hand, lightly closed over the loose turning material, must "feel" the fibers "biting" or twisting together.

-2. When the free end of the turning strand, which is against the loose material under the arm, takes in too thick a tuft of material, the left hand is closed, and so arrests the twist of the material between the left hand and the bundle. This allows teasing out the overall "bite," with the right hand, thus maintaining a uniform thickness of the spinning strand.

(c) Thickness of Strands. Equal thickness and twist for each of the strands throughout their length are important. The thickness should not be greater than is necessary with the material being used. For a grass rope, the strand should not be more than one-fourth inch diameter; for coarse bark or palm, not more than one- eighth or three-sixteenth inch; and for fine bark, hair, or sisal fiber, not more than one-eighth inch.

(d) Common Errors in Ropemaking:

-1. There is a tendency with beginners to feed unevenly. Thin wispy sections of strand are followed by thick portions. Such feeding degrades the quality of rope. Rope made from such strands will break with less than one-fourth of the tensile strain on the material.

-2. Beginners are wise to twist and feed slowly. Speed, with uniformity of twist and thickness, comes with practice.

-3. Thick strands do not help. It is useless to try and spin a rope from strands an inch or more in thickness. Such a rope will break with less than half the tensile strain on the material. Spinning "thick" strands does not save time in ropemaking.

(e) Lianas, Vines, and Canes. Lianas and ground vines are natural ropes, and grow in subtropical and tropical scrub and jungle. Many are of great strength and useful for braiding, tree climbing, and other purposes. The smaller ground vines, when "braided," give great strength and flexibility. Canes and stalks of palms provide excellent material if used properly. Only the outer skin is tough and strong, and this skin will split off easily if the main stalk is bent away from the skin. This principle also applies to the splitting of lawyer cane (calamus), palm leaf stalks, and all green material. If the split starts to run off, bend the material away from the thin side, and it will gradually gain in size and come back to an even thickness with the other split side.

(f) Bark Fibers:

-1. The fibers in many barks which are suitable for "ropemaking" are located near the innermost layers. This is the bark next to the sap wood. When seeking suitable barks of green timber, cut a small section about 3 inches long and 1 inch wide. Cut this portion from the wood to the outer skin of the bark.

-2. The specimen should be peeled and the different layers tested. Green bark fibers are generally difficult to spin because of "gum" and it is better to search around for windfall dead branches and try the inner bark of these. The gum probably has leached out, and the fibers should separate easily.

-3. Many shrubs have excellent bark fiber, and here it is advisable to cut the end of a branch and peel off a strip of bark for testing. Thin bark from green shrubs is sometimes difficult to spin into fine cord and is easier to use as braid for small cords.

-4. Where it is necessary to use green bark fiber for rope spinning the gum will generally wash out when the bark is teased and soaked in water for a day or so. After removing from the water, the bark strips should be allowed to dry before shredding and teasing into fiber.

(11) Braiding. One person may require a length of rope. If there is no help available to spin materials, it is necessary to find reasonably long material. With this material, one person can braid and make suitable rope. The usual three-strand braid makes a flat rope, and while quite good, it does not have finish or shape, nor is it as "tight" as the four-strand braid. On other occasions, it may be necessary to braid broad bands for belts or for shoulder straps. There are many fancy braids which can be developed from these, but these three are basic, and essential for practical woodcraft work. A general rule for all braids is to work from the outside into the center.

(a) Three Plait:

-1. The right-hand strand is passed over the strand to the left.

-2. The left-hand strand is passed over the strand to the right.

-3. This is repeated alternately from left to right (Figure 13-64).

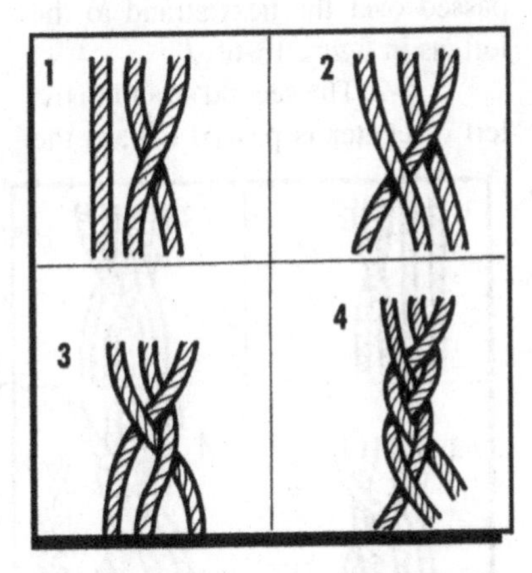

Figure 13-64. Three-Strand Braib.

(b) Flat Four-Strand Braid:

-1. The four strands are placed side by side. The right-hand strand is taken (Figure 13-65-1) and placed over the strand to the left.

-2. The outside left-hand strand (Figure 13-65-2) is laid under the next strand to itself and over what was the first strand.

-3. The outside right-hand strand is laid over the first strand to its left (Figure 13-65-3).

-4. The outside left strand is placed under and over the next two strands, respectively, moving toward the right.

-5. Thereafter, the right-hand strand goes over one strand to the left, and the left-hand strand under and over to the right (Figure 13-65-4).

(c) Broad Braid. Six or more strands are held flat and together.

-1. A strand in the center is passed over the next strand to the left, as in figure 13-66-1.

-2. The second strand to the left of center is passed toward the right and over the first strand so that it points toward the right (Figure 13-66-2).

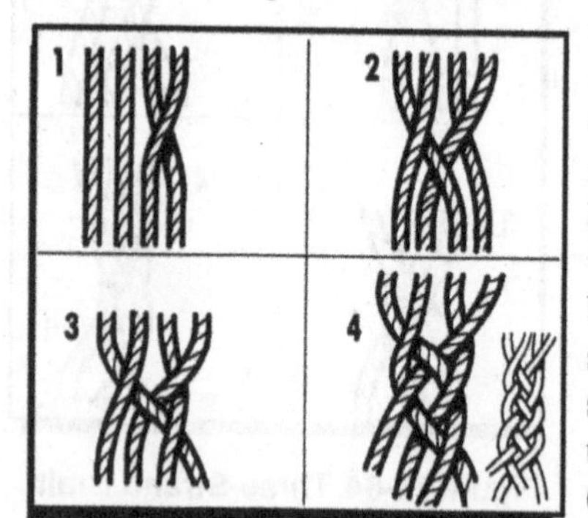

Figure 13-65. Four-Strand Braid.

-3. The strand next to the first one is taken and woven under and over (Figure 13-66-3).

-4. The next strands are woven from left and right alternately towards the center (Figure 13-66-4 through 6). The finished braid should be tight and close (Figure 13-66-7).

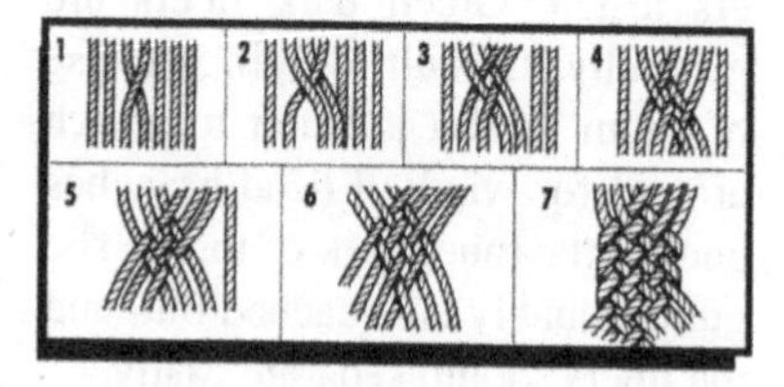

Figure 13-66. Broad Braid

-5. To finish the broad braid:

-a. One of the center strands is laid back upon itself (Figure 13-67-1).

-b. Now take the first strand which it enclosed in being folded back, and weave this back upon itself (Figure 13-67-2).

-c. Strand from the opposite side is laid back and woven between the strands already braided (Figure13-67-3).

-d. All the strands should be so woven back that no strands show an uneven pattern, and there should be a regular under-over-under of the alternating weaves (Figure 13-67-4).

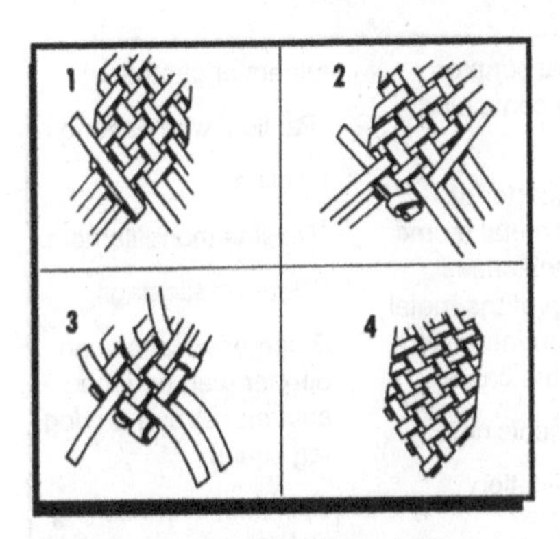

Figure 13-67. Finishing the Broad Braid.

-e. If the braid is tight, there may be a difficulty in working the loose ends between the plaited strands.

-f. This can be done easily by sharpening a thin piece of wood to a chisel edge to open the strands sufficiently to allow the ends being finished to pass between the woven strands.

-g. It should be rolled under a bottle or other round object and made smooth for final finishing.

13-7. Personal Survival Kit:

a. Even though a survival kit may be available, aircrew members should consider assembling and carrying personal survival kits. Survival experiences have occurred where survivors hit the ground running, and because of shock and fear left their survival kits behind. If survivors have a personal survival kit in a pocket, it may improve their survival chances considerably.

b. A great deal of thought should go into preparing personal survival kits. The potential needs of the survivors must be a consideration, such as the impact of the environmental elements, type of mission to be flown (tactical or nontactical), availability of rescue, and how far to friendly forces (Figure 13-68).

c. There are two basic ways to carry a personal survival kit. One way is to pack all items into one or two waterproof containers. The other way is to scatter the items throughout personal clothing. Any type of small container can be used to encase the contents of the personal survival kit. Plastic cigarette cases, soap dishes, and Band-Aid boxes are excellent containers.

d. Examples of items which can be packed into a small container are:

(1) Matches.
(2) Safety pins (varied sizes).
(3) Fishhooks.
(4) Knife (small, multibladed).
(5) Button compass.
(6) Prophylactic (for water container).
(7) Bouillon cubes.
(8) Salt.
(9) Snare wire.
(10) Water purification tablets.
(11) Signal mirror.
(12) Needles.

MINIMUM ESSENTIAL ITEMS

High quality pocket knife with at least two cutting blades.

Pocket compass.

Match safe with matches.

- Plastic or metallic container.
- Waterproof kitchen-type matches (cushion heads against friction), or
- Waterproof matches rolled in paraffin-soaked muslin in an easily opened container such as small soap box, toothbrush case, etc.

Needles — sailmakers, surgeons, and darning — at least one of each.

Assorted fishhooks in heavy foil, tin, or plastic holders.

Snare wire — small hank.

Needle-nosed pliers with side cutters; high quality.

Bar surgical soap or hand soap containing physohex.

Small fire starter of pyrophoric metal (some plastic match cases have a strip of the metal anchored on the bottom outside of the case).

Personal medicines.

Water purification tablets.

"Band-Aids."

Insect repellent stick.

Lip balm.

GOOD TO HAVE ITEMS

*Pen-gun and flares.

*Colored cloth or scarf for signaling.

Stick-type skin dye (for camouflage).

Plastic water bottle.

*Flexible saw (wire saw).

*Sharpening stone.

Safety pins (several sizes).

Travel razor.

Small steel mirror.

6" flat bastard file.

Aluminum foil.

ADDITIONAL SUGGESTIONS

Toothbrush— small type.

Surgical tape.

Prophylactics (make good waterproof containers or canteens).

*Penlight with batteries.

Fishline.

*Fishline monofilament.

*Clear plastic bags.

Emergency ration can opener (can be taped shut and strung on dog tag chain).

Split shot — for fishing sinkers.

Gill net.

Small, high quality candles.

INDIVIDUAL MEDICAL KIT

Sterile gauze compress bandage.

Antibiotic ointment (Neomycin polymycin bacitracin opthalmic ointment is good).

Tincture of zephrine — skin antiseptic.

Aspirin tablets.

Salt tablets.

Additional medications may be desirable, depending upon nature of the mission and an individual's particular personal needs.

This should be discussed with and procured from your local flight surgeon.

Especially valuable.

Figure 13-68. Personal Survival Kit Items.

(13) Band-aids.
(14) Aluminum Foil.
(15) Insect repellent stick.
(16) Chapstick.
(17) Soap (Antiseptic).

NOTE: All kits carried aboard the aircraft should be approved by the unit life support officer.

Part Five

SUSTENANCE

Chapter 14

FOOD

14.1. Introduction. Except for the water they drink and the oxygen they breathe, survivors must meet their body needs through the intake of food. This chapter will explore the relationship of proper nutrition to physical and mental efficiency. It is extremely important that survivors maintain a proper diet at all times. A nutritionally sound body stands a much better chance of surviving. Improper diet over a long period of time may lead to a lack of stamina, slower reactions, less resistance to illness, and reduced mental alertness, all of which can cost survivors their lives in a survival situation. A knowledge of the body's nutritional requirements will help survivors select foods to supplement their rations.

14-2. Nutrition. Survivors and evaders expend much more energy in survival situations than they would in the course of their normal everyday jobs and life. Basal metabolism is the amount of energy expended by the body when it is in a resting state. The rate of basal metabolism will vary slightly with regard to the sex, age, weight, height, and race of a person. The basic energy expended, or number of calories consumed by the hour will change as a person's activity level changes. A person who is simply sitting in a warm shelter, for example, may consume anywhere from 20 to 100 calories an hour, while that same person evading through thick undergrowth with a heavy pack, would expand a greater amount of energy. In a survival situation, proper food can make the difference between success and failure.

a. The three major constituents of food are carbohydrates, fats, and proteins. Vitamins and minerals are also important as they keep certain essential body processes in good working order. It is also necessary for survivors to maintain proper water and salt levels in their bodies, as they aid in preventing certain heat disorders.

(1) Carbohydrates. Carbohydrates are composed of very simple molecules which are easily digested. Carbohydrates lose little of their energy to the process of digestion and are therefore efficient energy suppliers. Because carbohydrates supply easily used energy, many nutritionists recommend that, if possible, survivors should try to use them for up to half of their calorie

intake. Examples of carbohydrates are: starches, sugars, and cellulose. These can be found in fruits, vegetables, candy, milk, cereals, legumes, and baked goods. Cellulose cannot be digested by humans, but it does provide needed roughage for the diet.

(2) Fats. Fats are more complex than carbohydrates. The energy contained in fats is more slowly released than the energy in carbohydrates. Because of this, it is a longer lasting form of energy. Fats supply certain fat-soluble vitamins. Sources of these fats and vitamins are butter, cheese, oils, nuts, egg yolks, margarine, and animal fats. If survivors eat fats before sleeping, they will sleep warmer. If fats aren't included in the diet of survivors, they can become run down and irritable. This can lead to both physical and psychological breakdown.

(3) Protein. The digestive process breaks protein down into various amino acids. These amino acids are formed into new body tissue protein, such as muscles. Some protein gives the body the exact amino acids required to rebuild itself. These proteins are referred to as "complete." Protein that lacks one or more of these essential amino acids is referred to as "incomplete." Incomplete protein examples are cheese, milk, cereal grains, and legumes. Incomplete protein, when eaten in combination with milk and beans for example, can supply an assortment of amino acids needed by the body. Some complete protein is found in fish, meat, poultry, and blood. No matter which type of protein is consumed, it will contain the most complex molecules of any food type listed.

(a) If possible, the recommended daily allowance of 2 ½ to 3 ounces complete protein should be consumed by each survivor each day. If only the incomplete protein is available, two, three, or even four types of foods may need to be eaten in combination so that enough amino acids are combined to form complete protein.

(b) If amino acids are introduced into the body in great numbers and some of them are not used for the rebuilding of muscle, they are changed into fuel or stored in the body as fat. Because protein contains the more complex molecules, over fats or carbohydrates, they supply energy after those forms of energy have been used up. A lack of protein causes malnutrition, skin and hair disorders, and muscle atrophy.

b. Vitamins occur in small quantities in many foods, and are essential for normal growth and health. Their chief function is to regulate the body processes. Vitamins can generally be placed into two groups: fat-soluble and water-soluble. The body only stores slight amounts of the water-soluble type. In a long survival episode where a routinely balanced diet is not available, survivors must overcome food aversions and eat as much of a variety of vitamin-rich foods as possible. Often one or more of the four basic food groups (meat, fish, poultry; vegetables and fruits; grain and cereal; milk and milk products) are not available in the form of familiar foods, and vitamin deficiencies such as beriberi or scurvy result. If the survivor can overcome aversions to local foods high in vitamins, these diseases as well as signs and symptoms such as depression and irritability can be warded off.

c. Adequate minerals can also be provided by a balanced diet. Minerals build and(or) repair the skeletal system and regulate normal body functions. Minerals needed by the body include iodine, calcium, iron, and salt, to name but a few. A lack of minerals can cause problems with muscle coordination, nerves, water retention, and the ability to form or maintain healthy red blood cells.

d. For survivors to maintain their efficiency, the following number of calories per day is recommended. These figures will change because of individual differences in basal metabolism, weight, etc. During warm weather survivors should consume anywhere from 3,000 to 5,000 calories per day. In cold weather the calorie intake should rise from 4,000 to 6,000 calories per day. A familiarity with the calorie and fat amounts in foods is important for survivors to meet their nutritional needs. For example, it would take quite a few mussels and dandelion greens to meet those requirements. Survivors should attempt to be familiar enough with foods that they can select or find foods that provide a high calorie intake (Figure 14-1).

FOOD	CALORIES	FAT
WHOLE LARGE DUCK EGG	177	12.0
SMALL OR LARGE MOUTH BASS – 3 TO 4 OZ.	109	3.6
CLAMS – 4 TO 5 LARGE	88	.2
FRESHWATER CRAYFISH – 3 TO 4 OZ.	75	.6
EEL – 3 TO 5 OZ.	240	20.0
OCTOPUS – 3 TO 4 OZ.	76	.9
ATLANTIC SALMON – 4 OZ.	220	14.0
RAINBOW TROUT – 4 OZ.	200	11.8
BANANA – ONE SMALL	87	.3
BREADFRUIT – 3 TO 4 OZ.	105	.5
GUAVA – ONE MEDIUM	64	.7
MANGO – ONE SMALL	68	.5
WILD DUCK – 4 OZ.	230	16.0
BAKED OPOSSUM – 4 OZ.	235	10.6
WILD RABBIT – 4 OZ.	124	4.0
VENISON – 4 OZ.	128	3.1
DANDELION GREENS – ONE CUP COOKED	70	1.4
POTATO – MEDIUM	78	.2
PRICKLY PEAR – 4 OZ.	43	.2

Figure 14-1. Food and Calorie Diagram.

(1) Survivors should also be familiar with the number of calories supplied by the food in issued rations. In most situations, rations will have to be supplemented with other foods procured by survivors. If possible, survivors should limit their activities to save energy. Rationing food is a good idea since

survivors never know when their ordeal will end. They should eat when they can, keeping in mind that they should maintain at least a minimum calorie intake to satisfy their basic activity needs.

(2) Caloric and fat values of selected foods are shown in the chart, and unless otherwise specified, the foods listed are raw. Depending on how survivors cook the food, the usable food value can be increased or decreased.

14-3. Food. Survivors should be able to find something to eat wherever they are. One of the best places to find food is along the seacoast, between the high and low watermark. Other likely spots are the areas between the beach and a coral reef; the marshes, mud flats, or mangrove swamps where a river flows into the ocean or into a larger river; riverbanks, inland waterholes, shores of ponds and lakes, margins of forests, natural meadows, protected mountain slopes, and abandoned cultivated fields.

a. Rations placed in survival kits have been developed especially to provide some of the proper sustenance needed during survival emergencies. When eaten as directed on the package, it will keep the survivor relatively efficient. If enough other food can be found, rations should be conserved for emergency use.

b. Consideration must be given to available food and water and how long the survival episode may last. Environmental conditions must also be considered. If a survivor is in a cold environment, more of the *proper* food will be required to provide necessary body heat. Rescue may vary from a few hours to several months, depending on the environment, operational commitments, and availability of rescue resources in that area. Available food must be rationed based on the estimated time which will elapse before being able to supplement issued rations with natural foods. If it is decided that some of the survivors should go for help, each traveler should be given twice as much food as those remaining behind. In this way, the survivors resting at the encampment and those walking out will stay in about the same physical condition for about the same length of time.

c. If available water is less than a quart a day, avoid dry, starchy, and highly seasoned foods and meat. Keep in mind that eating increases thirst. For water conservation, the best foods to eat are those with high carbohydrate content, such as hard candy and fruit. All work requires additional food and water. When work is being performed, the survivor must increase food and water consumption to maintain physical efficiency. If food is available, it is alright to nibble throughout the day. It is preferable though to have at least two meals a day, with one being hot. Cooking usually makes food safer, more digestible, and palatable. The time spent cooking will provide a good rest period. On the other hand, some food such as sapodilla, star apple, and soursop, are not palatable unless eaten raw.

d. Native foods may be more appetizing if they are eaten by themselves. Rations and native foods usually do not mix well. In many countries, vegetables are often contaminated by human feces which the natives use as fertilizer. Dysentary is transmitted in this way. If possible, survivors should try to select and prepare their own meals. If necessary to avoid offending the natives, indicate that religious beliefs or taboos require self-preparation of food.

e. Learn to overcome food prejudices. Foods that may not look good to the survivor are often a part of the natives' regular diet. Wild foods are high in mineral and vitamin content. With a few exceptions, all animals are edible when freshly killed. Avoid strange looking fish and fish with flesh that remains indented when depressed as it is probably becoming spoiled and should not be eaten. With knowledge and the ability to overcome food prejudices, a survivor can eat and sustain life in strange or hostile environment.

14-4. Animal Food. Animal food gives the most food value per pound. Anything that creeps, crawls, swims, or flies is a possible source of food. People eat grasshoppers, hairless caterpillars, wood-boring bettle larvae and pupae, ant eggs, spider bodies, and termites. Such insects are high in fat and should be cooked until dried. Everyone has probably eaten insects contained in flour, cornmeal, rice, beans, fruits, and greens in their daily foods.

a. Man as a Predator. To become successful in hunting, the hunter must go through a behavioral change and reorganize personal priorities. This means the one and only goal for the present is to kill an animal to eat. To kill this animal, the hunter must mentally become a predator. The hunter must be prepared to undergo stress in order to hunt down and kill an animal. Because of the type of weapons survivors are likely to have, it will be necessary to get very close to the animal to immobilize or kill it. This is going to require all the stealth and cunning survivors can muster. In addition to stealth and cunning, knowledge of the animal being hunted is very important. If in an unfamiliar area, survivors may learn much about the animal life of the area by studying signs such as trails, droppings, and bedding areas.

b. Animal Sign. The survivor should establish the general characteristics of the animals. The size of the tracks will give a good idea of the size of the animal. The depth of the tracks will indicate the weight of the animal. The animal dung can tell the hunter much.

For example, if it is still warm or slimy, it was made very recently; if there is a large amount scattered around the area, it could well be a feeding or bedding area. The droppings may indicate what the animal feeds upon. Carnivores often have hair and bone in the dung; herbivores have coarse portions of the plants they have eaten. Many animals mark their territory by urinating or scraping areas on the ground or trees. These signs could indicate good trap or ambush sites. Following the signs (tracks, droppings, etc.) may reveal the feeding, watering, and resting areas. Well worn trails will often lead to the animal's watering place. Having made a careful study of all the signs of the animal, the hunter is in a much better position to procure it, whether electing to stalk, trap, or snare it, or lie in wait to shoot it.

c. Hunting. If survivors elect to hunt, there are some basic techniques which will be helpful and improve chances of success. Wild animals rely entirely upon their senses for their preservation. These senses are smell, vision, and hearing. Humans have lost the keenness of some of their senses like smelling, hearing, etc. To overcome this disadvantage, they have the ability to reason. As an example, some animals have a fantastic sense of smell, but this can be overcome by approaching the quarry from a downwind direction. The best times to hunt are at dawn and dusk as animals are either leaving or returning to their bedding areas. Both diurnal and nocturnal animals are active at this time. There are five basic methods of hunting:

(1) Still or Stand. This is the best method for inexperienced hunters as it involves less skill. The main principle of this method is to wait in ambush along a well-used game trail, until the quarry approaches within killing range. Morning and evening are usually the best times to still hunt. Care should be taken not to disturb the area; always wait downwind. Patience and self-control are necessary to remain motionless for long periods of time.

(2) Stalking. "Stalking" refers to the stealthily approach toward game. This method is normally used when an animal has been sighted and the hunter then proceeds to close the distance using all available cover. Stalking must be done slowly so that minimum noise is made; quick movement is easily detected by the animal. Always approach from the downwind side and move when the animal's head is down eating, drinking, or looking in another direction. The same techniques are used in blind stalking as in the regular stalk, the main difference being that the hunter is stalking a position where the animal is expected to be while the animal is not in sight.

(3) Tracking. Tracking is very difficult unless conditions are ideal.

This method involves reading all of the signs left behind by the animal, interpreting what the animal is doing, and how it can best be killed. The most common signs are trails, beds, urine, droppings, blood, tracks, and feeding signs.

(4) Driving. Some wild animals can be scared or driven in a direction where other hunters or traps have been set. This method is normally used where the game can be funneled; a valley or canyon is a good place to make a drive. More than one person is usually necessary to make a drive.

(5) Calling. Small predators may be called in by imitating an injured animal. Ducks and geese can be attracted by imitating their feeding calls. These noises can be made by sucking on the hand, blowing on a blade of grass or paper, sucking the lip, or using specially designed devices. Survivors should not call animals unless they know what they are doing as strange noises may "spook" the animal.

d. Killing Implements. It is difficult to kill animals of any size without using some type of tool or weapon. As our technology has increased in complexity, so have our killing tools. If a firearm is available, a basic knowledge of shooting and hunting techniques is necessary.

(1) Learning to become proficient with primitive weapons is important. Many primitive tribes of the world are still effectively using spears, clubs, bows and arrows, sling shots, etc., to provide food for their families. One of the limiting factors in the use of firearms is the amount of ammunition on hand. Therefore, a survivor cannot afford to waste ammunition on moving game or game which is beyond the effective range of the firearm being used. Wait for a pause in the animal's motions. The shot must be placed in a vital area with any firearm. Aim for the brain, spine, lungs, or heart (Figure 14-2). A hit in these areas is usually fatal.

(2) A full-jacketed bullet often won't immediately down a larger animal hit in a vital area such as the lungs or heart. The alternative to losing the animal is tracking it to where it falls. Often it's better to wait awhile before pursuing the animal. If not pursued, it may lay down and stiffen or perhaps bleed to death. Follow the blood trail to where the game has gone down and kill it if it is still alive. Even though ammunition might be limited, small game may be more productive than large game. Although they present smaller targets and have less meat, they are less wary, more numerous, and travel less distance to escape if wounded. A large amount of edible meat on small game can be destroyed from a bullet wound. On rodents, most of the meat is on the hindquarters and frontquarters; birds, it is the breast and legs. The survivor should try to hit a vital spot that spoils the least meat.

Figure 14-2. Shooting Game.

(3) Night hunting is usually best, since most animals move at night. A flashlight or torch may be used to shine in the animal's eyes. It will be partly blinded by the light and a survivor can get much closer than in the daytime. If no gun is available, the animal can be killed with a club or a sharpened stick used as a spear.

(4) Remember that large animals, when wounded, cornered, or with their young, can be dangerous. Be sure the animal is dead, not just wounded, unconscious, or playing "possum." Animals usually die with their eyes open and glazed-over. Poke all "dead" animals in the eye with a long sharp stick before approaching them.

(5) Small freshwater turtles can often be found sunning themselves along rivers and lakeshores. If they dash into shallow water, they can still be procured with nets, clubs etc; watch out for mouth and claws. Frogs and snakes also sun and feed along streams. Use both hands to catch a frog—one to attract it and keep it busy while grabbing it with the other. Bright cloth on a fish-hook also works. All snakes are good eating and can be killed with a long stick. Both marine and dry-land lizards are edible. A noose, small fishhook baited with a bright cloth lure, slingshot, or club can be used. A slingshot can be made with a forked stick and the elastic from the parachute pack or surgical

tubing found in some survival kits (Figure 14-3). With practice, the slingshot can be very effective for killing small animals.

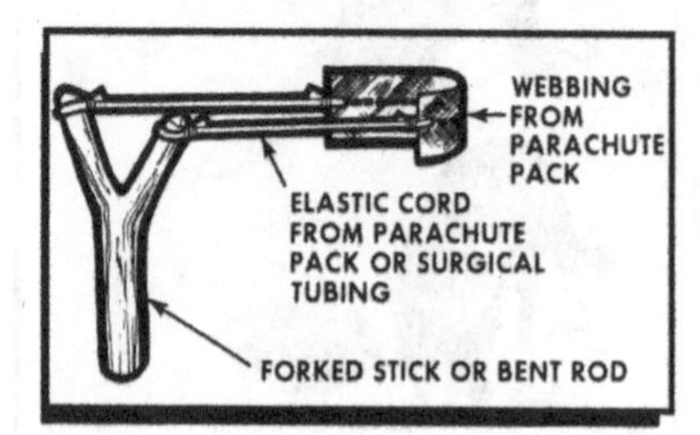

Figure 14-3. Slingshot.

e. Snaring and Trapping. Snaring and trapping animals are ways survivors can procure animal food to supplement issued rations. Since small animals are usually more abundant than large animals, they will probably be the survivor's main source of food. Snares should be set out on a 15:1 ratio; 15 snares should be set out for every one animal expected to be caught.

(1) Using traps and snares are more advantageous then going out on foot and physically hunting the animal. The most important advantage being that traps work 24 hours a day with no assistance from the hunter. A large area can be effectively strapped with the possibility of catching many animals within the same period of time. Survivors (generally) use much less energy maintaining a trapline than is used by hunting. This means less food is required because less energy is expended.

(2) The traps or snares should be set in areas where the game is known to live or travel. Look for signs such as tracks, droppings, feeding signs, or actual sightings of the animal. If snares are used, they should be set up to catch the animal around the neck. Therefore, the loop must allow the head to pass through but not the body. Loops will vary in size from one animal to another. When placing snares, try to find a narrow area of the game trail where the animal has no choice but to enter the loop. If a narrow area cannot be found, brush or other obstacles can be arranged to funnel the animal into the snare (Figure 14-4). Do not overdo the funneling; use as little as possible. Avoid disturbing the natural surroundings if possible. Do not walk on game trails, but approach 90 degrees to the trail, set the snare, and back away. Snares may also be set over holes or burrows. All

Figure 14-4. Funneling.

snares and traps should be set during the midday because most animals are nocturnal in nature. Check snares and traps twice daily. If possible, check after sunup and before sunset. The checks should be made from a distance so any animals moving at the time of checking will not be disturbed or frightened away.

(3) There are three ways to immobilize or trap animals.

(a) Strangle. This is done by simply using a free-sliding noose which, when tightened around the neck, will restrict circulation of air and blood. The materials should be strong enough to hold the animal; for example, suspension line, string, wire, cable, or rawhide.

(b) Mangle. Mangle traps use a weight which is suspended over the animal's trail or over bait. When the animal trips the trigger, the weight (log) will descend and mangle the animal (Figure 14-5).

(c) Hold. Any means of impeding the animal and detaining

Figure 14-5. Mangle.

Figure 14-6. Apache Foot Snare.

its progress would be considered a hold-type trap.

(4) The apache foot snare is an example of a hold-type trap. It is used for large browsers and grazers like deer (Figure 14-6). It should be located along game trails where an obstruction, such as a log, blocks the trail. When animals jump over this obstruction, a very shallow depression is formed where their hooves land. The apache foot snare should be placed at this depression. The box trap for birds is another example of hold-type traps (Figure 14-7). The simple loop is the quickest snare to construct. All snares and traps should be simple in construction with as few moving parts as possible. This loop can be constructed from any type of bare wire, suspension line, inner core, vines, long strips of green bark, clothing strips or belt, and any other material that will not break under the strain of holding the animal. If wire is being used for snares, a figure "8" or locking loop should be used (Figure 14-8). Once tightened around the animal, the wire is locked into place by the figure "8" which prevents the loop from opening again. A simple loop snare is generally placed in the opening of a den, with the end of the snare anchored to a stake or similar object (Figure 14-8). The simple loop snare can also be used when making a squirrel pole (Figure 14-9) or with some types of trigger devices.

Figure 14-7. Box Trap.

f. Triggers. Triggers may be used with traps. The purpose of the trigger is to set the device in motion, which will eventually

Figure 14-8. Locking Loop and Setting Noose.

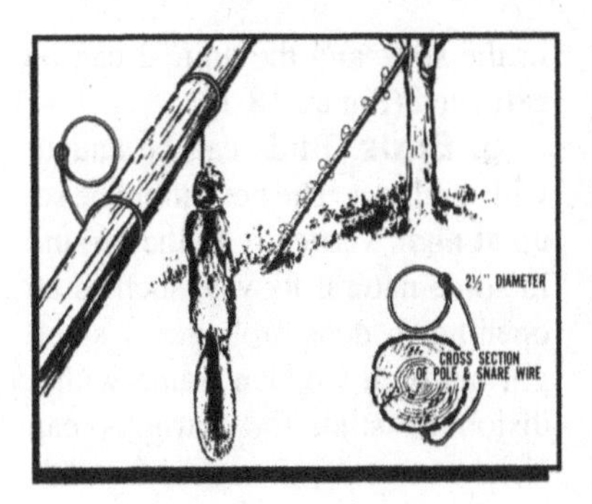

Figure 14-9. Squirrel Pole.

strangle, mangle, or hold the animal. There are many triggers. Some of the more common ones are:

(1) Two-pin toggle with a counterweight for small to medium animals which are lifted out of the reach of predators (Figure 14-10).

Figure 14-10. Two-Pin Toggle.

(2) Figure "H" with wire snare for small mammals and rodents (Figure 14-11).

(3) Canadian ace for predators such as bobcat, coyote, etc., (Figure 14-12).

Figure 14-11. Figure H.

Figure 14-12. Canadian Ace.

(4) Three-pin toggle with deadfall for medium to large animals (Figure 14-13). Medium and

Figure 14-13. Three-Pin Toggle.

large animals can be captured using deadfalls, but this type of trap is recommended only when big game exists in large quantities to justify the great expense of time and effort spent in constructing the trap.

(5) The twitch-up snare which incorporates the simple loop, can be used to catch small animals (Figure14-14). When the animal is

Figure 14-14. Twitch Up.

caught, the sapling jerks it up into the air and keeps the carcass out of the reach of predators. This type of snare will not work well in cold climates, since the bent sapling will freeze in position and not spring up when released.

(6) A long forked stick can be used as a twist stick to procure ground squirrels, rabbits, etc. A den that has signs of activity must be located. Using the long forked stick, the survivor probes the hole with the forked end until something soft is felt then twisting the stick will entangle the animal's hide in the stick and the animal can be extracted (Figure 18-18).

g. Birds. Birds can be caught with a gill net. The net should be set up at night vertically to the ground in some natural flyway, such as an opening in dense foliage. A small gill net on a wooden frame with a disjointed stick for a trigger can also be used. A gill net can be made by using inner core from parachute suspension line (Figure 18-20).

(1) Birds can be caught on baited fishhooks (Figure 14-15) or simple slipping loop snares. Bird's nest can be a source of food. All bird eggs are edible when fresh. Large wading birds such as cranes and herons often nest in mangrove swamps or in high trees near water.

Figure 14-15. Baited Fishhook.

(2) During molting season, birds cannot fly because of the loss of their "flight" feathers; they can be procured by clubbing or netting.

(3) Birds can be also caught in an Ojibway snare. This snare is made by cutting a 1- or 2-inch thick

sapling at a height of 4½ to 5 feet above the ground (Figure 14-16). A springy branch is then whittled flat at the butt end and a rectangular hole is cut through the flattened end. One end of a ½-inch thick stick, 15 inches long, is then whittled to fit slightly loose in the hole and the top corner of the whittled end is rounded off so the stick will easily drop away from the hole. The branch is then tied by its butt end to the top of the sapling. A length of

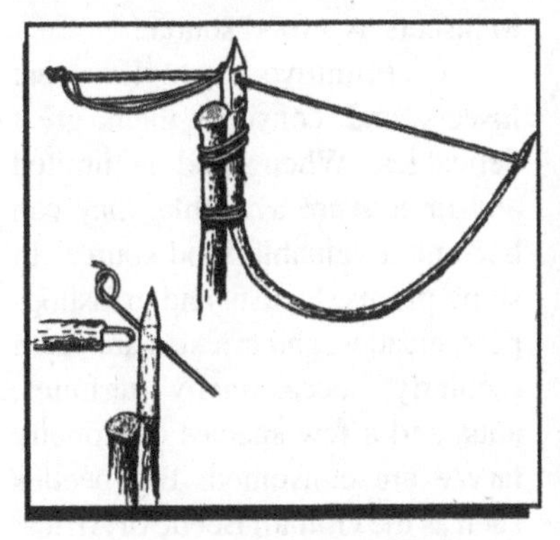

Figure 14-16. Obibwa Bird Snare.

inner core from suspension line is tied to the bottom end of the branch and the branch is bent into a bow with the line passing through the hole in the butt end. A knot is tied in the line and the 15-inch stick is then placed in the hole to lock the line in place (just behind the knot). An 8-inch loop is made at the end of the line and laid out on the 15-inch stick (spread out as well as possible). A piece of bait is placed on top of the sapling, and when a bird comes to settle on the 15-inch stick, the stick drops from the hole causing the loop to tighten around the bird's legs.

(4) When many birds frequent a particular type of bush, some simple loop snares may be set up throughout the bush. Make the snares as large as necessary for the particular type of birds that come to perch, feed, or roost there (Figure 18-17).

Figure 14-17. Ptarmigan or Small Game Snare.

(5) In wild, wooded areas, many larger species of birds such as spruce grouse and ptarmigan may be approached. The spruce grouse, "which has merited the name of "fools hen," can be approached and killed with a stick with little trouble. It often sits on the lower branches of trees and can be easily caught with a long stick with a loop at the end (Figure 14-18).

(6) Ground feeding birds (Quail, Hungarian Partridge, Chukar) can be trapped in a trench dug into the ground. The trench

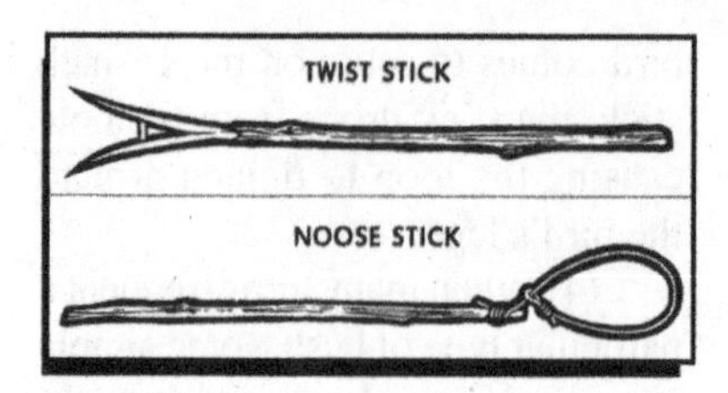

Figure 14-18. Twist Stick and Noose Stick.

should be just wide enough for the bird to walk into, so survivors must first observe the type of ground feeding birds in the area. The trench should be 2 to 3 feet long and about 10 to 12 inches deep at the deep end. The other end of the trench should be ramped down from the surface level. Bait is scattered along the surface into the pit, and after having pecked the last piece of bait the bird will not be able to get out of the pit because it can't fly out or climb out, its feathers keep it from backing out, and it can't turn around to walk out.

(7) Perching birds may be captured by using bird lime. Bird lime is a term applied to any sticky or gluey substance which is rubbed on a branch to prevent the flight of a bird which has landed on it or has flapped a wing against it. Bird lime is usually made from the sap of plants in the Euphorbia family. The common names of some of these plants are spotted spurge, cypress spurge, snow-on-the-mountain, and poinsettias. The Euphorbias have a wide range in North and Central America. The milky sap is poisonous and may cause blisters on the skin and should be handled with care. Bird lime is most effective in the desert and jungle, but it will not work in cold weather. Dust will make bird lime ineffective, so it should be used in spots where dust is not prevalent. The sap of the breadfruit tree makes excellent bird lime as it swells and become glutinous upon contact with air.

h. Insects. If there ever is a time when food aversions must be overcome, it is when survivors turn to insects as a food source.

(1) Primitive peoples eat insects and consider them great delicacies. When food is limited and insects are available, they can become a valuable food source. In some places, locusts and grasshoppers, cicadas, and crickets are eaten regularly; occasionally termites, ants, and a few species of stonefly larvae are consumed. Big beetles such as the Goliath Beetle of Africa, the Giant Water Beetles, and the big Long Horns are relished the world over. Clusters, like those of the Snipefly Atherix (that overhang the water), and the windrows of Brinefly puparia are eaten. Aquatic water bugs of Mexico are grown especially for food. All stages of growth can be eaten, including the eggs but, the large insects must be cooked to kill internal parasites.

(2) Termites and white ants are also an important food source. Strangely enough, these are closely related to cockroaches. The reason

they are eaten so extensively in Africa is the fact that they occur in enormous numbers and are easily collected both from their nests and during flight. They are sometimes attracted to light in unbelievable numbers and the natives become greatly excited when the large species appear.

(3) Many American Indian tribes made a habit of eating the large carpenter ants that are sometimes pests in houses. These were eaten both raw and cooked. Even today the practice of eating them has not entirely disappeared, although they do not form an essential part of the diet of any of the inhabitants of this country (Figure 14-19).

(4) It is not at all unnatural that the American Indians should have relished the honey ants in all parts of the continent where they occur. These ants are peculiar in that some of the workers become veritable storehouses for honey, their abdomens becoming more or less spherical and so greatly enlarged that they are scarcely able to move. They cluster on the ceilings and walls of their nests and disgorge part of their stored food to other inhabitants. The Indians discovered the sweetness stored in these insects and made full use of it. At first they ate the ants alive, later gathering them in quantity and crushing them so that they formed an enticing

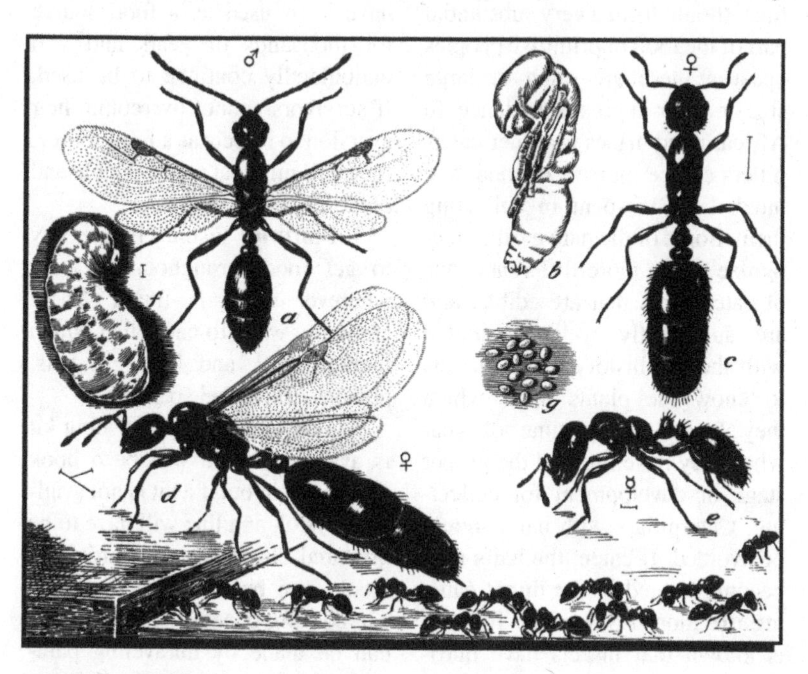

Figure 14-19. Ants.

dish—one which was considered a delicacy and served to guests of distinction as a special favor. The next step in the use of the honey ant was the extraction of the pure honey by crushing the insects and straining the juices. After the honey was extracted, it was allowed to ferment, forming what is said to be a highly flavored wine.

(5) Indians of the American tropics, with a much larger ant fauna from which to choose, select the queens of the famous leafcutting or so-called umbrella ants upon which to feed, eating only the abdomens, either raw or cooked.

(6) It is natural that caterpillers, the larvae of moths and butterflies, should form a very substantial part of the food of primitive peoples because these are often of large size or occur in great abundance. In Africa, many tribes consider caterpillers choice morsels of food, and much time is spent in collecting them. Some of the native tribes recognize 20 or more different kinds of caterpillars that are edible, and are sufficiently well acquainted with the life history of the insects to know the plants upon which they feed and the time of year when they have reached the proper stage of development for collecting. Caterpillars with hairs should be avoided. If eaten, the hairs may become lodged in the throat causing irritation or infection. Today it is known that insects have nutritional or medicinal value. The praying mantis, for example, contains 58 percent protein, 12 percent fat, 3 percent ash, vitamin B complex, and vitamin A. The insect's outer skeleton is an interesting compound of sugar and amino acids.

(7) Bee larvae were eaten by the ancient Chinese. Some Chinese today eat locusts, dragonflies, and bumblebees. Cockroaches and locusts are a favorite dish in Szechuan. In Kwangtun, grasshoppers, golden June beetles, crickets, wasp larvae, and silkworm larvae are used for food.

(8) Stinging insects should have their stinging apparatus removed before they are eaten.

(9) As can be seen, insects have been used as a food source for thousands of years and will undoubtedly continue to be used. If survivors cannot overcome their aversion to insects as a food source, they will miss out on a valuable and plentiful supply of food.

i. Fishing. Fishing is one way to get food throughout the year wherever water is found. There are many ways to catch fish which include hook and line, gill nets, poisons, traps, and spearing.

(1) If an emergency fishing kit is available, there will be a hook and line in it, but if a kit is not available a hook and line will have to be procured elsewhere or improvised. Hooks can be made from wire or carved from bone or wood. The line can be made by unraveling parachute suspension line or by twisting

Figure 14-20. Fishing Places.

threads from clothing or plant fibers. A piece of wire between the fishing line and the hook will help prevent the fish from biting through the line. Insects, smaller fish, shell-fish, worms, or meat can be used as bait. Bait can be selected by observing what the fish are eating. Artificial lures can be made from pieces of brightly colored cloth, feathers, or bits of bright metal or foil tied to a hook. If the fish will not take the bait, try to snag or hook them in any part of the body as they swim by. In freshwater, the deep-est water is usually the best place to fish. In shallow streams, the best places are pools below falls, at the foot of rapids, or behind rocks. The best time to fish is usually early morning or late evening (Figure 14-20). Sometimes fishing is best at night, especially in moonlight or if a light is available to attract the fish. The survivor should be patient and fish at different depths in all kinds of water. Fishing at different times of the day and changing bait often is rewarding.

(2) The most effective fishing method is a net because it will catch fish without having to be attended (Figures 14-21 and 14-22). If a gill net is used, stones can be used for anchors and wood for floats. The net should be set at a slight angle to the current to clear itself of any floating refuse that comes down the stream. The net should be checked at least twice daily (Figure 14-23). A net with poles attached to each end works effectively if moved up or down a stream as rapidly as possible while moving stones and threshing the bottom or edges of the streambanks. The net should be

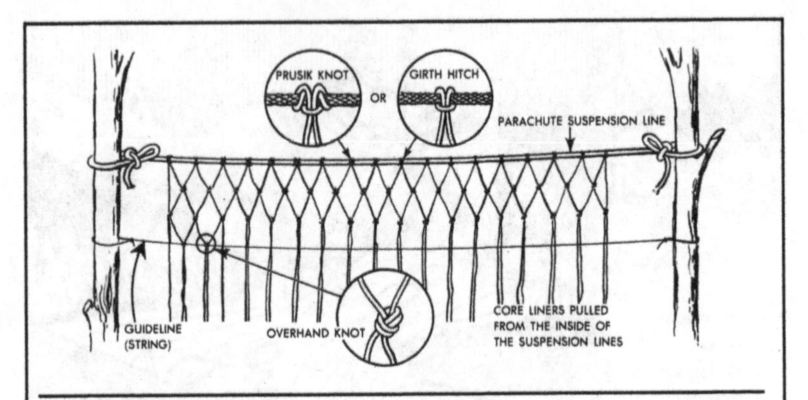

1. Suspend a suspension line casing (form which the core liners have been pulled) between two uprights, approximately at eye level.

2. Hang core liners (an even number) from the line suspended as in 1, above. These lines should be attached with a Prusik knot or girth hitch and spaced in accordance with the mesh you desire. One-inch spacing will result in a 1-inch mesh, etc. The number of lines used will be in accord with the width of the net desired. If more than one man is going to work on the net, the length of the net should be stretched between the uprights, thus providing room for more than one man to work. If only one man is to make up the net, the depth of the net should be stretched between the uprights and step 8, below, followed.

3. Start at left or right. Skip the first line and tie the second and third lines together with an overhand knot. Space according to mesh desired. Then tie fourth and fifth, sixth and seventh, ect. One line will remain at the end.

4. On the second row, tie the first and second, third and fourth, fifth and sixth, etc., to the end.

5. Third row, skip the first line and repeat step 3 above.

6. Repeat step 4, and so on.

7. You may want to use a guide line which can be moved down for each row of knots to ensure equal mesh. Guide line should run across the net on the side opposite the one you are working from so that it will be out of your way.

8. When you have stretched the depth between the uprights and get close to ground level, move the net up by rolling it on a stick and continue until the net is the desired length.

9. String suspension line casing along the sides when net is completed to strengthen it and make the net easier to set.

Figure 14-21. Making a Gill Net.

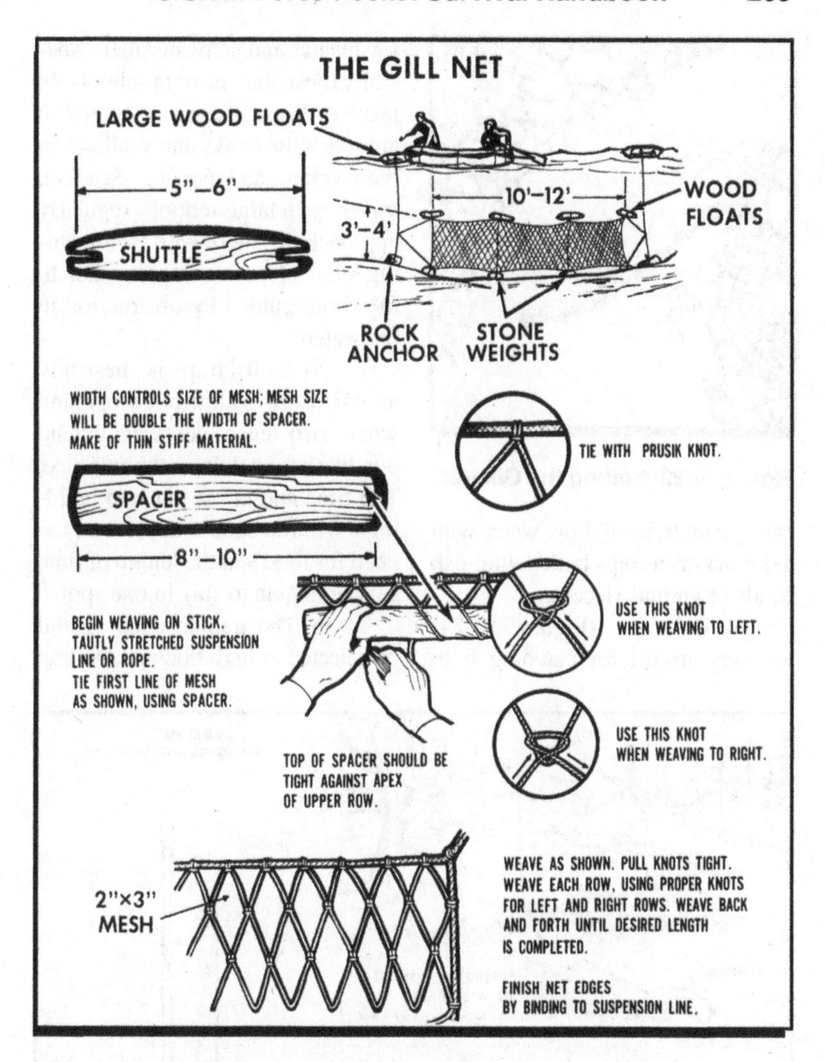

Figure 14-22. Making a Gill Net With Shuttle and Spacer.

checked every few moments so the fish cannot escape.

(3) Shrimp (prawns) live on or near the sea bottom and may be scraped up. They may be lured to the surface with light at night. A hand net made from parachute cloth or other material is excellent for catching shrimp. Lobsters are creeping crustaceans found on or near the sea bottom. A lobster trap, jig, baited hook, or dip net can be used to catch lobster. Crabs will creep, climb, and burrow and are

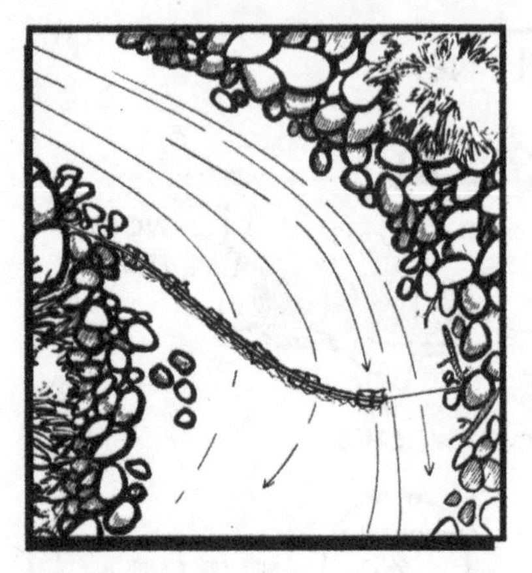

Figure 14-23. Setting the Gill Net.

easily caught in shallow water with a dip net or in traps baited with fish heads or animal viscera.

(4) Fishtraps (Figure 14-24) are very useful for catching both freshwater and saltwater fish, especially those that move in schools. In lakes or large streams, fish tend to approach the banks and shallows in the morning and evening. Sea fish, traveling in large schools, regularly approach the shore with the incoming tide, often moving parallel to the shore guided by obstruction in the water.

(a) A fishtrap is basically an enclosure with a blind opening where two fence-like walls extend out, like a funnel, from the entrance. The time and effort put into building a fishtrap should depend on the need for food and the length of time survivors plan to stay in one spot.

(b) The trap location should be selected at high tide and the trap

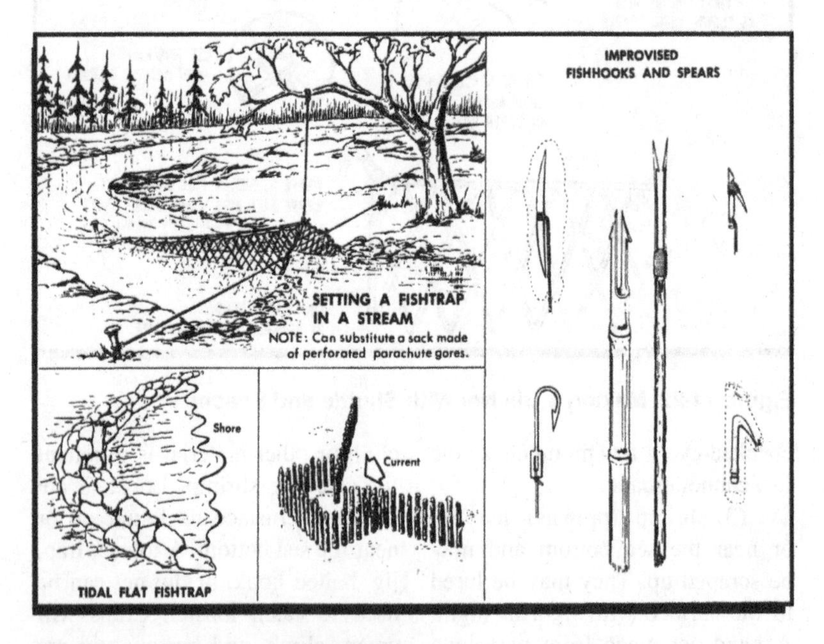

Figure 14-24. Maze-type Fishtraps.

built at low tide. One to 2 hours of work should do the job. Consider the location, and try to adapt natural features to reduce the labors. Natural rock pools should be used on rock shores. Natural pools on the surface of reefs should be used on coral islands by blocking the opening as the tide recedes. Sandbars, and the ditches they enclose, can be used on sandy shores. The best fishing off sandy beaches is the lee side of offshore sandbars. By watching the swimming habits of fish, a simple dam can be built which extends out into the water forming an angle with the shore. This will trap fish as they swim in their natural path. When planning a more complex brush dam, select protected bays or inlets using the narrowest area and extending one arm almost to the shore.

(c) In small, shallow streams, the fishtraps can be made with stakes or brush set into the stream bottom or weighted down with stones so that the stream is blocked except for a small narrow opening into a stone or brush pen or shallow water. Wade into the stream, herding the fish into the trap, and catch or club them when they get in shallow water. Mud-bottom streams can be trampled until cloudy and then netted. The fish are blinded and cannot avoid the nets. Freshwater crawfish and snails can be found under rocks, logs, overhanging bushes, or in mud bottoms.

(5) Fish may be confined in properly built enclosures and kept for days. In many cases, it may be advantageous to keep them alive until needed and thus ensure there is a fresh supply without danger of spoilage. Mangrove swamps are often good fishing grounds. At low tide, clusters of oysters and mussles are exposed on the mangrove "knees" or lower branches. Clams can be found in the mud at the base of trees. Crabs are very active among branches or roots and in the mud. Fish can be caught at high tide. Snails are found on mud and clinging to roots. Shellfish which are not covered at high tide or those from a colony containing diseased members should not be eaten. Some indications of diseased shellfish are shells gaping open at low tide, foul odor, and(or) milky juice.

(6) Throughout the warm regions of the world, there are various plants which the natives use for poisoning fish. The active poison in these plants is harmful only to cold-blooded animals. Survivors can eat fish killed by this poison without ill effects.

(a) In Southeast Asia, the derris plant is widely used as a source of fish poison. The derris plant, a large woody vine, is also used to produce a commercial fish poison called rotenone. Commercial rotenone can be used in the same manner as crushed derris roots; it causes respiratory failure in fish, but has no ill effects on humans. However, rotenone has no effect if dusted over the surface of a pond. It should be mixed to a malted-milk consistency with a little water, and then distributed in the water. If the concentration is strong, it takes effect within 2 minutes in warm water, or it may take an hour in colder water. Fish sick enough to turn over on their backs will eventually die. An ounce of 12 percent rotenone can kill every fish for a half mile down a slow-moving stream that is about 25 feet wide. A few facts to remember about the use of rotenone are:

-1. It is very swift acting in warm water at 70°F and above.

-2. It works more slowly in cold water and is not practical in water below 55 °F.

-3. It is best applied in small ponds, streams, or tidal pools.

-4. Excess usage will be wasted. However, too little will not be effective.

(b) A small container of 12 percent rotenone (one-half ounce) is a valuable addition to any emergency kit. Do not expose it unnecessarily to air or light; it retains its toxicity best if kept in a dark-colored vial. Lime thrown in a small pond or tidal pool will kill fish in the pool. Lime can be obtained by burning coral and seashells.

(c) The most common method of using fish-poison plants is to crush the plant parts (most often the roots) and mix them with water. Drop large quantities of the crushed plant into pools or the headwaters of small streams containing fish. Within a short time, the fish will rise in a helpless state to the surface. After putting in the poison, follow slowly down stream and pick up the fish as they come to the surface, sink to the bottom, or swim crazily to the bank. A stick dam or obstruction will aid in collecting fish as they float downstream. The husk of "green" black walnuts can be crushed and sprinkled into small sluggish streams and pools to act as a fish stupefying agent. In the southwest Pacific, the seeds and bark from the barringtonia tree (Figure 14-25) are commonly used as a source of fish poison. The barringtonia tree usually grows along the seashore.

(7) Tickling can be effective in small streams with undercut banks or in shallow ponds left by receding flood waters. Place hands in the water and reach under the bank slowly, keeping the hands close to the bottom if possible. Move the fingers slightly until they make contact with a fish. Then work hands gently along its belly until reaching its gills. Grasp the fish firmly just behind the gills and scoop it onto land. In the tropics, this type of fishing can be dangerous due to hazardous marine life in the water such as piranhas, eels, and snakes.

14-5. Plant Food. The thought of having a diet consisting only of plant food is often distressing to stranded aircrew members. This is

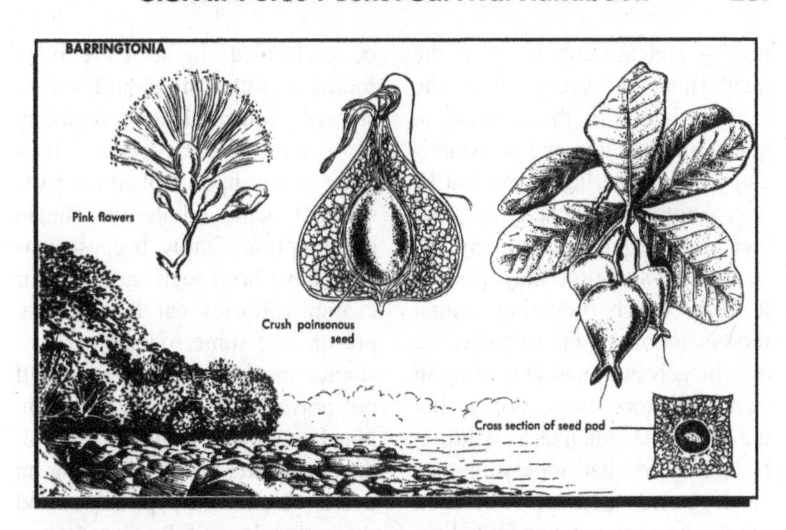

Figure 14-25. Barringtonia Plant for Poisoning Fish.

not the case if the survival episode is entered into with the confidence and intelligence based on knowledge or experience. If the survivors know what to look for, can identify it, and know how to prepare it properly for eating, there is no reason why they can't find sustenance. In many isolated regions, survivors who have had some previous training in plant identification can enjoy wild plant food.

a. Plants provide carbohydrates, which provide body energy and calories. Carbohydrates keep weight and energy up, and include important starches and sugars.

b. A documented and authoritative example of the value of a strictly plant diet in survival can be cited in the case of a Chinese botanist who had been drafted into the Japanese Army during World War II. Isolated with his company in a remote section of the Philippines, the Chinese botanist kept 60 of his fellow soldiers alive for 16 months by finding wild plants and preparing them properly. He selected six men to assist him, and then found 25 examples of edible plants in the vicinity of their camp. He acquainted the men with these samples, showing them what parts of the plants could be used for food. He then sent the men out to look for similar plants and had them separate the new plants according to the original examples to avoid any poisonous plant mingling with the edible ones. The result of this effort was impressive. Though all the men had a natural desire for ordinary food, none suffered physically from the plant food diet. The report was especially valuable because the botanist kept a careful record of all the food used, the results, and the comments of the men. This case history reflects the same opinions as those found in questionnaires directed to American survivors during World War II.

c. Another advantage of a plant diet is availability. In many instances, a situation may present itself in which procuring animal food is out of the question because of injury, being unarmed, being in enemy territory, exhaustion, or being in an area which lacks wildlife. If convinced that vegetation can be depended upon for daily food needs, the next question is "where to get what and how."

(1) Experts estimate there are about 300,000 classified plants growing on the surface of the earth, including many which thrive on mountain tops and on the floors of the oceans. There are two considerations that survivors must keep in mind when procuring plant foods. The first consideration, of course, is the plant be edible, and preferably, palatable. Next, it must be fairly abundant in the areas in which it is found. If it includes an inedible or poisonous variety in its family, the edible plant must be distinguishable to the average eye from the poisonous one. Usually a plant is selected because one special part is edible, such as the stalk, the fruit, or the nut.

(2) To aid in determining plant edibility, there are general rules which should be observed and an edibility test that should be performed. In selecting plant foods, the following should be considered. Select plants resembling those cultivated by people. It is risky to rely upon a plant (or parts thereof) being edible for human consumption simply because animals have been seen eating it (for example, horses eat leaves from poison ivy; some rodents eat poisonous mushroom). Monkeys will put poisonous plants and fruits in pouches of their mouths and spit them out later. When selecting an unknown plant as a possible food source, apply the following general rules:

(a) Mushrooms and fungi should not be selected. Fungi have toxic peptides, a protein-base poison which has no taste. There is no field test other than eating to determine whether an unknown mushroom is edible. Anyone gathering wild mushrooms for eating must be absolutely certain of the identity of every specimen picked. Some species of wild mushrooms are difficult for an expert to identify. Because of the potential for poisoning, relying on mushrooms as a viable food source is not worth the risk.

(b) Plants with umbrella-shaped flowers are to be completely avoided, although carrots, celery, dill, and parsley are members of this family. One of the most poisonous plants, poison water hemlock, is also a member of this family (Figure 14-26).

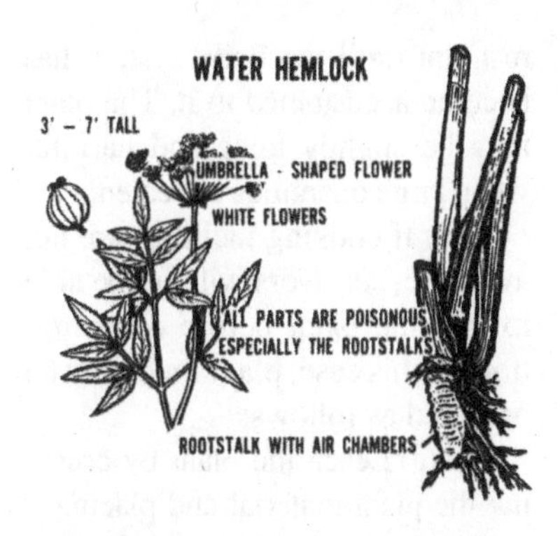

Figure 14-26. Water Hemlock.

(c) All of the legume family should be avoided (beans and peas). They absorb minerals from the soil and cause problems. The most common mineral absorbed is selenium. Selenium is what has given locoweed its fame. (Locoweed is a vetch.)

(d) As a general rule, all bulbs should be avoided. Examples of poisonous bulbs are tulips and death camas.

(e) White and yellow berries are to be avoided as they are almost always poisonous. Approximately one-half of all red berries are poisonous. Blue or black berries are generally safe for consumption.

(f) Aggregated fruits and berries are always edible (for example, thimbleberry, raspberry, salmonberry, and blackberry).

(g) Single fruits on a stem are generally considered safe to eat.

(h) Plants with shiny leaves are considered to be poisonous and caution should be used.

(i) A milky sap indicates a poisonous plant.

(j) Plants that are irritants to the skin should not be eaten, such as poison ivy.

(k) A plant that grows in sufficient quantity within the local area should be selected to justify the edibility test and provide a lasting source of food if the plant proves edible.

(l) Plants growing in the water or moist soil are often the most palatable.

Plants are less bitter when growing in shaded areas.

(3) The previously mentioned information concerning plants is general. There are exceptions to every rule, but when selecting unknown plants for consumption, plants with these characteristics should be avoided. Plants that do not have these characteristics should be considered as possible food sources. Apply the edibility test to only one plant at a time so if some abnormality does occur, it will be obvious which plant caused the problem. Once a plant has been selected for the edibility test, proceed as follows:

(a) Crush or break part of the plant to determine the color of its sap. If the sap is clear, proceed to the next step.

(b) Touch the plant's sap or juice to the inner forearm or tip of the tongue. (A small taste of a poisonous plant will not do serious harm.) If there are no ill effects, such as a rash or burning sensation to the skin, bitterness to the taste, or numbing sensation of the tongue or lips, then proceed with the rest of the steps. (NOTE: Sometimes heavy smokers are unable to taste various poisons, such as alkaloids).

(c) Prepare the plant or plant part for consumption by boiling in two changes of water. The toxic properties of many plants are water soluble or destroyed by heat; cooking and discarding in two changes of water lessens the amount of poisonous material or removes it completely. Parboiling is a process of boiling the individual plant parts in repeated changes of water to remove bitter elements. This boiling period should last about 5 minutes.

(d) Place about 1 teaspoonful of the prepared plant food in the mouth for 5 minutes and chew but do not swallow it. A burning, nauseating, or bitter taste is a warning of possible danger. If any of these ill effects occur, remove the material from the mouth at once and discard that plant as a food source. However, if no burning sensation or other unpleasant effect occurs, swallow the plant material and wait 8 hours.

(e) If after this 8 hours there are no ill effects, such as nausea, cramps, or diarrhea, eat about 2 tablespoonfuls and wait an additional 8 hours.

(f) If no ill effects occur at the end of this 8-hour period, the plant may be considered edible.

(g) Keep in mind that any new or strange food should be eaten with restraint until the body system has become accustomed to it. The plant may be slightly toxic and harmful when large quantities are eaten.

(4) If cooking facilities are not available, survivors will not be able to boil the plant before consumption. In this case, plant food may be prepared as follows:

(a) Leach the plant by crushing the plant material and placing it in a container. Pour large quantities of cold water over it (rinse the plant parts). Leaching removes some of the bitter elements of nontoxic plants.

(b) If leaching is not possible, survivors should follow the steps they can in the edibility test.

d. The survivor will find some plants which are completely edible, but many plants which they may find will have only one or more identifiable parts having food and thirst-quenching value. The variety of plant component parts which might contain substance of food value is shown in figure 14-27.

(1) Underground Parts:

(a) Tubers. The potato is an example of an edible tuber. Many other kinds of plants produce tubers such as the tropical yam, the Eskimo potato, and tropical water lilies. Tubers are usually found below the ground. Tubers are rich in starch and should be cooked by roasting in an earth oven or by boiling to break down the starch for ease in digestion. The following are some of the plants with edible tubers.

EDIBLE PARTS OF PLANTS	
Underground Parts	Tubers Roots and Rootstalks Bulbs
Stems and Leaves (potherbs)	Shoots and Stems Leaves Pith Bark
Flower Parts	Flowers Pollen
Fruits	Fleshy Fruits (dessert and vegetable) Seeds and Grains Nuts Seed Pods Pulps
Gums and Resins	
Saps	

Figure 18-27. Edible Parts of Plants.

-1.Arrowroot, East Indian.
-2.Taro.
-3.Cassava (Tapioca).
-4.Bean, Yam.
-5.Chufa (Nut Grass).
-6. Water Lily (Tropical).
-7.Sweet Potato (Kamote).
-8.Yam Tropical.

(b) Roots and Rootstalks. Many plants produce roots which may be eaten. Edible roots are often several feet in length. In comparison, edible rootstalks are underground portions of the plant which have become thickened, and are relatively short and jointed. Both true roots and rootstalks are storage organs rich in stored starch. The following are some of the plants with edible roots or rootstalks (rhizomes):

-1.Baobab.
-2.Pine, Screw.
-3.Bean, Goa.
-4.Plantain, Water.
-5.Bracken.
-6.Reindeer Moss.
-7.Calla, Wild (Water Arum).
-8.Rock Tripe.
-9.Pollypody.
-lO.Canna Lily.
-11. Rush, Flowering.
-12.Cattail.
-13.Spinach, Ceylon.
-14.Chicory.
-15.Ti Plant.
-16. Horseradish.
-17.Tree Fern.
-18.Lotus Lily.
-19.Water Lily (Temperate Zone).
-20.Manioc.

(c) Bulbs. The most common edible bulb is the wild onion, which can easily be detected by its characteristic odor. Wild onions may be eaten uncooked, but other kinds of bulbs are more palatable if cooked. In Turkey and Central Asia, the bulb of the wild tulip may be eaten. All bulbs contain a high percentage of starch. (Some bulbs are poisonous, such as the death camas which has white or yellow flowers.) The following are some of the plants with edible bulbs:

-1 .Lily, Wild.
-2.Tulip, Wild.
-3,Onion, Wild.
-4.Blue Camas.
-5.Tiger Lily.

(2) Shoots and Leaves:

(a) Shoots (Stems). All edible shoots grow in much the same fashion as asparagus. The young shoots of ferns (fiddleheads) and especially those of bamboo and numerous kinds of palms are desirable for food. Some kinds of shoots may be eaten raw, but most are better if first boiled for

5 to 10 minutes, the water drained off, and the shoots reboiled until they are sufficiently cooked for eating (parboiled). (See figure 14-28).

-1. Agave (Century Plant).
-2. Palm, Coconut.
-3. Purslane.
-4. Reindeer Moss.
-5. Bamboo.
-6. Palm, Fishtail.
-7. Bean, Goa.
-8. Palm, Nipa.
-9. Bracken.
-lO. Palm, Rattan.
-11. Rhubarb, Wild.
-12. Cattail.
-13. Palm, Sago.
-14. Spinach, Ceylon.
-15. Rock Tripe.
-16. Colocynth.
-17. Palm, Sugar.
-18. Papaya.
-19. Sugar Cane.
-20. Lotus Lily.
-21. Pokeweed (poisonous roots).
-22. Sweet Potato-Kamote.
-23. Luffa Sponge.
-24. Water Lily (Tropical).
-25. Polypody.
-26. Palm, Buri.
-27. Willow, Arctic.

(b) Leaves. The leaves of spinach-type plants (potherbs), such as wild mustard, wild lettuce, and lamb quarters, may be eaten either raw or cooked. Prolonged cooking, however, destroys most of the vitamins. Plants which produce edible leaves are perhaps the most numerous of all edible plants. The young tender leaves of nearly all nonpoisonous plants are edible. The following are only some of the plants with edible leaves:

-l. Amarath.
-2. Luffa Sponge.
-3. Rock Tripe.
-4. Avocado.
-5. Mango.
-6. Sorrel, Wild.

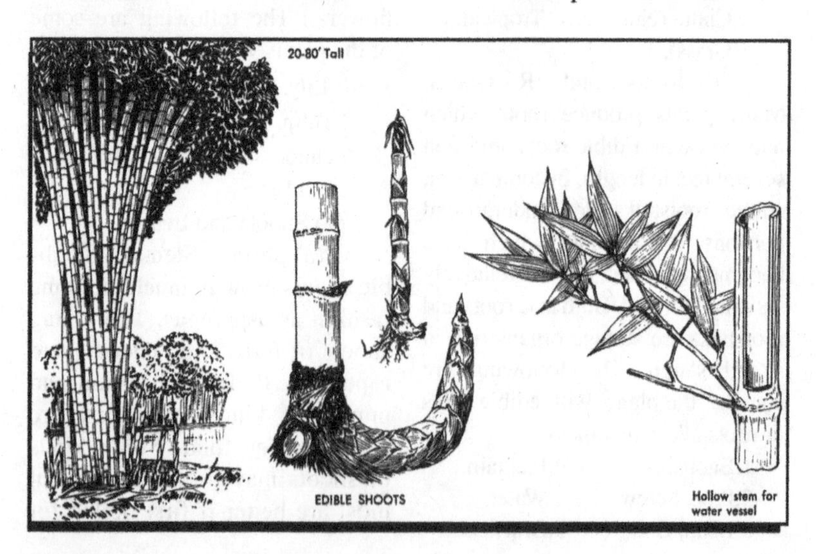

Figure 18-28. Bamboo.

-7. Baobab.
-8. Orach, Sea.
-9. Bean, Goa.
-10. Papaya.
-11. Spinach, Ceylon.
-12. Cassava.
-13. Chickory.
-14. Pine, Screw.
-15. Spreading Wood Fern.
-16. Dock.
-17. Plantain.
-18. Pokeweed (poisonous roots).
-19. Sweet Potato-Kamote.
-20. Tamarind.
-21. Horseradish.
-22. Prickly Pear.
-23. Taro (only after cooking).
-24. Lettuce, Water.
-25. Purslane.
-26. Ti Plant.
-27. Willow, Arctic.
-28. Lotus Lily.
-29. Reindeer Moss.

(c) Pith. Some plants have an edible pith in the center of the stem. The pith of some kinds of tropical plants is quite large. Pith of the sago palm is particularly valuable because of its high food value. The following are some of the palms with edible pith (starch):

-1. Buri.
-2. Fishtail.
-3. Sago.
-4. Coconut.
-5. Rattan.
-6. Sugar

(d) Bark. The inner bark of a tree—the layer next to the wood—may be eaten raw or cooked. It is possible in northern areas to make flour from the inner bark of such trees as the cottonwood, aspen, birch, willow, and pine. The outer bark should be avoided in all cases because this part contains large amounts of bitter tannin. Pine bark is high in vitamin C. The outer bark of pines can be cut away and the inner bark stripped from the trunk and eaten fresh, dried, or cooked, or it may be pulverized into flour. Bark is most palatable when newly formed in spring. As food, bark is most useful in the arctic regions, where plant food is often scarce.

(3) Flower Parts:

(a) Flowers and Buds. Fresh flowers may be eaten as part of a salad or to supplement a stew. The hibiscus flower is commonly eaten throughout the southwest Pacific area. In South America, the people of the Andes eat nasturtium flowers. In India, it is common to eat the flowers of many kinds of plants as part of a vegetable curry. Flowers of desert plants may also be eaten. The following are plants with edible flowers:

-1. Abal.
-2. Colocynth.
-3. Papaya.
-4. Banana.
-5. Horseradish.
-6. Caper, Wild.
-7. Luffa Sponge.

(b) Pollen. Pollen looks like yellow dust. All pollen is high in food value and in some plants, especially the cattail. Quantities of pollen may easily be collected and eaten as a kind of gruel.

(4) Fruits. Edible fruits can be divided into sweet and nonsweet (vegetable) types. Both are the seed bearing parts of the plant. Sweet fruits are often plentiful in all areas of the world where plants grow. For instance, in

the far north, there are blueberries and crowberries; in the temperate zones, cherries, plums, and apples; and in the American deserts, fleshy cactus fruits. Tropical areas have more kinds of edible fruit than other areas, and a list would be endless. Sweet fruits may be cooked, or for maximum vitamin content, left uncooked. Common vegetable fruits include the tomato, cucumber, and pepper.

(a) Fleshy Fruits (Sweet). The following are plants with edible fruits:

-1. Apple, Wild.
-2. Bael Fruit.
-3. Banana.
-4. Bignay.
-5. Blueberry, Wild.
-6. Bullocks Heart.
-7. Cloudberry.
-8. Crabapple.
-9. Cranberry.
-lO. Fig, Wild.
-11. Grape, Wild.
-12. Huckleberry.
-13. Jackfruit.
-14. Jujube, Common.
-15. Mango.
-16. Mulberry.
-17. Papaya.
-18. Plum, Batako.
-19. Pokeberry.
-20. Prickly Pear.
-21. Rose Apple.
-22. Soursop.
-23. Sweetsop.

(b) Fleshy Fruits (Vegetables). The following are plants with edible fruits (vegetables):

-1. Breadfruit.
-2. Horseradish.
-3. Plantain.
-4. Caper, Wild.
-5. Luffa Sponge.

(c) Seeds and Grains. Seeds of many plants, such as buckwheat, ragweed, amaranth, and goosefoot, contain oils and are rich in protein. The grains of all cereals and many other grasses, including millet, are also extremely valuable sources of plant protein. They may either be ground between stones, mixed with water and cooked to make porridge, parched or roasted over hot stones. In this state, they are still wholesome and may be kept for long periods without further preparation (Figure 14-29). The following are some of the plants with edible seeds and grains:

-1. Amaranth. (Tropical).
-2. Millet, Italian.
-3. Rice.
-4. Bamboo.
-5. Millet, Pearl.
-6. Palm, Nipa.
-7. Tamarind.
-8. Pine, Screw.
-9. Coloynth.
-10. Water Lily
-11. Sterculia.
-12. Baobab.
-13. Orach, Sea.
-14. St. John's Bread.
-15. Bean, Goa.
-16. Lotus Lily.
-17. Purslane.
-18. Water Lily (Temperate).
-19. Luffa Sponge.

(d) Nuts. Nuts are among the most nutritious of all raw plant foods and contain an abundance of valuable protein. Plants bearing edible nuts occur in all the climatic zones of the world and in all continents except in the arctic regions. Inhabitants of the temperate zones are familiar with walnuts, filberts, almonds, hickory nuts, acorns, hazelnuts, beechnuts, and pine nuts, to mention just a few. Tropical zones produce coconuts and other palm nuts, brazil nuts, cashew nuts, and macadamia nuts (Figure 14-30). Most nuts can be eaten raw

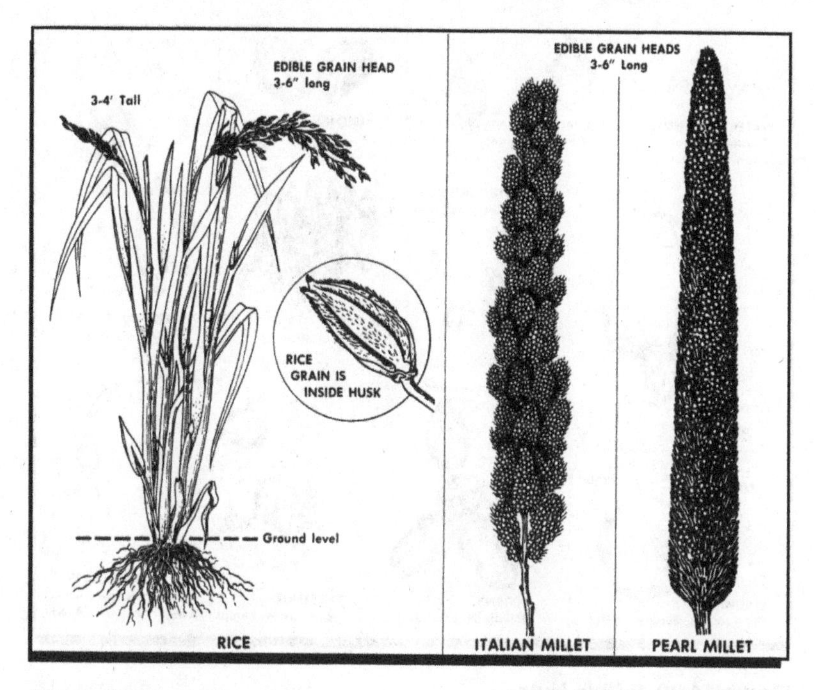

Figure 14-29. Grains.

but some such as acorns, are better when cooked. The following are some of the plants with edible nuts:

-1. Almond.
-2. Chestnut, Water (TrapaNut).
-3. Palm, Buri.
-4. Almond, Indian or Tropical.
-5. Chestnut, Mountain.
-6. Palm, Coconut.
-7. Beechnut.
-8. Filbert (Hazelnut).
-9. Palm, Fishtail.
-10. Jackfruit Seeds.
-11. Oak, English (Acorn).
-12. Palm, Sago.
-13. Palm, Sugar.
-14. Pine.
-15. Pistachio, Wild.
-16. Walnut.

(e) Pulps. The pulp around the seeds of many fruits is the only part that can be eaten. Some fruits produce sweet pulp; others have a tasteless or even bitter pulp. Plants that produce edible pulp include the custard apple, inga pod, breadfruit, and tamarind. The pulp of breadfruit must be cooked, whereas in other plants, the pulp may be eaten uncooked. Use the edibility rules in all cases of doubt.

(5) Gums and Resins. Gum and resin are sap that collects and hardens on the outside surface of the plant. It is called gum if it is soft and soluble, and resin if it is hard and not soluble. Most people are familiar with the gum which

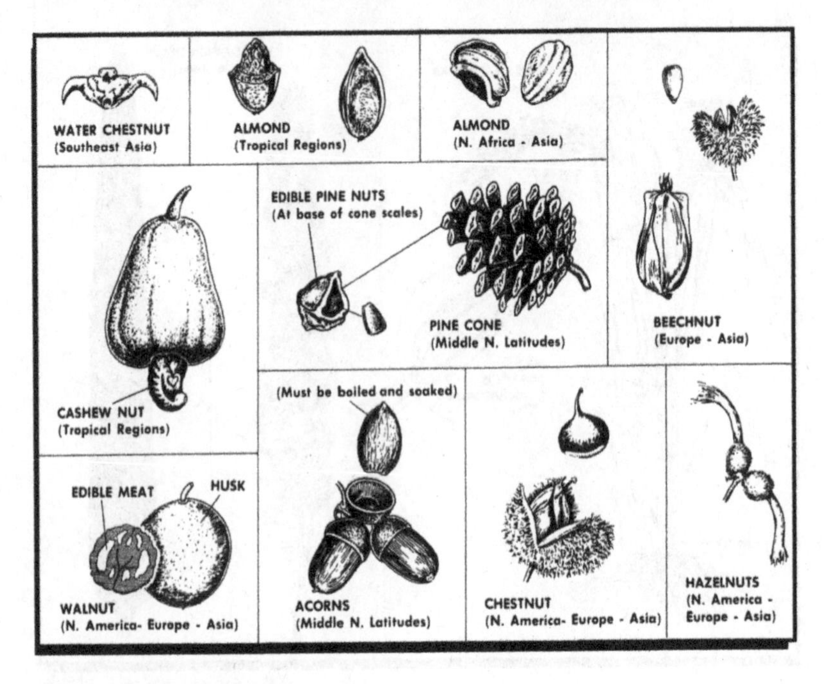

Figure 14-30. Edible Nuts.

exudes from cherry trees and the resin which seeps from the pine trees. These plant byproducts are edible and are a good source of nutritious food which should not be overlooked.

(6) Saps. Vines or other plant parts may be tapped as potential sources of usable liquid. The liquid is obtained by cutting the flower stalk and letting the fluid drain into some sort of container such as a bamboo section. Palm sap with its high-sugar content is highly nutritious. The following are some plants with edible sap and drinking water:

(a)Acacia, Sweet (water),
(b)Colocynth (water).
(c)Palm, Coconut (sap),
(d)Palm, Fishtail (sap).
(e)Agave (water).
(f)Cuipo Tree (water).
(g)Saxual (water).
(h)Palm, Nipa (sap).
(i)Palm, Rattan (water).
(j)Cactus (water).
(k)Grape (water).
(l)Banana (water).
(m)Palm, Sago (sap).
(n)Palm, Sugar (sap).
(o)Palm, Buri (sap).

18-6. Food in Tropical Climate. There are more types of animals in the jungles of the world than in any other region. A jungle visitor who is unaware of the life style and eating habits of these animals would not observe the presence of a large number of the animals.

a. Game trails are the normal routes along which animals travel through a jungle. Some of the animals used as food are hedgehogs, porcupines, anteaters, mice, wild pigs, deer, wild cattle, bats, squirrels, rats, monkeys, snakes, and lizards.

(1) Reptiles are located in all jungles and should not be overlooked as a food source. All snakes should be considered poisonous and extreme caution used when killing the animal for a food source. All cobras should be avoided since the spitting cobra aims for the eyes; the venom can blind if not washed out immediately. Lizards are good food, but may be difficult to capture since they can be extremely fast. A good blow to the head of a reptile will usually kill it. Crocodiles and caimans are extremely dangerous on land as well as in the water.

(2) Frogs can be poisonous; all brilliantly colored frogs should be totally avoided. Some frogs and toads in the tropics secrete substances through the skin which has a pungent odor. These frogs are often poisonous.

(3) The larger, more dangerous animals such as tigers, rhinocerous, water buffalo, and elephants are rarely seen and should be left alone. These larger animals are usually located in the open grasslands.

b. Seafood such as fish, crabs, lobsters, crayfish, and small octopi can be poked out of holes, crevices, or rock pools (Figure 14-31). Survivors should be ready to spear them before they move off into deep water. If they are in deeper water, they can be teased shoreward with a baited hook, or a stick.

(1) A small heap of empty oysters shells near a hole may indicate the presence of an octopus. A baited hook placed in the hole will often catch the octopus. The survivor should allow the octopus to surround the hook and line before lifting. Octopi are not scavengers like sharks, but they are hunters, fond of spiny lobster and other crab-like fish. At night, they come into shallow water and can be easily seen and speared.

(2) Snails and limpets cling to rocks and seaweed from the low-water mark up. Large snails called chitons adhere tightly to rocks just above the surf line.

(3) Mussels usually form dense colonies in rock pools, on logs, or at the bases of boulders. Mussels are poisonous in tropical zones during the summer,especially when seas are highly phosphorescent or reddish.

(4) Sluggish sea cucumbers and conchs (large snails) live in deep water. The sea cucumber will

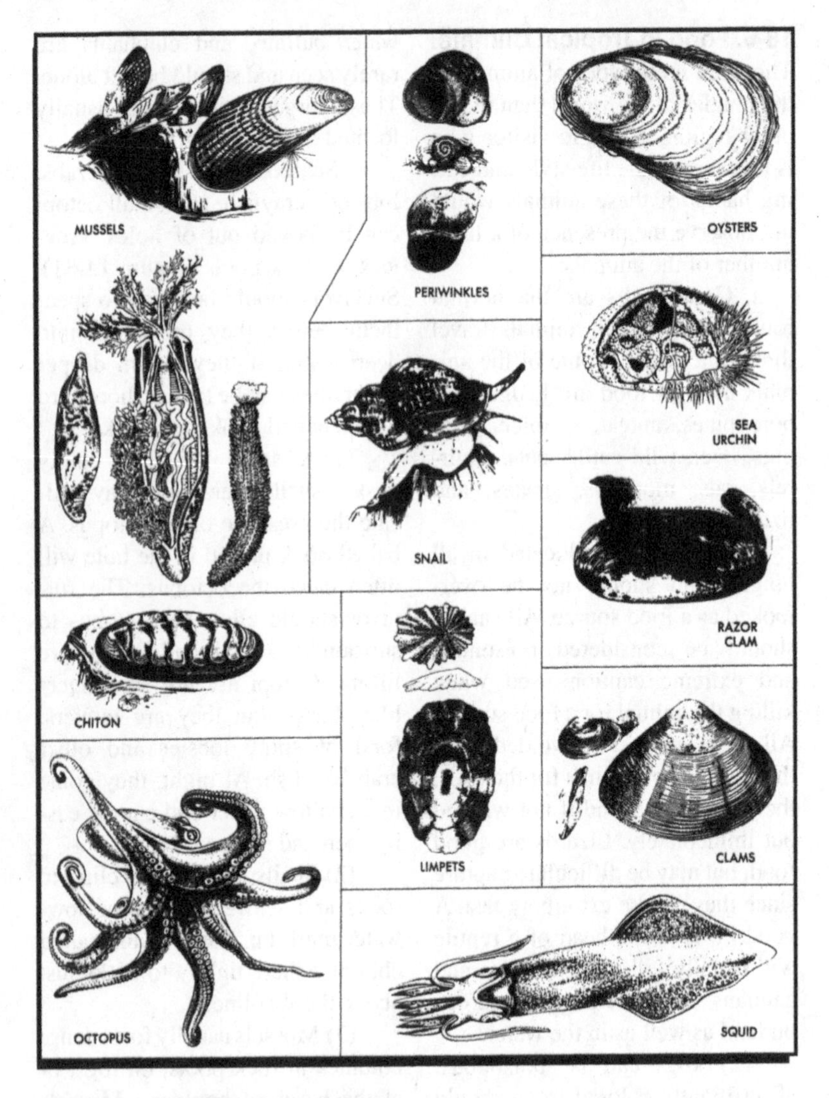

Figure 14-31. Edible Invertebrates.

shoot out its stomach when excited. The stomach is not edible. The skin and the five strips of muscle can be eaten after boiling. Conches can be boiled out of their shells and have very firm flesh. Use care when picking conches up. The bottom of their "foot" has a boney covering which can severely cut the survivor who procures it.

(5) The safest fish to eat are those from the open sea or deep

water beyond the reef. Silvery fishes, river eels, butterfly fishes, and flounders from bays and rivers are good to eat.

(6) Land crabs are common on tropical islands and are often found in coconut groves. An open coconut can be used for bait.

(7) A number of methods can be used for procuring fish.

(a) Hook-and-Line Fishing. This type of fishing on a rocky coast requires a lot of care to keep the line from becoming entangled or cut on sharp edges. Most shallow-water fish are nibblers. Unless the bait is well placed and hooked and the barb of the hook offset by bending, the bait may be lost without catching a fish. Use hermit crabs, snails, or the tough muscle of a shellfish as bait. Take the cracked shells and any other animal remains and drop them into the area to be fished. This brings the fish to the area and provides a better procurement opportunity. Examine stomach contents of the first fish caught to determine what the fish are feeding on.

(b) Jigging. A baited or spooned hook dipped repeatedly beneath the surface of the water is sometimes effective. This method may be used at night.

(c) Spearing. This method is difficult except when the stream is small and the fish are large and numerous during the spawning season, or when the fish congregate in pools. Make a spear by sharpening a long piece of wood, lashing two long thorns on a stick, or fashioning a bone spear point, and take a position on a rock over a fish run. Wait patiently and quietly for a fish to swim by.

(d) Chop Fishing. Chop fishing is effective at night during low tide. This method requires a torch and a machete. The fish are attracted by the light of the torch, and then they may be stunned by slashing at them with the back of the machete blade. Care should be taken when swinging the machete (Figure 14-32).

c. The jungle environment has a uniquely favorable condition for plant and animal life. The variety and richness of plant growth in these areas are paralleled nowhere else on the earth. Because the rainfall is distributed throughout the

Figure 14-32. Chop Fishing.

year and there is a lack of cold seasons, plants in the humid regions can grow, produce leaves, and flower the year round. Some plants grow very rapidly. For example, the stem of the giant bamboo may grow more than 22 inches in a single day.

(1) A survivor in search of plant food should apply some basic principles to the search. A survivor is lucky to find a plant that can readily be identified as edible. If a plant resembles a known plant, it is very likely to be of the same family and can be used. If a plant cannot be identified, the edibility test should be applied. A survivor will find many edible plants in the tropical forest, but chances of finding them in abundance are better in an area that has been cultivated in the past (secondary growth).

(2) Some plants a survivor might find:

(a) Citrus fruit trees may be found in uncultivated areas, but are primarily limited to areas of secondary growth. The many varieties of citrus fruit trees and shrubs have leaves 2 to 4 inches long alternately arranged. The leaves are leathery, shiny, and evergreen. The leaf stem is often winged. Small (usually green) spines are often present by the side of the bud. The flowers are small and white to purple in color. The fruit has a leathery rind with numerous glands and is round and fleshy with several cells (fruit sections or slices) and many seeds. The great number of wild and cultivated fruits (oranges, limes, lemons, etc.) native to the tropics are eaten raw or used in beverages.

(b) Taro can be found in both secondary growth and in virgin areas. It is usually found in the damp, swampy areas in the wild, but certain varieties can be found in the forest. It can be identified by its large heart-shaped or arrowhead-shaped leaves growing at the top of a vertical stem. The stem and leaves are usually green and rise a foot or more from a tuber at the base of the stem. Taro leaf tips point down; poisonous elephant ear points up. All varieties of taro must be cooked to break down the irritating crystals in the plant.

(c) Wild pineapple can be found in the wild, and common pineapples may be found in secondary growth areas. The wild pineapple is a coarse plant with long clustered, sword-shaped leaves with sawtoothed edges. The leaves are spirally arranged in a rosette. Flowers are violet or reddish. The wild pineapple fruit will not be as fully developed in the wild state as when cultivated. The seeds from the flower of the plant are edible as well as the fruit. The ripe fruit may be eaten raw, but the green fruit must be cooked to avoid irritation. (The leaf fibers make excellent lashing material and ropes can be manufactured from it.)

(d) Yams may be found cultivated or wild. There are many varieties of yam, but the most common has a vine with square-shaped cross section and two rows of heart-shaped leaves growing on opposite sides of the vine. The vine can be followed to the ground to locate the tuber. The tubers should be cooked to destroy the poisonous properties of the plant (Figure 14-33).

(e) Ginger grows in the tropical forest and is a good source of flavoring for food. It is found in shaded areas of the primary forest. The ginger plant grows 5 to 6 feet high. It has seasonal white snapdragon-type flowers, some variations have red flowers. The leaves when crushed produce a very sweet odor and are used for seasoning or tea. The tea is used by primitive people to treat colds and fever.

(f) The coconut palm is found wild on the seacoast and in farmed areas inland. It is a tree 50 to 100 feet high, either straight or curved, marked with ringlike leaf scars. The base of the tree is swollen and surrounded by a mass of rootlets. The leaves are leathery and reach a length of 15 to 20 feet. (The leaves make excellent sheathing for shelter.) The fruit grows in clusters at the top of the tree. Each nut is covered with a fibered hard shell. The "heart" of the coconut palm is edible and is found at the top. (The new leaves grow out of the heart.) Cut the tree down and remove the leaves to gain access to the heart. The flower of the coconut tree is also edible and is best used as a cooked vegetable. The germinat-

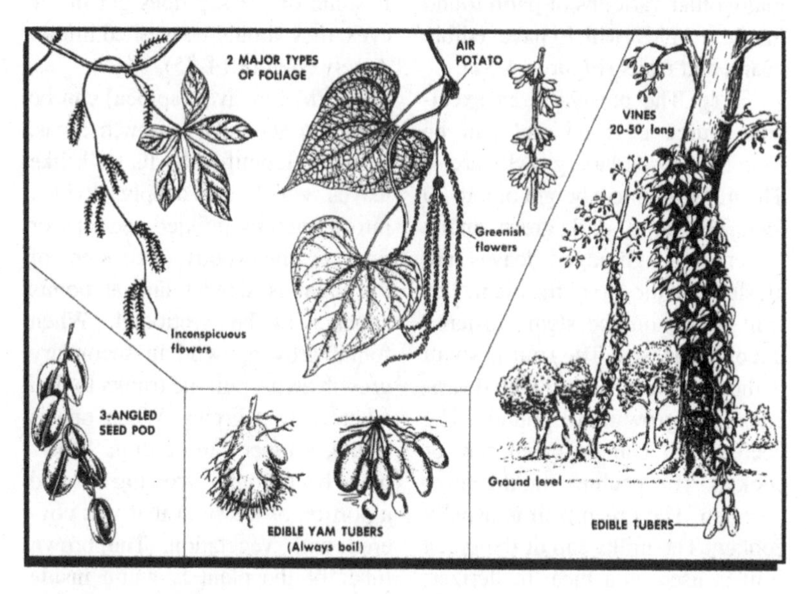

Figure 14-33. Yams.

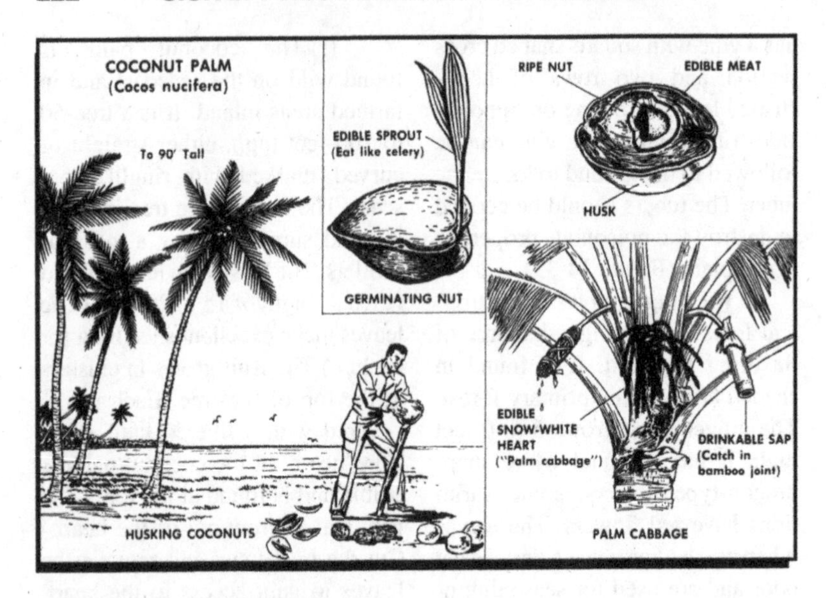

Figure 14-34. Coconut Palm.

ing nut is filled with a meat that can be eaten raw or cooked. There are many other varieties of palm found in the tropics which have edible hearts and fruits (Figure 14-34).

(g) The papaya is an excellent source of food and can be found in secondary growth areas. The tree grows to a height of 6 to 20 feet. The large, dark green, many fingered, rough-edged leaves are clustered at the top of the plant. The fruit grows on the stem clustered under the leaves. The fruit is small in the wild state, but cultivated varieties may grow to 15 pounds. The peeled fruit can be eaten raw or cooked. The peeling should never be eaten. The green fruit is usually cooked. The milky sap of the green fruit is used as a meat tenderizer; care should be taken not to get it in the eyes. Always wash the hands after handling fresh green papayas. If some of the sap does get in the eyes, they should be washed immediately (Figure 14-35).

(h) Cassava (tapioca) can be found in secondary growth areas. It can be identified by its stalk-like leaves which are deeply divided into numerous pointed sections or fingers. The woody (red) stem of the plant is slender and at points appears to be sectioned. When found growing wild in secondary growth areas, pull the trunks to find where a root grows. When one is found, a tuber can be dug. Tubers have been found growing around a portion of the stem that was covered with vegetation. The brown tuber of the plant is white inside and must be boiled or roasted. The

Figure 14-35. Papaya.

tuber must also be peeled before boiling. (The green-stemmed species of cassava is poisonous and must be cooked in several changes of water before eating it.)

(i) Ferns can be found in the virgin tropical forest or in secondary growth areas. The new leaves (fiddle heads) at the top are the edible parts. They are covered with fuzzy hair which is easily removed by rubbing or washing. Some can be eaten raw, but as a rule, should be cooked as a vegetable (Figure 14-36).

(j) Sweet sops can be found in the tropical forest. It is a small tree with simple, oblong leaves. The fruit is shaped like a blunt pine cone with thick grey-green or yellow, brittle spines. The fruit is easily split or broken when ripe, exposing numerous dark brown seeds imbedded in the cream colored, very sweet pulp.

(k) The star apple is common in the tropical forests. The tree grows up to a height of 60 feet and can be identified by the leaves which have shiny, silky, brown hairs on the bottom. The fruit looks like a small apple or plum with a smooth greenish or purple skin. The meat is greenish in color and milky in texture. When cut through the center, the brown, elongated seeds make a figure like a 6- or 10-pointed star. The fruit is sweet and eaten only when fresh. When cut, the rind will, like other

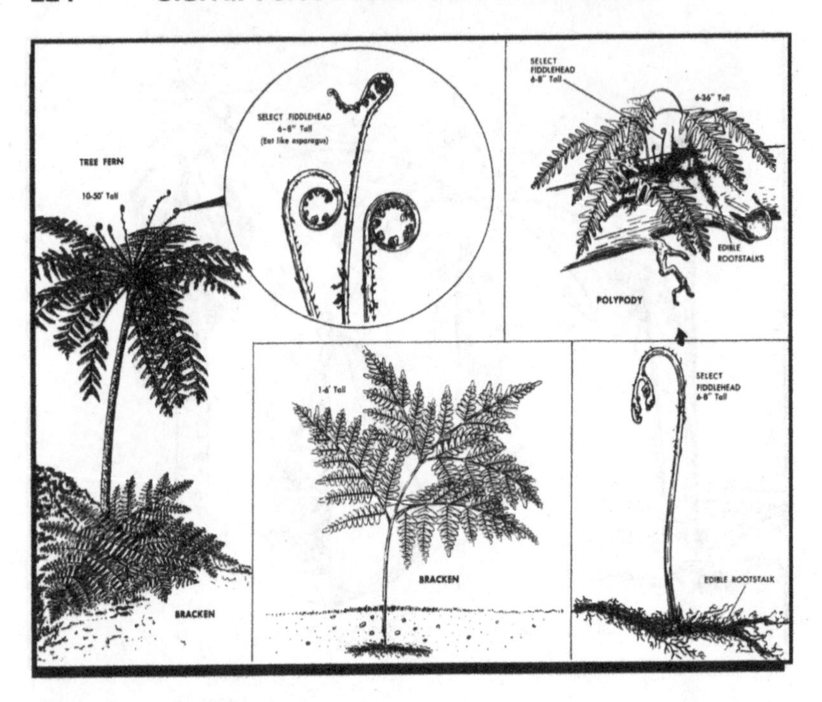

Figure 14-36. Edible Ferns.

parts of the tree, emit a white sticky juice or latex which is not poisonous (an exception to the milky sap rule).

(3) Of the 300,000 different kinds of wild plants in the world, a large number of them are found in the tropics and many of them are potentially edible. Very few are deadly when eaten in small quantities. Those which are poisonous may be detected by using the edibility rules. Only a small number of jungle plants have been discussed. It would be of great benefit to anyone flying over or passing through a tropical environment to study the plant foods available in this type of environment.

14-7. Food in Dry Climates.

Although not as readily available as in the tropical climate, food is available and obtainable.

a. Plant life in the desert is varied due to the different geographical areas. It must be remembered, therefore, that available plants will depend on the actual desert, the time of year, and if there has been any recent rainfall. The aircrew member should be familiar with plants in the area to be flown over.

(1) Date palms are located in most deserts and are cultivated by the native people around oases and irrigation ditches. They bear

a nutritious, oblong, black fruit (when ripe).

(2) Fig trees are normally located in tropical and subtropical zones; however, a few species can be found in the deserts of Syria and Europe. Many kinds are cultivated. The fruit can be eaten when ripe. Most figs resemble a top or a small pear somewhat squashed in shape. Ripe figs vary greatly as to palatability. Many are hard, woody, covered with irritating hairs, and worthless as survival food. The edible varieties are soft, delectable, and almost hairless. They are green, red, or black when ripe.

(3) Millet, a grain bearing plant, is grown by natives around oases and other water sources in the Middle East deserts.

(4) The fruit of all cacti are edible. Some fruits are red, some yellow, but all are soft when ripe. Any of the flat leaf variety, such as the prickly pear, can be boiled and eaten as greens (like spinach) if the spines are first removed. During severe droughts, cattlemen burn off the spines and use the thick leaves for fodder. Although the cactus originates in the American deserts, the prickly pear has been introduced to the desert edges in Asia, Africa, the Near East, and Australia, where it grows profusely. Natives eat the fruit as fast as it ripens.

(5) There are two types of onions in the Gobi desert. A hot, strong, scallion-type grows in the late summer. It will improve the taste of food, but should not be used as a primary food. The highland onions grow 2 to 2.5 inches in diameter. These can be eaten like apples and the greens can also be eaten raw or cooked.

(6) All desert flowers can be eaten except those with milky or colored sap.

(7) All grasses are edible. Usually the best part is the whitish tender end that shows when the grass stalk is pulled from the ground. All grass seeds are edible.

b. Animal food sources may be used to supplement diets and provide needed protein and fats. When looking at a desert area, it is sometimes difficult to visualize an abundance of animal life existing in it. There is, however, a great quantity of animal life present. Most are edible, but some may be hazardous to a survivor during the procurement stage. Some of the abundant animal life includes:

(1) At the peak of seasonal plant growth, the desert crawls and buzzes with an enormous number and variety of beetles, ants, wasps, moths, and bugs. They appear with the first good rains and generally feed during nighttime. The Ute Indians of North America have harvested crickets, and peoples of the Middle East have roasted locusts. The human

diet in Mexico and the American Indians of the Southwest frequently includes grasshoppers and caterpillars.

(2) On the Playas of the Sonora and Chichuahua deserts, several species of freshwater shrimp appear every summer in warm temporary ponds. In the Mohave Desert, where summer rains are rare, they may appear only a few times in a century.

(3) Snakes, lizards, tortoises, etc., have adapted well to the desert environment. Care must be observed when procuring them as some are hazardous, such as the Gila monster and rattlesnake. The desert tortoise, about a foot long when full grown, lives in some of the harshest regions of the Mohave and Sonora deserts. It is club footed, herbivorous, and can crawl about 20 feet per minute. The tortoise converts some of its food into water which is stored for the hot months in two sacs under the upper shell. A pint of water lasts the dry season. In spring and fall, the tortoise browses in broad daylight, becoming livelier as the day warms up. In the heat of the summer, it comes out of its shallow burrow in the early morning, the late evening, or not at all.

(4) In general, desert birds stay in areas of heavier vegetation and many need water daily; therefore, most will be found within short flights of some type of water source. Many birds will migrate during the drought season. If an abundance of birds is seen, insects, vegetation, and a water source will normally be nearby.

(5) Rabbits, prairie dogs, and rats have learned to live in deserts. They remain in the shade or burrow into the ground protecting themselves from the direct sun and heated air as well as from the hot desert surface.

(6) Larger mammals are also found in the desert. This group consists of gazelles, antelope, deer, foxes, small cats, badgers, dingos, hyenas, etc., and are amazingly abundant. Most are nocturnal and generally avoid humans. They roam at night eating smaller game and insects; a few eat plants; and a few can be hazardous to a survivor. Any of these mammals should be approached with caution.

(7) Only a few of the available animals and plants have been discussed. If the possibility of having to survive in a desert area exists, the aircrew member should try to become familiar with the food source available in that area.

14-8. Food in Snow and Ice Climates. In the snow and ice climates, food is more difficult to find than water. Animal life is normally more abundant during the warm months, but it can still be found in the cold months. Fish are available in most waters during the warmer months but they congregate in deep waters, large rivers, and lakes during the cold months. Some edible plant life can be found throughout the year in most areas of the arctic.

a. All animals in the arctic regions are edible, but the livers of seals and polar bears must not be eaten because of the high concentration of vitamin A. Death could result from ingesting large quantities of the liver. On the open sea ice, game animals such as seal, walrus, polar bear, and fox are available. Many types of birds can be found during the warmer months. Fish can be caught throughout the year.

(1) Seal will probably be the main source of food when stranded on the open sea ice. They can be found in open leads, areas of thin ice, or where snow has drifted over a pressure ridge forming a cave which could have open water or very thin ice. These areas may also house polar bears which feed primarily on seals. Polar bears should be avoided.

(a) Newborn seals have trouble staying afloat or swimming and will be found on the ice in the early summer. The seal cubs can be easily killed with a club, spear, knife, or firearm and make an excellent source of food. The meat, blubber (fat), and coagulated milk in their stomachs are edible. When killing a cub, it is best to keep a lookout for the mother. She tends to protect offspring in any way possible.

(b) Seals must surface periodically to breathe. When the icepack is thin, the seals poke their noses through the ice and take a breath of air in a lead or in open water. In thick ice, the seal will chew and(or) claw a breathing hole through the ice. Normally most seals will have more than one breathing hole. In hunting seals, it is best to take a position beside a breathing hole and wait until a seal comes up to breathe, then spear or strike it on the head with a club. Seals are very sensitive to blows on or about the nose. They will often lose consciousness but not die. A hook can be suspended through the breathing hole so it hangs down at least 6 inches below the ice. When a seal comes to breathe, it can become hooked when it tries to depart the breathing hole. Seals can be recovered by gaffing or grabbing by hand, but in some cases, the breathing hole might have to be enlarged to pull the body through. If the seal is killed in open water, a "manak" or "grapple hook" can be used to retrieve it. All seals killed in open water or those that fall into open water should be recovered immediately. During the cold months, they will float for quite awhile, but during the warm months or when a female is nursing young, they sink rapidly. This is due to the loss of body fat.

(2) Birds are plentiful during the summer months and can be procured by spearing, clubbing,

catching with a baited fishhook, or use of a weapon.

(3) On tundra areas, there are large game, small game, and birds available as a food source.

(a) The large game consists of caribou, musk oxen, sheep, wolf, and bears. Even though the large game animals can be a food source, they will be difficult to procure if a firearm is not available. Therefore, they should be considered a hazard to a survivor without a firearm. In the spring, bears tend to congregate along rivers and streams due to the amount of food available—normally salmon. During the fall, bears will be found feeding at berry patches. During certain seasons of the year, these areas should be avoided.

(b) Small game animals of the tundra include hares, lemmings, mice, ground squirrels, marmots, and foxes. They may be trapped or killed the entire year. When snaring, it is best to use a simple loop made of strong line or wire. The wire must be a two-strand twisted wire since metal becomes brittle in the cold and breaks very easily. Other snares and triggers will be less effective in the cold climate. A gill net can be used as a snare by spreading it across a trail so that the animal will entangle itself.

(c) Surface water is generally plentiful due to the number of lakes, ponds, bogs, and marshes. Water fowl and birds are very abundant during the warm months and include ducks, terns, geese, gulls, owls, and ptarmigan. The eggs and young birds are an excellent food source and can be easily procured.

(4) As in the tundra areas, the forested areas in the arctic and arctic-like areas abound in wildlife.

(a) The large game species include moose, deer, caribou, and bear.

(b) Small game of the forests includes hares, squirrels, porcupine, muskrat, and beaver. They can be snared or trapped easily in winter or summer. Small animal trails can be found in the winter with great ease. Most animals do not like to travel in deep snow so they tend to travel the same trail most of the time and this trail will look like a small superhighway — the snow packed down well below the normal snow level. Most trails will also be located in heavy cover or undergrowth or parallel to roads and open areas. The same trails will normally be used during the summer.

(5) During the summer months, the open water provides an excellent opportunity to procure all types of fish, both freshwater and saltwater, and freshwater mussels. The mussels can be handpicked off the bottom, while the fish can be netted, speared, clubbed, or caught with a hook and line. After freezeup, fishing is still possible through the ice. Shallow lakes, rivers, or ponds can freeze completely killing off all fish life. Fish tend to congregate in

the deepest water possible. A hole should be cut through the ice at the estimated deepest point. Other good locations are at outlets or where tributaries flow into lakes or ponds. The ice is normally thinner over rapid moving water and at the edges of deep streams or rivers with snowdrifts extending out from the banks. Open water is often marked by a mist or fog formed over the area by vaporizing water. All methods of procuring fish in the summer will work in the winter.

(6) The ocean shores are rich hunting grounds for edible sea life such as clams, mussels, scallops, snails, limpets, sea urchins, chitons, and sea cucumbers. They can be procured most of the year wherever there is open water. Tidal pools usually contain a great number of both fish and mollusks. The fish can be netted, speared, or hand caught. All sea life can be eaten raw, but cooking usually makes it more palatable (Figure 14-37).

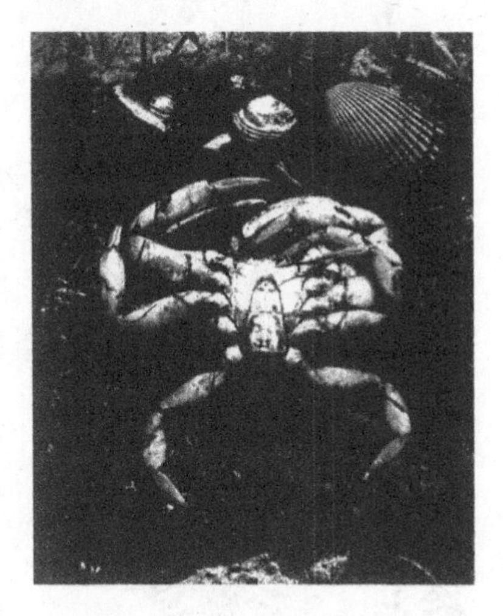

Figure 14-37. Shell Fish.

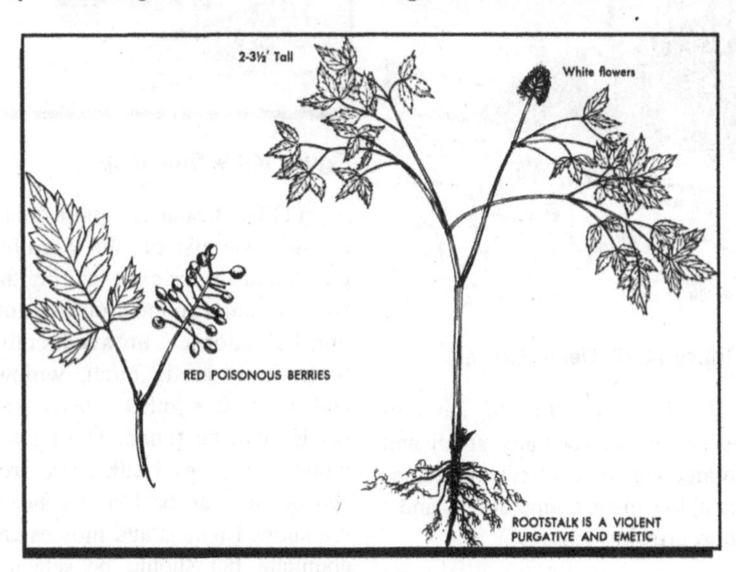

Figure 14-38. Baneberry.

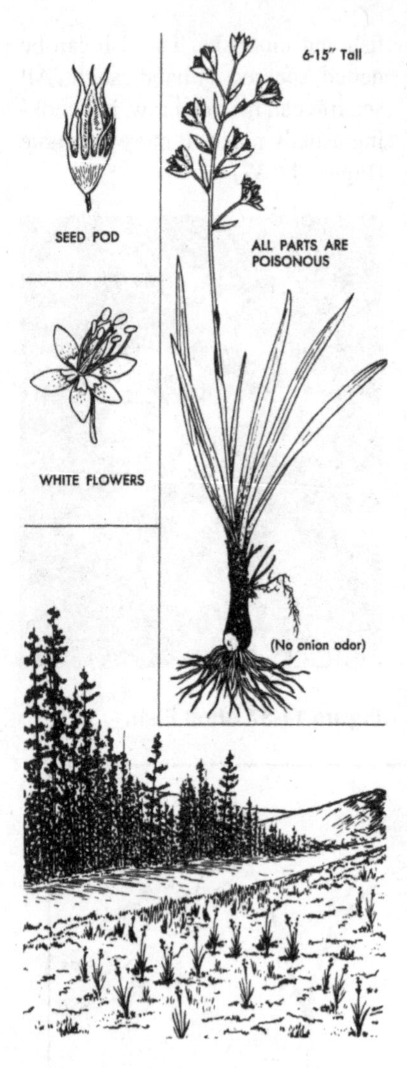

Figure 14-40. Death Camas.

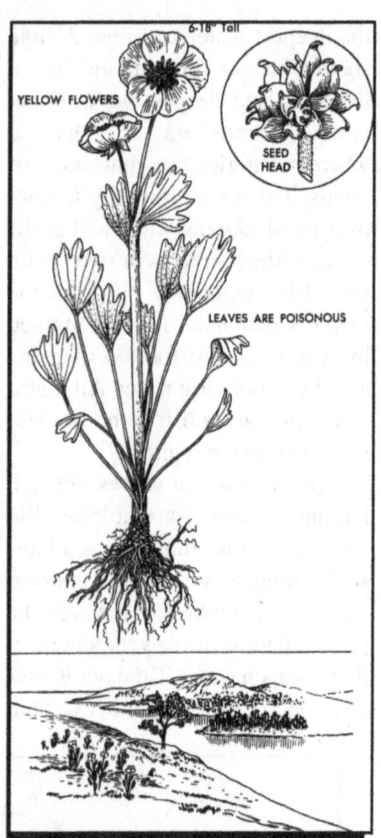

Figure 14-39. Buttercup.

b. The plant life of the arctic regions is generally small and stunted due to the effects of permafrost, low mean temperatures, and a short growing season.

(1) On the barren tundra areas, a wide variety of small edible plants and shrubs exist. During the short summer months on the tundra, Labrador tea, fireweed, coltsfoot, dwarf arctic birch, willow, and numerous other plants and berries can be found. During the winter, roots, rootstalks, and frozen berries can be found beneath the snow. Lichens and mosses are abundant but should be selected

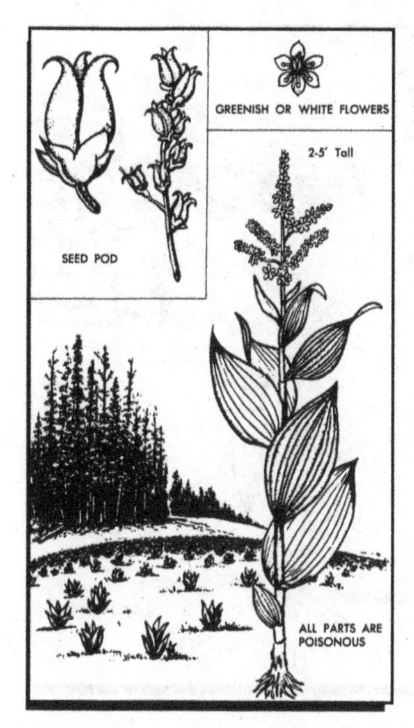

Figure 14-41. False Hellebore

carefully as some species are poisonous.

(2) In bog or swamp areas, many types of water sedge, cattail, dwarf birch, and berries are available. During spring and summer, many young shoots from these plants are easily collected.

(3) The wooded areas of the arctic contain a variety of trees (birch, spruce, poplar, aspen, and others). Many berry-producing plants can be found, such as blueberries, cranberries, raspberries, cloud berries, and crow berries. Wild rose hips, Labrador tea, alder, and other shrubs are very abundant. Many wild edible plants are highly nutritious. Greens are particularly rich in carotene (vitamin A). Leafy greens, many berries, and rose hips are all rich in ascorbic acid (vitamin C). Many roots and rootstalks contain starch and can be used as a potato substitute in stews and soups.

(4) Although there are several types of edible mushrooms, fungi, and puff-balls in the arctic, a person should avoid ingesting them because it is difficult to identify the poisonous and nonpoisonous species. During the growing season, the physical characteristics can change considerably making positive identification even more difficult.

(5) There are many poisonous plants and a few poisonous berries in the arctic. Very few cause death; many will cause extreme nausea, dizziness, abdominal pain, and diarrhea. Contact poisonous plants, such as poison ivy, are not found in the arctic. The more common poisonous plants are shown in figures 14-38 through 14-45.

(6) When selecting edible plants, select young shoots when possible as these will be the most tender. Plants should be eaten raw to obtain the most nutritive value. Some of the more common edible plants are:

(a) Dandelions generally grow with grasses but may be scattered over rather barren areas. Both leaves and roots are edible

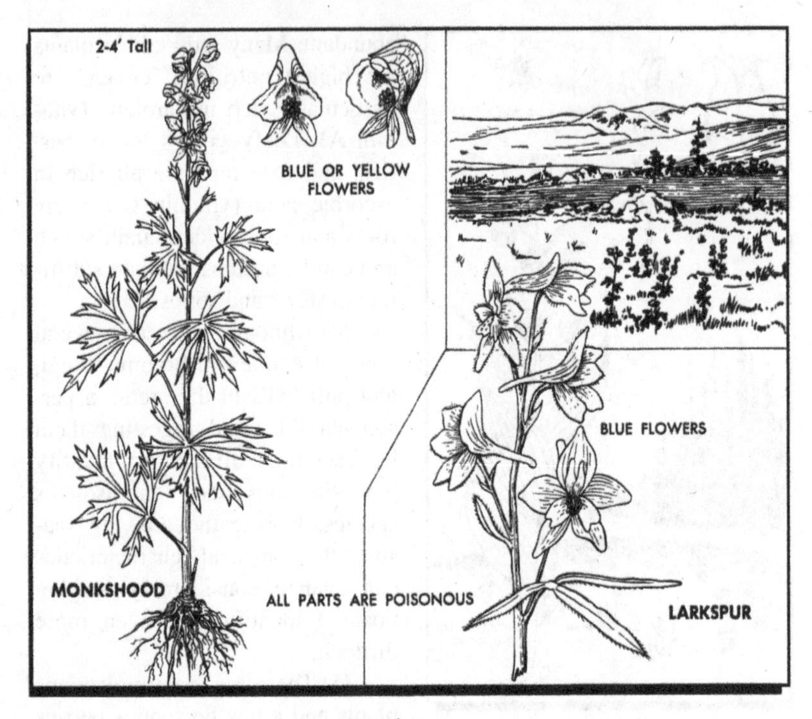

Figure 14-42. Monkshood and Larkspur.

raw or cooked. The young leaves make good greens; the roots (when roasted) are used as a substitute for coffee.

(b) Black and white spruce are generally the northern most evergreens. These trees have short, stiff needles that grow singularly rather than in clusters like pine needles. The cones are small and have thin scales. Although the buds, needles, and stems have a strong resinous flavor, they provide essential vitamin C by chewing them raw. In spring and early summer, the inner bark can be used for food.

(c) The dwarf arctic birch is a shrub with thin tooth-edged leaves and bark which peels off in sheets. The fresh green leaves and buds are rich in vitamin C. The inner bark may also be eaten.

(d) There are many different species of willow in the arctic. Young tender shoots may be eaten as greens and the bark of the roots is also edible. They have a decidedly sour taste but contain a large amount of vitamin C (Figure 14-46).

(7) Lichens are abundant and widespread in the far North and can be used as a source of emer-

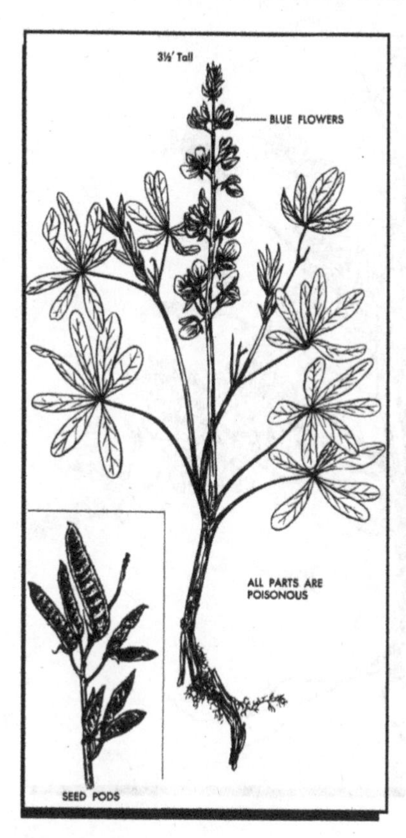

Figure 14-43. Lupine.

gency food. Many species are edible and rich in starch-like substances, including Iceland moss, peat moss and reindeer lichen. Beard lichen growing on trees has been used as food by Indians. However, some of it contains a bitter acid which causes irritation of the digestive tract. If lichens are boiled, dried, and powdered, this acid is removed and the powder can then be used as flour or made into a thick soup.

14-9. Food on Open Seas. Almost all sea life is not only edible, but is also an excellent source of nutrients essential to humans. The protein is complete because it contains all the essential amino acids, and the fats are similar to those of vegetables. Seafoods are high in minerals and vitamins. The majority of life in the sea (fish, birds, plants, and aquatic animals) is edible.

a. Most seaweeds are edible and are a good source of food, especially for vitamins and minerals. Some seaweeds contain as much as 25 percent protein, while others are composed of over 50 percent carbohydrates. At least 75 different species are used for food by seacoast residents around the world. For many people, especially the Japanese, seaweeds are an essential part of the diet, and the most popular varieties have been successfully farmed for hundreds of years. The high cellulose content may require gradual adaptation because of their laxative quality if they comprise a large part of the diet. As with vegetables, some species are more flavorful than others. Generally, leafy green, brown, or red seaweeds can be washed and eaten raw or dried. The following list of edible seaweeds gives a description of the plant, tells where it may be found, and in many cases, suggests a method of preparation:

(1) Common green seaweeds (Figure 14-47), often called sea

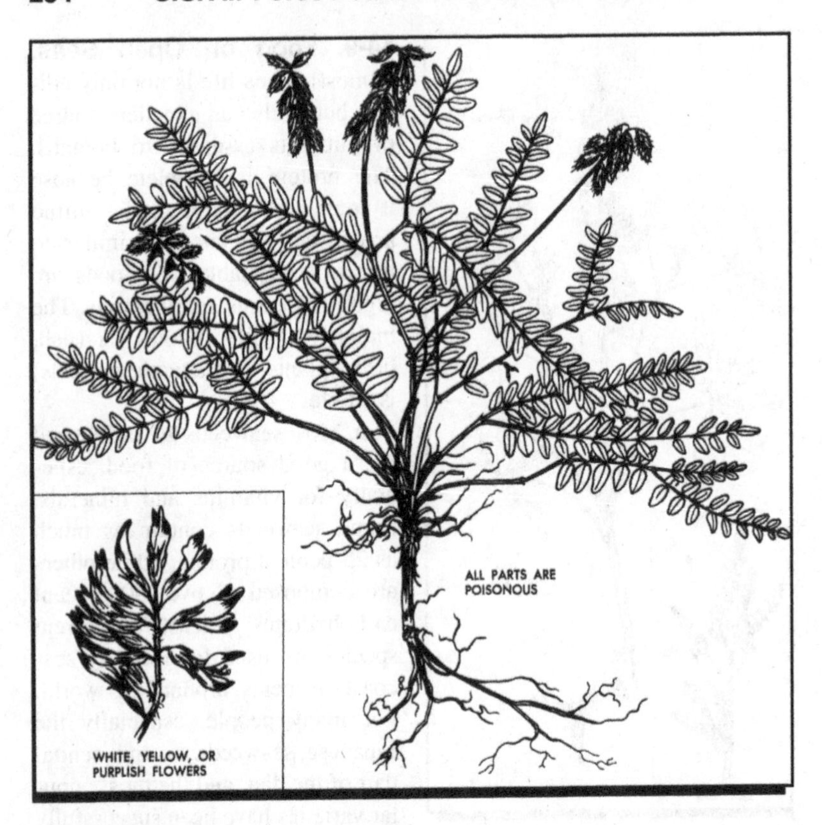

Figure 14-44. Vetch and Locoweed.

lettuce (Ulva lactuca), are in abundance on both sides of the Pacific and North Atlantic oceans. After washing it in clean water, it can be used as a garden lettuce.

(2) The most common edible brown seaweeds are the sugar wrack, kelp, and Irish moss (Figure 14-48).

(a) The young stalks of the sugar wrack are sweet to taste. This seaweed is found on both sides of the Atlantic and on the coasts of China and Japan.

(b) Edible kelp has a short cylindrical stem and thin, wavy olive-green or brown fronds one to several feet in length. It is found in the Atlantic and Pacific oceans, usually below the high-tide line on submerged ledges and rocky bottoms. Kelp should be boiled before eating. It can be mixed with vegetables or soup.

(c) Irish moss, a variety of brown seaweed, is quite edible, and is often sold in market places. It is found on both sides of the Atlantic Ocean and can be identified by its tough, elastic, and leathery texture; however, when dried, it becomes crisp and shrunken. It should be

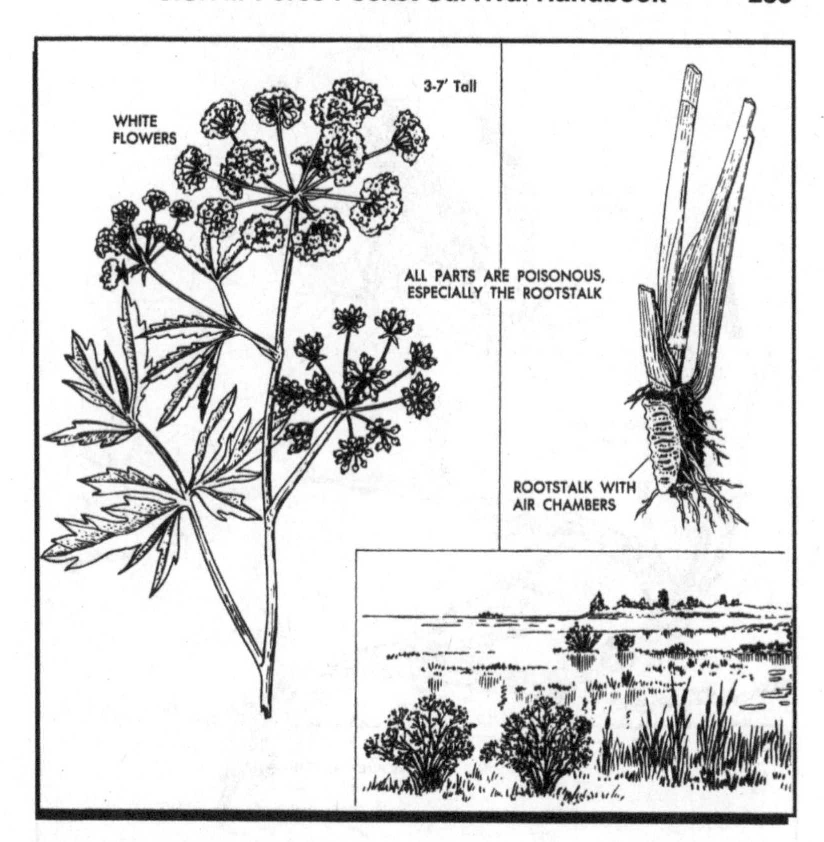

Figure 14-45. Water Hemlock.

boiled before eating. It can be found at or just below the high-tide line. It is sometimes found cast upon the shore.

(3) Red seaweeds can usually be identified by their characteristic reddish tint, especially the edible varieties. The most common and edible red seaweeds include the dulse, laver, and other warm-water varieties (Figure 14-49).

(a) Dulse has a very short stem which quickly broadens into a thin, broad, fan-shaped expanse which is dark red and divided by several clefts into short, round-tipped lobes. The entire plant is from a few inches to a foot in length. It is found attached to rocks or coarser seaweeds, usually at the low-tide level, on both sides of the Atlantic Ocean and in the Mediterranean. Dulse is leathery in consistency and is sweet to the taste. If dried and rolled, it can be used as a substitute for tobacco.

(b) Laver is usually red, dark purple, or purplish-brown, and has a satiny sheen or filmy luster.

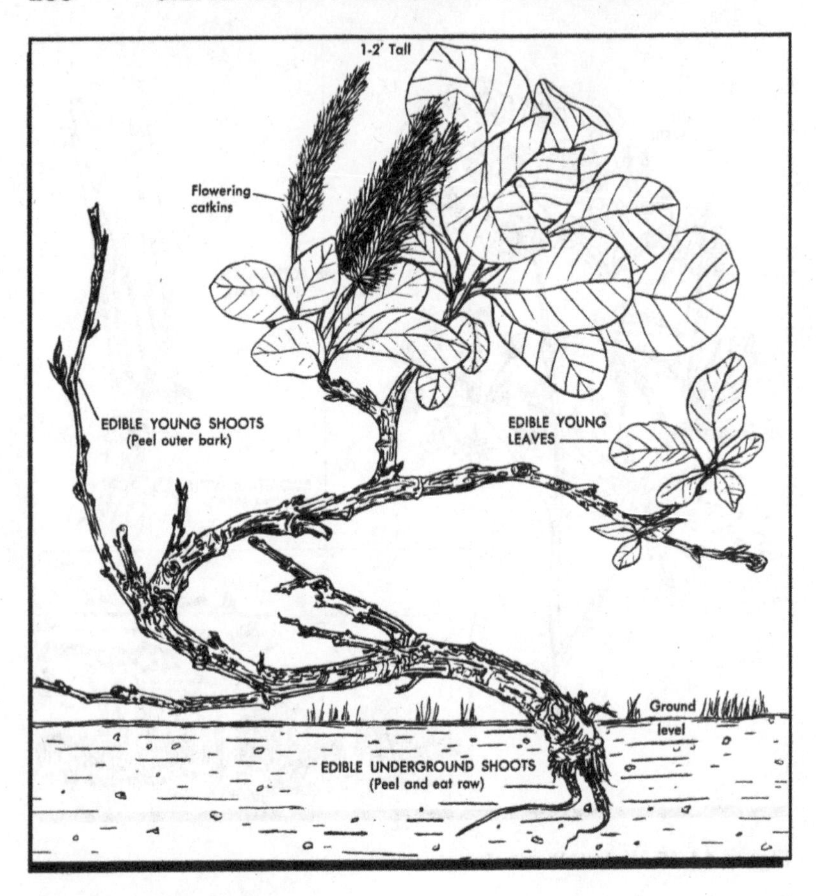

Figure 14-46. Arctic Willow.

Common to both the Atlantic and Pacific oceans, it has been used as food for centuries. This seaweed is used as a relish, or is cleaned and then boiled gently until tender. It can also be pulverized and added to crushed grains and fried in the form of flatcakes. During World War II, laver was chewed for its thirst-quenching value by New Zealand troops. Laver is usually found on the beach at the low-tide level.

(c) A great variety of red, warm-water seaweed is found in the South Pacific area. This seaweed accounts for a large portion of the native diet. When found on the open sea, bits of floating seaweed may not only be edible but often contains tiny animals that can be used for food. The small fish and crabs can be dislodged by shaking the clump of seaweed over a container.

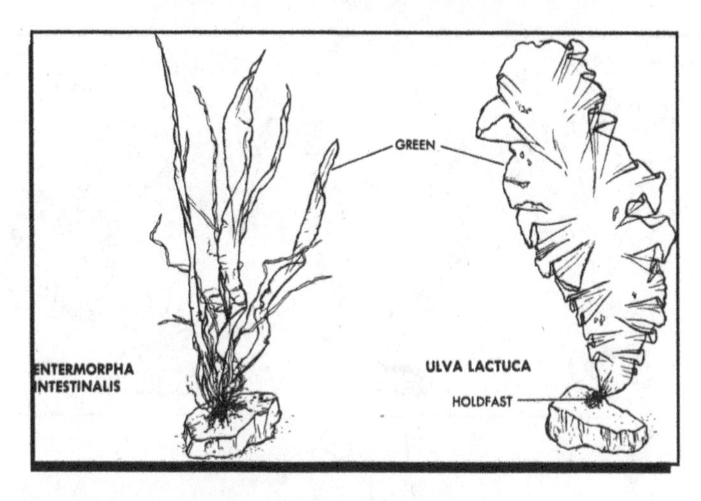

Figure 14-47. Edible Green Seaweeds.

b. Plankton includes both minute plants and animals that drift about or swim weakly in the ocean. These basic organisms in the marine food chain are generally more common near land since their occurrence depends upon the nutrients dissolved in the water. Plankton can be caught by dragging a net through the water. The taste of the plankton will depend upon the types of organisms predominant in the area. If the population is mostly fish larvae, the plankton will taste like fish. If the population is mostly crab or shellfish larvae, the plankton will taste like crab or

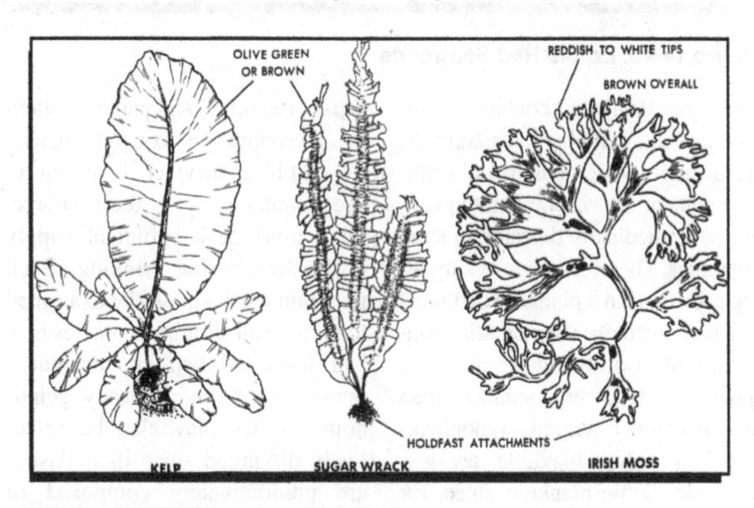

Figure 14-48. Edible Brown Seaweeds.

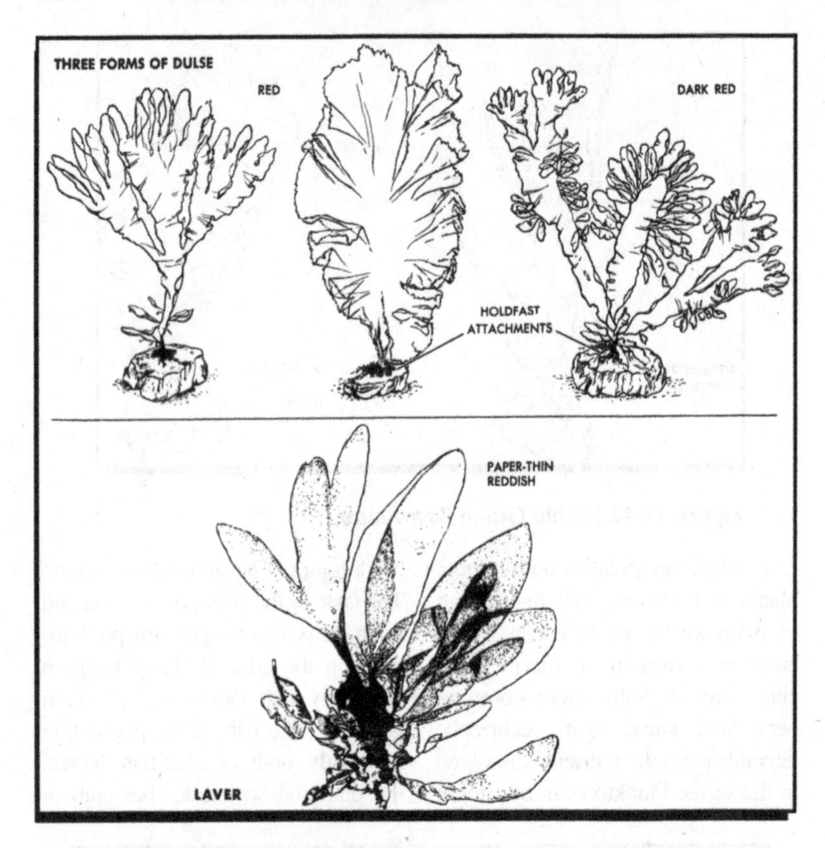

Figure 14-49. Edible Red Seaweeds.

shellfish. Plankton contains valuable protein, carbohydrates and fats. Because of its high chiton and cellulose content, however, plankton cannot be immediately digested in large quantities. Therefore, anyone subsisting primarily on a plankton diet must gradually increase the quantities consumed. Most of the planktonic algae (phytoplankton) are smaller than the planktonic animals (zooplankton) and, although edible, are less palatable. Some plankton algae, for example, those dinoflagellates that cause "red tides" and paralytic shellfish poisoning, are toxic to humans.

(1) If a survivor is going to use plankton as a food source, there must be a sufficient supply of freshwater for drinking. Each plankton catch should be examined to remove all stinging tentacles broken from jellyfish or Portuguese man-of-war. The primarily gelatinous species may also be selectively discarded since their tissues are predominately composed of saltwater. When the plankton is

found in subtropical waters during the summer months, and the presence of poisonous dinoflagellates is suspected (due to discoloration or high luminescence of the ocean), the edibility test should be applied before eating.

(2) The final precaution which a survivor may wish to take before ingesting plankton is to feel or touch the plankton to check for species that are especially spiney. The catch should be sorted (visually) or dried and crushed before eating if it contains large numbers of these spiney species.

c. If a fishing kit is available, the task of fishing will be made much easier. Small fish will usually gather under the shadow of the raft or in clumps of floating seaweed. These fish can be eaten or used as bait for larger fish. A net can also be used to procure most all sea life. Light attracts some types of fish. A flashlight or reflected moonlight can be used. It is not advisable to secure fishing lines to the body or the raft because a large fish may pull a person out of the raft or damage the raft. Fish, bait, or bright objects dangling in the water can attract large dangerous fish. All large fish should be killed outside the raft by a blow to the head or by cutting off the head.

d. Sea birds have proven to be a useful food source which may be more easily caught than fish. Survivors have reported capturing birds by using baited hooks, by grabbing, and by shooting. Freshly killed birds should be skinned, rather than plucked, to remove the oil glands. They can be eaten raw or cooked. The gullet contents can be a good food source. The flesh should be eaten or preserved immediately after cleaning. The viscera, along with any other unused parts, make good fish bait.

e. Marine mammals are rarely encountered by a person in the water, although they may be seen from a distance. Any large mammal is capable of inflicting injuries, but unless such mammals are pursued, they will generally avoid people. The killer whale (Orca) is rarely seen and, although large enough to feed on humans, has never been known to do so. Almost all sea mammals are a good source of food but difficult to obtain. The liver, especially that of any arctic or cold-water mammal, should not be eaten because of toxic concentrations of vitamin A.

f. All sea life must be cleaned, cut up, and eaten as soon as possible to avoid spoilage. Any meat left over can be preserved by sun-drying or smoking. The internal parts can be used as bait.

14-10. Preparing Animal Food. Survivors must know how to use the meat of game and fish to their advantage and how to do this with the least effort and physical exertion. Many people have died from starvation because they had failed to take full advantage of a game carcass. They abandoned the carcass on the mistaken theory that they could get more game when needed.

a. If the animal is large, the first impulse is usually to pack the meat to camp. In some cases, it might be easier to move the camp to the meat. A procedure often advocated for transporting the kill is to use the skin as a sled for dragging the meat. When the entire animal is dragged, this method may prove satisfactory only on frozen lakes or rivers or over very smooth snow-covered terrain. In rough or brush-covered country, however, it is generally more difficult to use this method, although it will work. Large mountain animals can sometimes be dragged down a snow-filled gully to the base of the mountain. If meat is the only consideration, and the survivors do not care about the condition of the skin, mountain game can sometimes be rolled for long distances. Before transporting a whole animal, it should be gutted and the incision closed. Once the bottom of the hill is reached, almost invariably the method is either to backpack the meat to camp, making several trips if no other survivors are present, or to pack the camp to the animal. Under survival conditions, home is on the back. When the weight of the meat proves excessive and moving the camp is not practical, some of the meat could be eaten at the scene. The heart, liver, and kidneys should be eaten as soon as possible to avoid spoilage.

(1) Under survival conditions, skinning and butchering must be done carefully so that all edible meat can be saved. When the decision is made to discard the skin, a rough job can be done. However, considerations should be given to possible uses of the skin. A square of fresh skin, long enough to reach from the head to the knees, will not weigh much less when it is dried, and is an excellent ground cloth for use under a sleeping bag on frozen ground or snow. The best time to skin and butcher an animal is immediately after the kill. However, if an animal is killed late in the day, it can be gutted immediately and the other work done the next morning. An effort to keep the carcass secure from predators should be made.

(2) When preparing meat under survival conditions, all edible fat should be saved. This is especially important in cold climates, as the diet may consist almost entirely of lean meat. Fat must be eaten in order to provide a complete diet. Rabbits lack fat, and the fact that a person will die after an extended diet consisting only of rabbit meat indicates the importance of fat in a primitive diet. The same is true of birds, such as the ptarmigan.

(3) Birds should be handled in the same manner as other animals. They should be cleaned after killing and protected from flies. Birds, with the exception of sea birds, should be plucked and cooked with the skin on. Carrion-eating birds, such as vultures, must be boiled for at least 20 minutes to kill parasites before further cooking and eating. Fish-eating birds have a strong, fish- oil flavor. This may be

lessened by baking them in mud or by skinning them before cooking.

b. There are two general ways to skin animals depending upon the size: the big game method, or the glove skinning method.

(1) Survivors should use the big game method when skinning and butchering large game.

(a) The first step in skinning is to turn the animal on its back and with a sharp knife, cut through the skin on a straight line from the tail bone to a point under its neck as illustrated in figure 14-50. In making this cut, pass around the anus and, with great care, press the skin open until the first two fingers can be inserted between the skin and the thin membrane enclosing the entrails. When the fingers can be forced forward, place the blade of the knife between the fingers, blade up, with knife held firmly. While forcing the fingers forward, palm upward, follow with the knife blade, cutting the skin but not cutting the membrane.

(b) If the animal is a male, cut the skin parallel to, but not touching the penis. If the tube leading from the bladder is accidentally cut, a messy job and unclean meat will result. If the gall or urine bladders are broken, washing will help clean the meat. Otherwise, it is best not to wash the meat but to allow it to form a protective glaze.

(c) On reaching the ribs, it is no longer possible to force the fingers forward, because the skin adheres more strongly to flesh and bone. Furthermore, care is no longer

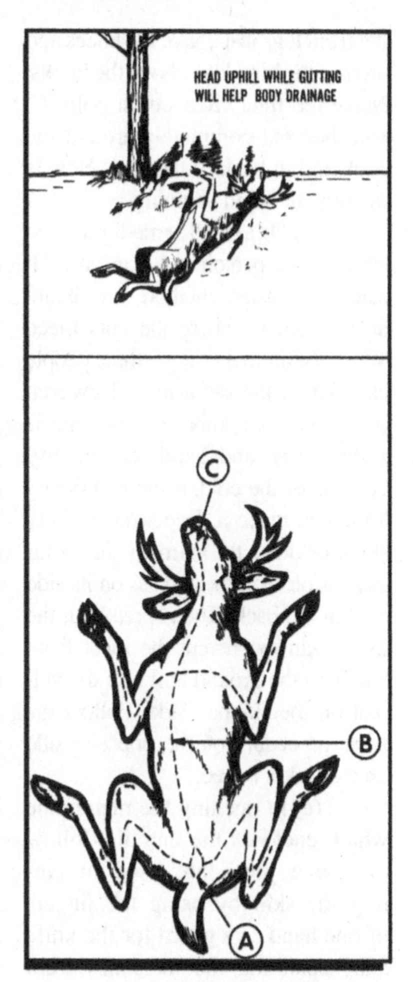

Figure 14-50. Big Game Skinning

necessary. The cut to point C can be quickly completed by alternately forcing the knife under the skin and lifting it. With the central cut completed, make side cuts consisting of incisions through the skin, running from the central cut (A-C) up the inside of each leg to the knee and hock joints. Then make cuts around the front legs just above the knees and around the hind legs above the hocks. Make the final cross cut at point C, and then cut completely around the neck and in back of the ears. Now is the time to begin skinning.

(d) On a small or medium-sized animal, one person can skin on each side. The easiest method is to begin at the corners where the cuts meet. When the animal is large, three people can skin at the same time. However, one should remember that when it is getting dark and hands are clumsy because of the cold, a sharp skinning knife can make a deep wound. After skinning down the animal's side as far as possible, roll the carcass on its side to skin the back. Then spread out the loose skin to prevent the meat from touching the ground and turn the animal on the skinned side. Follow the same procedure on the opposite side until the skin is free.

(e) In opening the membrane which encloses the entrails, follow the same procedure used in cutting the skin by using the fingers of one hand as a guard for the knife and separating the intestines from membrane. This thin membrane along the ribs and sides can be cut away in order to see better. Be careful to avoid cutting the intestines or bladder. The large intestine passes through an aperture in the pelvis. This tube must be separated from the bone surrounding it with a knife. Tie a knot in the bladder tube to prevent the escape of urine. With these steps completed, the entrails can be easily disengaged from the back and removed from the carcass. Another method of gutting or field dressing is shown in figure 14-51. After gutting is completed, it may be advisable to hang the animal. Figure 14-52 shows two methods. (NOTE: If it is hot, gut the animal before skinning it.)

(f) The intestines of a well-conditioned animal are covered with a lace-like layer of fat which can be lifted off and placed on nearby bushes to dry for later use. The gall bladder which is attached to the liver of some animals should be carefully removed. If it should happen to rupture, the bile will taint anything it touches. Be sure to clean the knife if necessary. The kidneys are imbedded in the back, forward of the pelvis, and are covered with fat. Running forward from the kidneys on each side of the backbone are two long strips of chop-meat or muscle called tenderloin or backstrap. Eat this after the liver, heart, and kidneys as it is usually very tender. Edible meat can also be removed from the head, brisket, ribs, backbone, and pelvis.

(g) Large animals should be quartered. To do this, cut down between the first and second rib and then sever the backbone with an axe or machete. Cut through the brisket of the front half and then chop lengthwise through the backbone to produce the front quarters. On the rear half, cut through the pelvic bone and lengthwise through the backbone. To make the load lighter and easier to transport, a knife could be used to

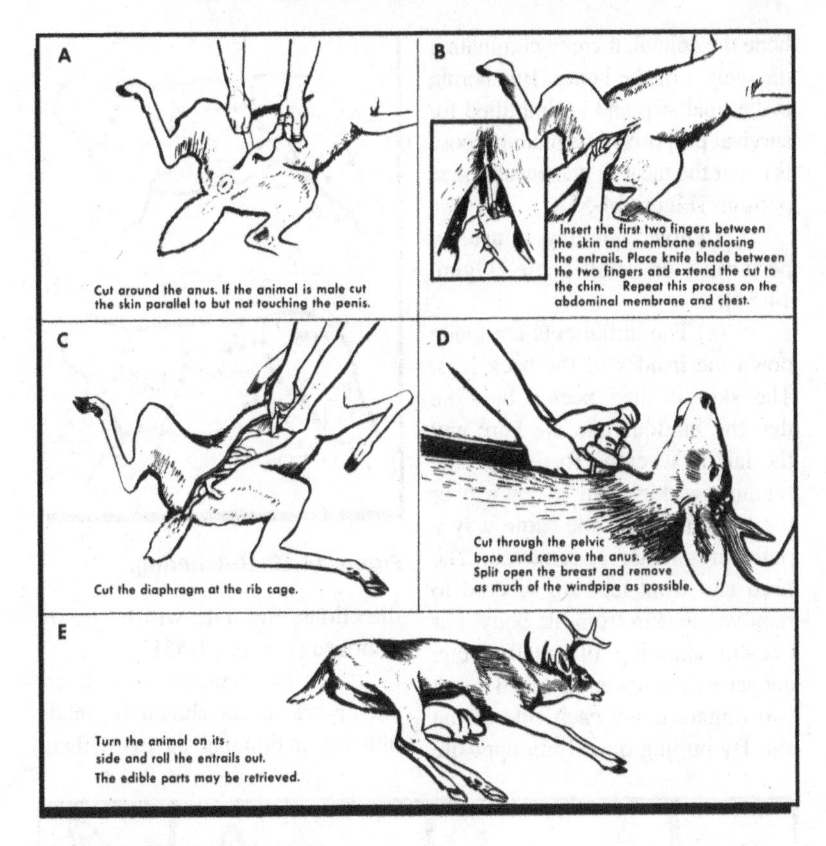

Figure 14-51. Field Dressing.

Figure 14-52. Hanging Game.

bone the animal, thereby eliminating the weight of the bones. Butchering is the final step and is simplified for survival purposes. The main purpose is to cut the meat in manageable size portions (Figure 14-53).

(2) Glove skinning is usually performed on small game (Figure 14-54).

(a) The initial cuts are made down the insides of the back legs. The skin is then peeled back so that the hindquarters are bare and the tail is severed. To remove the remaining skin, pull it down over the body in much the same way a pullover sweater is removed. The head and front feet are severed to remove the skin from the body. For one-cut skinning of small game, cut across the lower back and insert two fingers under each side of the slit. By pulling quickly in opposite directions, the hide will be easily removed (Figure 14-55).

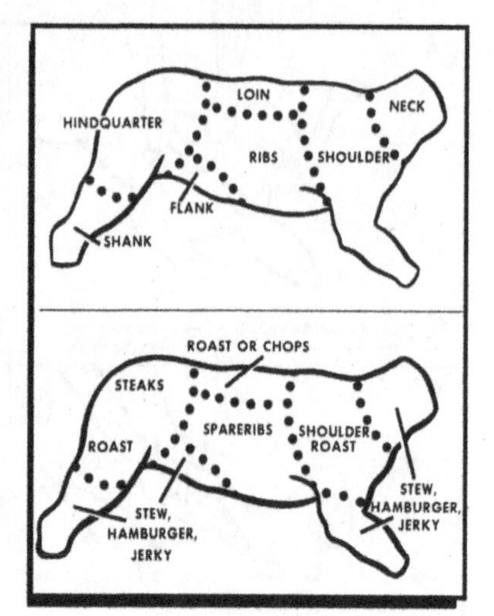

Figure 14-53. Butchering.

(b) To remove the internal organs, a cut should be made into the abdominal cavity without

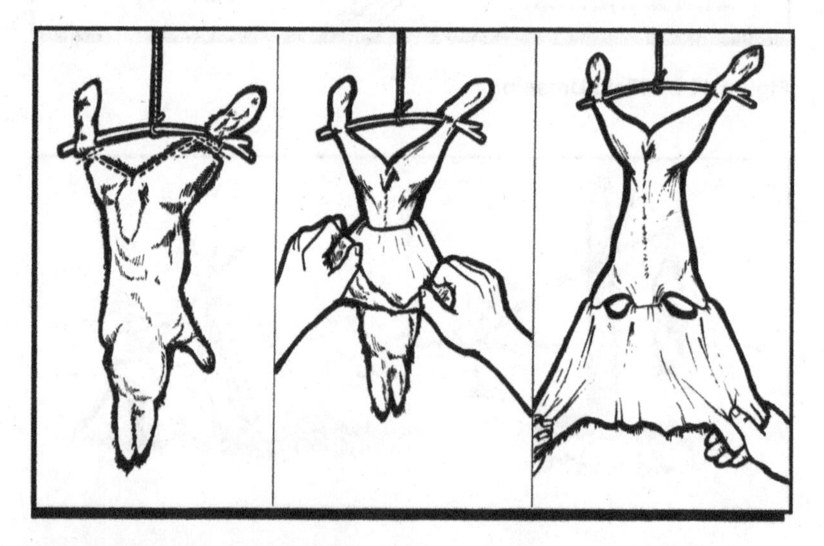

Figure 14-54. Glove Skinning.

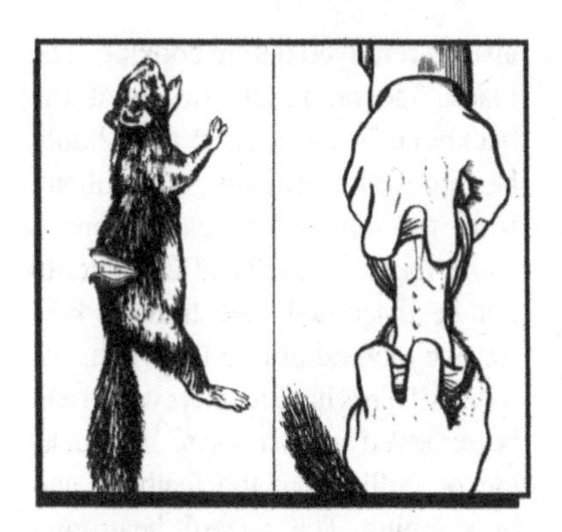

Figure 14-55. Small Animal Skinning.

puncturing the organs. This cut must run from the anus to the neck. There are muscles which connect the internal organs to the trunk and they must be severed to allow the viscera to be removed. A rabbit may be gutted by using a knife-less method with no mess and little time lost. Squeeze the entrails toward the rear resulting in a tight bulging abdomen. Raise the rabbit over the head and sling it down hard striking the forearms against the thighs. The momentum will expel the entrails through a tear in the vent (Figure 14-56). Save the internal organs such as heart, liver, and kidneys, as they are nutritious. The liver should be checked for any white blotches and discarded if affected as these indicate tularemia (also known as rabbit fever). The disease is transmitted by rodents but also infects humans.

c. Cold-blooded animals are generally easy to clean and prepare.

(1) Snakes and lizards are very similar in taste and they have similar skin. Like the mammals, the skin and viscera should be removed. The easiest way to do this is to sever the head and(or) legs. In the case of a lizard, peel back enough skin so that it may be grasped securely and simply pull it down the length of the body turning the skin inside out as it goes. If the skin does not come away easily, a cut down the length of the animal can be made. This will allow the skin to part from the body more easily. The entrails are then removed and the animal is ready to cook.

Figure 14-56. Dressing a Rabbit Without a Knife.

(2) Except for the larger amphibians such as the bullfrog, the hind legs are the largest portion of the animal worth saving. To remove the hindquarters, simply cut through the backbone with a knife, leaving the abdomen and upper body. Pull the skin from the legs and they are ready to cook. With the bullfrogs and larger amphibians, the whole body can be eaten. The head, the skin, and viscera should be removed and discarded (use as bait to catch something else).

d. Most fish need little preparation before they are eaten. Scaling the fish before cooking is not necessary. A cut from the anus to the gills will expose the internal organs which should be removed. The gills should also be removed before cooking. The black line along the inside of the backbone is the kidney and should be removed by running a thumbnail from the tail to the head. There is some meat on the head and should not be discarded. See figure 14-57 for one method of filleting a fish.

e. All birds have feathers which can be removed in two ways: by plucking or pulling out the feathers, and by skinning. The gizzard, heart, and liver should be retained. The gizzard should be split open as it contains partially digested food and stones which must be discarded before being eaten.

f. Insects are an excellent food source and they require little or no preparation. The main point to remember is to remove all hard portions such as the hind legs of a

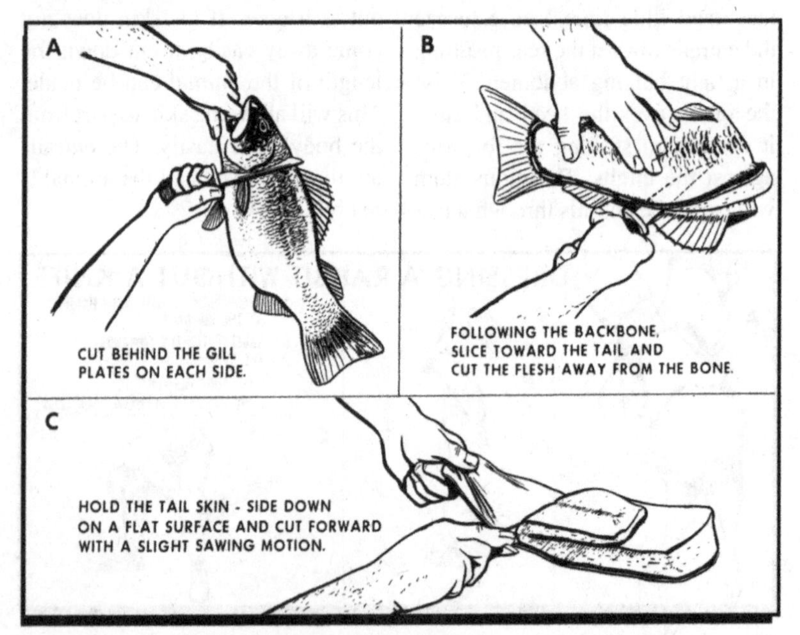

Figure 14-57. Filleting a Fish.

grasshopper and the hard wing covers of beetles. The rest is edible.

14-11. Cooking. All wild game, large insects (grasshoppers), freshwater fish, clams, mussels, snails, and crawfish must be thoroughly cooked to kill internal parasites. Mussels and large snails may have to be minced to make them tender.

a. Boiling is the most nutritious, simplest, and safest method of cooking. Numerous containers can be used for boiling; for example, a metal container suspended above, or set beside, a heat source to boil foods. Green bamboo makes an excellent cooking container. Stone boiling is a method of boiling using super-heated rocks and a container that holds water but cannot be suspended over an open flame. Example of containers are survival kit containers, flying helmet, a hole in the ground lined with waterproof material, or a hollow log. The container is filled with food and water and then heated with super-hot stones until the water boils. Stones from a stream or damp area should not be used. The moisture in the stones may turn to steam and cause the stone to explode while the stones are being heated in the fire. The container should be covered and new stones added as the water stops boiling. The rocks can be removed with the aid of a wire secured to the rock before being put into the container or two sticks used in a chopstick fashion.

b. Baking is a good method of cooking as it is slow and is usually done by putting food into a container and cooking it slowly. Baking is often used with various types of ovens. Foods may be wrapped in wet leaves (Figure 14-58) (avoid using a type of plant that will give an unpleasant flavor to what is being cooked), placed inside a metal container, or they may be packed with mud or clay and placed directly on the coals. Fish and birds packed in mud and baked must not be skinned because the scales, skin, or feathers will come off the animal when the mud or clay is removed.

Figure 14-58. Baking.

Figure 14-59. Clam Baking.

Clambake-style baking is done by heating a number of stones in a fire and allowing the fire to burn down to coals. A layer of wet seaweed or leaves is then placed over the hot rocks. Food such as mussels and clams in their shells are placed on the wet seaweed and(or) leaves (Figure 14-59). More wet seaweed and(or) leaves and soil is used as a cover. When thoroughly steamed in their own juices, clam, oyster, and mussel shells will open and may be eaten without further preparation.

c. Any type of food can be cooked in the ground in a rock oven (Figure 14-60). First, a hole is dug about 2 feet deep and 2 or 3 feet square, depending on the amount of food to be cooked. The sides and bottom are then lined with rock. Next, procure several green trees about 6 inches in diameter and long enough to bridge the hole. Firewood and grass or leaves for insulation should also be gathered. A fire is started in the hole. Two or three green trees are placed over the hole and several rocks are placed on the trees. The fire must be maintained until the green trees burn through. This indicates the fire has burned long enough to thoroughly heat the rocks and the oven is ready. The fallen rocks, fire, and ash are removed from the hole and a thin layer of dirt is spread over the bottom. The insulating material (grass, leaves, moss, etc.) is placed over the soil, then the food more insulating material on top and around the food, another thin layer of soil, and the extra hot rocks are placed on top. The hole is then filled with soil up to ground level. Small pieces of meat (steaks, chops, etc.) cook in 1½ to 2 hours and large pieces take 5 to 6 hours.

d. Roasting is less desirable as it involves exposing the food to

Figure 14-60. Rock Oven.

direct heat which quickly destroys the nutritional properties. Putting a piece of meat on a stick and holding it over the fire is considered roasting.

e. Broiling is the quickest way to prepare fish. A rock broiler may be made by placing a layer of small stones on top of hot coals, and laying the fish on the top. Scaling the fish before cooking is not necessary, and small fish need not be cleaned. Cooked in this manner, fish have a moist and delicious flavor. Crabs and lobsters may also be placed on the stones and broiled.

f. Meat may be cooked by laying it on a flat board or stone (planking) which is propped up close to the fire (Figure 14-61). The meat will have to be turned over at least once to allow thorough cooking. The cooking time depends on how close the meat is to the fire.

g. Frying is by far the least favorable method of preparing food. It tends to make the meat tough because most all of the natural juices are cooked out of the meat. Some of the nutritional value of the meat will also be destroyed.

Figure 14-61. Planking.

Frying can be done on any nonporous surface which can be heated. Examples are unpainted aircraft parts, turtle shells, large seashells, flat rocks, and some survivial kit parts.

14-12. Preparing Food in Enemy Areas. The problem of preparing food in a hostile area becomes acute when a fire, even a small cooking fire, can bring about capture. After finding food in a hostile area, the problem of preparing the food in a manner which will not compromise the survivors presence must be resolved. Of course, it would be simple to state that the best solution would be to eat the food without cooking it.

a. In some respects, this would be a more reasonable solution than it might initially seem to be. From the standpoint of palatability, it is mostly a matter of adjusting the "frame of mind." Animal foods are recognized as being palatable when cooked to a very minor degree. The need for food cannot be ignored and the situation may demand that it be eaten partially cooked or even uncooked.

b. With regard to the health considerations involved, many of the reasons for cooking are recognized as a means of destroying organisms that may be present in the food and can cause sickness or ill effects if they enter into the body. Under survival conditions in a hostile area, one may be forced to forego thorough cooking and accept the risk involved until their return to friendly forces where professional treatment is available.

c. Assuming that there will have to be a way to prepare food under hostile conditions, a survivor should be aware of some of the ways in order to achieve some degree of safety, and at the same time, improve palatability. Parasites and other organisms living in the flesh of the animals depend upon the body temperature of the animals, the moisture within the flesh of the animals, and other factors to support their life. Any action that modifies these conditions (for example, freezing or thorough drying of the meat) and kills some parasites may improve the palatability.

d. If cooking is considered necessary, use extreme care in selecting the site for a fire and ensure that security considerations are favorable. The food should be prepared in very small quantities in order to keep the size of the fire as small as possible. The use of the "Dakota Hole" configuration is more appropriate for cooking food during a tactical situation.

14-13. Preserving Food. Finding natural foods is an uncertain aspect of survival. The survivor must make the best use of the available food. Food, especially meat, has a tendency to spoil within a short period of time unless it is preserved. There are many ways to preserve food; some of the most common are cooking, refrigeration, freezing, and dehydration.

a. Cooking will slow down the decomposition of food but will not eliminate it. This is because many bacteria are present which work to break it down. Cooking methods which are the best for immediate consumption, such as boiling, are the least effective for preserving food. Food should be recooked every day until all is consumed.

b. Cooling is an effective method of storing food for short periods of time. Heat tends to accelerate the decomposition process where cooling retards decomposition. The colder food becomes, the less the likelihood of deterioration until freezing eliminates decomposition. Cooling devices available to a survivor are:

(1) Food items buried in snow will maintain a temperature of approximately 32 °F.

(2) Food wrapped in waterproof material and placed in streams will remain cool in summer months. Care should be taken to ensure food is secured.

(3) Earth, below the surface, particularly in shady areas or along streams, remains cooler than the surface. A hole may be dug, lined with grass, and covered to form an effective cool storage area much the same as a root celler.

(4) When water evaporates, it tends to cool down the surrounding area. Using this fact, articles of food may be wrapped in an absorbent material such as cotton or burlap and rewetted as the water evaporates.

c. Once food is frozen, it will not decompose. Food should be frozen in meal-sized portions so refreezing is avoided.

d. Drying removes all moisture and preserves the food. Drying is done by sunning, smoking, or burying it in hot sand.

(1) For sun-drying, the food should be sliced very thin and placed in direct sunlight. Meat should be cut across the grain to improve tenderness and decrease drying time. If salt is available, it should be added to improve flavor and accelerate the drying process.

(2) Smoking is a process done through the use of nonresinous wood such as willow or aspen and is used to produce smoke which adds flavor and dries the meat. A smoke rack is also necessary to contain the smoke (Figure 14-62). The following are the procedures for drying meat using smoke:

(a) Cut meat very thin and across the grain. If the meat is

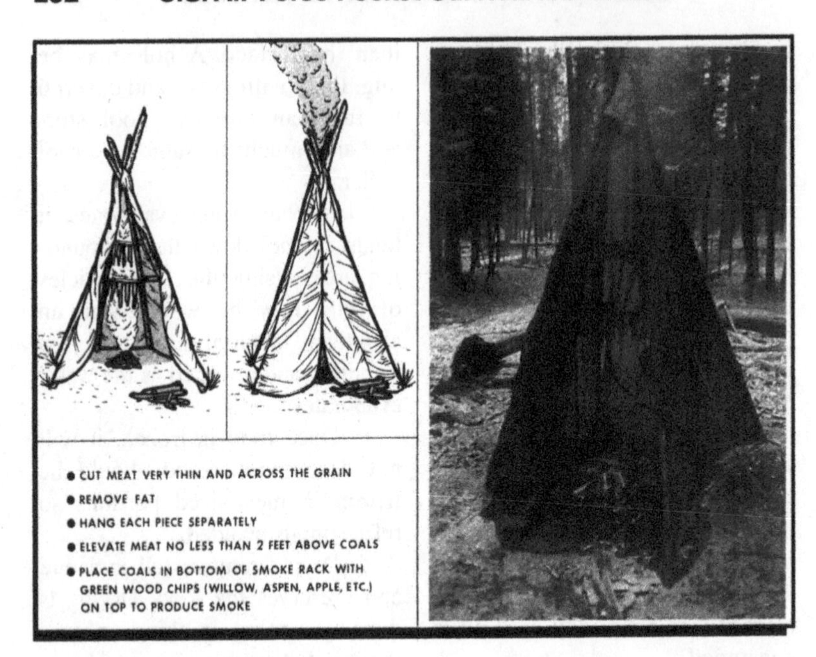

Figure 14-62. Smoke-Drying.

warm and difficult to slice thin, cut the meat in 1 or 2-inch cubes and beat it thin with a clean wooden mallet (improvised).

(b) Remove fat.

(c) Hang the meat on a rack so each piece is separate.

(d) Elevate meat no less than 2 feet above coals.

(e) Coals are placed in the bottom of a smoke rack with green woodchips on top to produce smoke.

e. The method used to preserve fish through warm weather is similar to that used in preserving meat (Figure 14-63). When there is no danger of predatory animals disturbing the fish, the fish should be placed on available fabric and allowed to cool during the night. Early the next morning, before the air gets warm, the fish should be rolled in moist fabric (and leaves). This bundle can be placed inside the survivor's pack. During the rest periods, or when the pack is removed, it should be placed in a cool location out of the sun's rays.

(1) Fish may be dried in the same manner described for smoking meat. To prepare fish for smoking, the heads and backbone are removed and the fish are spread flat on a grill. Thin willow branches with bark removed make skewers.

(2) Fish may also be dried in the sun. They can be suspended

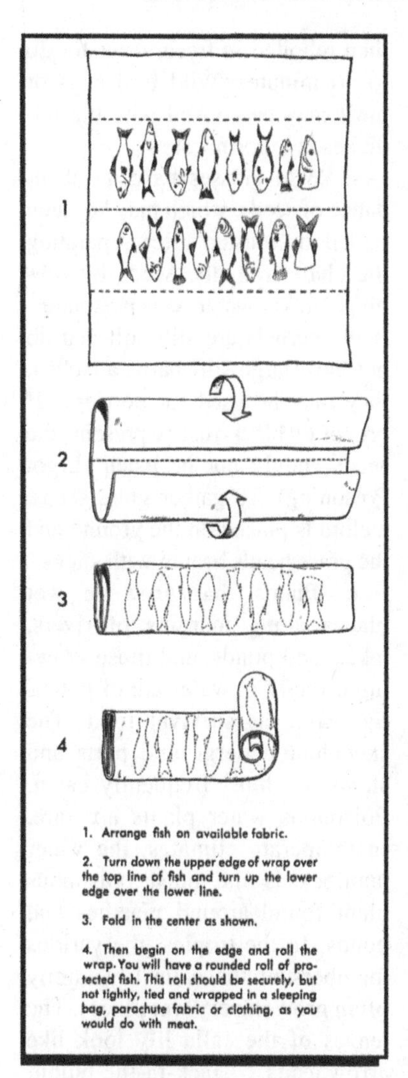

Figure 14-63. Preserving Fish.

from branches or spread on hot rocks. When the meat has dried, sea water or salt should be used on the outside, if available.

f. In survival environments, there are many animals and insects that will devour a survivor's food if it is not correctly stored. Protecting food from insects and birds is done by wrapping it in parachute material, wrapping and tying brush around the bundle, and finally, wrapping it with another layer of material. This creates "dead air" space making it more difficult for insects and birds to get to the food. If the outer layer is wetted, evaporation will also cool the food to some degree. In most cases, if the food is stored several feet off the ground, it will be out of reach of most animals. This can be done by hanging the food or putting it into a "cache." If the food is dehydrated, the container must be completely waterproof to prevent reabsorption. Frozen food will remain frozen only if the outside temperature remains below freezing. Burying food is a good way to store as long as scavengers are not in the area to uncover it. Insects and small animals should also be remembered when burying the food. Food should never be stored in the shelter as this may attract wild animals and could be hazardous to the survivors.

14-14. Preparing Plant Food.

Preparing plant foods can be more involved than preparing animal life.

a. Some plant foods, such as acorns and tree bark may be bitter because of tannin. These plants will require leaching by chopping up the plant parts, and pouring several changes of fresh water over them. This will help wash out the tannin, making the plant more palatable. Other plants such as cassava and green papaya must be cooked before eating to break down the harmful enzymes and chemical crystals within them and make them safe to eat. Plants such as skunk cabbage must undergo this cooking process several times before it is safe to eat.

b. All starchy foods must be cooked since raw starch is difficult to digest. They are boiled, steamed, roasted, or fried and are eaten plain, or mixed with other wild foods. The manioc (cassava) is best cooked, because the bitter form (green stem) is poisonous when eaten raw. Starch is removed from sago palm, cycads, and other starch-producing trunks by splitting the trunk and pounding the soft, whitish inner parts with a pointed club. This pulp is washed with water and the white sago (pure starch) is drained into a container. It is washed a second time, and then it may be used directly as a flour. One trunk of the sago palm will supply a survivor's starch needs for many weeks.

c. The fiddleheads of all ferns are the curled, young succulent fronds which have the same food value as cabbage or asparagus. Practically all types of fiddleheads are covered with hair which makes them bitter. The hair can be removed by washing the fiddleheads in water. If fiddleheads are especially bitter, they should be boiled for 10 minutes and then reboiled in fresh water for 30 to 40 minutes. Wild bird eggs or meat may be cooked with the fiddleheads to form a stew.

d. Wild grasses have an abundance of seeds, which may be eaten boiled or roasted after separating the chaff from the seeds by rubbing. No known grass is poisonous. If the kernels are still soft and do not have large stiff barbs attached, they may be used for porridge. If brown or black rust is present, the seeds should not be eaten (Ergot Poisoning). To gather grass seeds, a cloth is placed on the ground and the grass heads beaten with sticks.

e. Plants that grow in wet places along margins of rivers, lakes, and ponds, and those growing directly in water are of potential value as survival food. The succulent underground parts and stems are most frequently eaten. Poisonous water plants are rare. In temperate climates, the water hemlock is the most poisonous plant found around marshes and ponds. In the tropics, the various members of the calla lily family often grow in very wet places. The leaves of the calla lily look like arrowheads. Jack-in-the-pulpit, calla lily, and sweet flag are members of the Arum family. To be eaten, the members of this plant family must be cooked in frequent changes of water to destroy the irritant crystals in the stems. Two kinds of marsh and water plants are the cattail and the water lily.

(1) The cattail (Typha) is found worldwide except in tundra regions of the far north (Figure 14-64). Cattails can be found in the more moist places in desert areas of all continents as well as in the moist tropic and temperate zones of both hemispheres. The young shoots taste like asparagus. The spikes can be boiled or steamed when green and then eaten. The rootstalks, without the outer covering, are eaten boiled or raw. Cattail roots can be cut into thin strips, dried, and then ground into flour. They are 46 percent starch, 11 percent sugar, and the rest is fiber. While the plant is in flower, the yellow pollen is very abundant; this may be mixed with water and made into small cakes and steamed as a substitute for bread.

(2) Water lilies (Nymphaea and Nuphar) occur on all the continents, but principally in southern

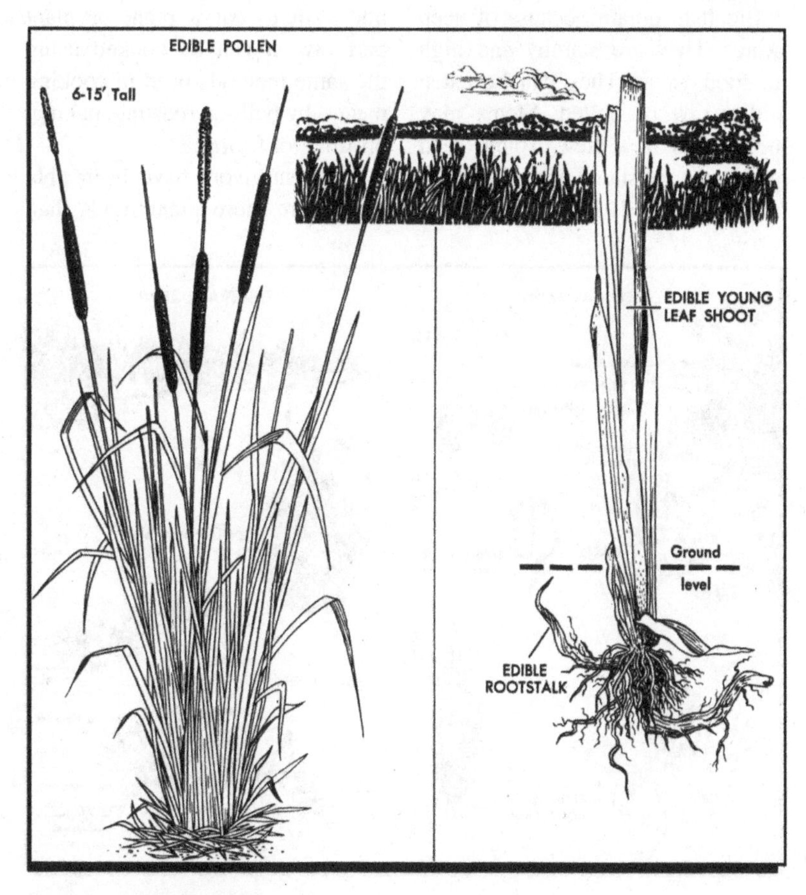

Figure 14-64. Cattails.

Asia, Africa, North America, and South America (Figure 14-65). Two main types are:

(a) Temperate water lilies produce enormous rootstalks and yellow or white flowers which float on the water.

(b) Tropical water lilies produce large edible tubers and flowers which are elevated above the water surface.

(3) Rootstalks or tubers may be difficult to obtain because of deep water. They are starchy and high in food value. They can be eaten either raw or boiled. Stems may be cooked in a stew. Young seed pods may be sliced and eaten as a vegetable. Seeds may be bitter, but are very nourishing. They may be parched and rubbed between stones as flour. The water lily is considered an important food source by native peoples in many parts of the world.

f. Nuts are very high in nutritional value and usually can be eaten raw. Nuts may be roasted in the fire or roasted by shaking them in a container with hot coals from the fire. They may then be ground to make a flour. If a survivor does not wish to eat a plant or plant part raw, it can be cooked using the same methods used in cooking meat—by boiling, roasting, baking, broiling, or frying.

g. If survivors have been able to procure more plant foods than

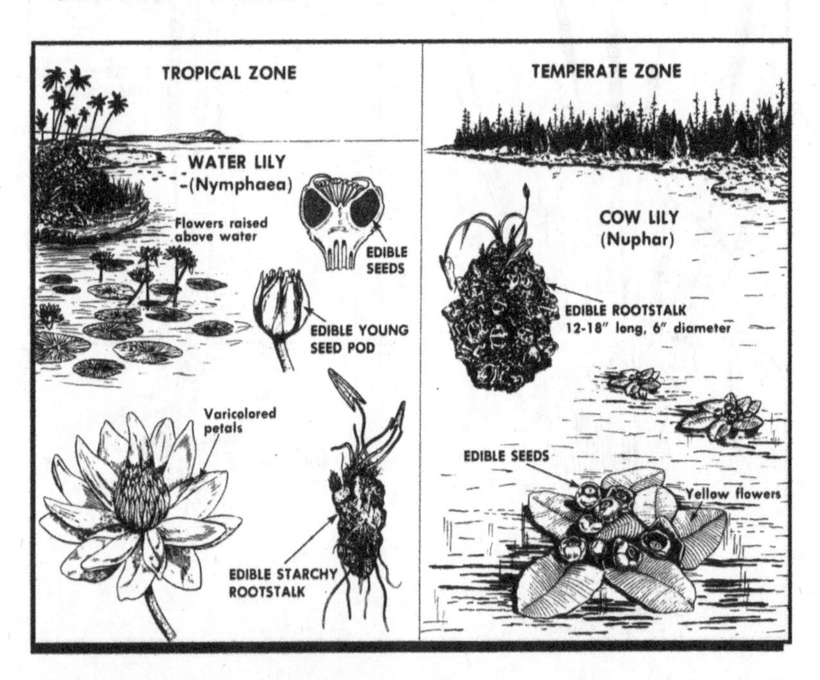

Figure 14-65. Water Lilies.

can be eaten, the excess can be preserved in the same manner as animal foods. Plant foods can be dried by wind, air, sun, or fire, with or without smoke. A combination of these methods can be used. The main object is to remove the moisture. Most wild fruits can be dried. If the plant part is large, such as some tubers, it should be sliced, and then dried. Some type of protection may be necessary to prevent consumption and(or) contamination by insects. Extra fruits or berries can be carried with the survivor by wrapping them in leaves or moss.

Chapter 15

WATER

15-1. Introduction. Nearly every survival account details the need survivors had for water. Many ingenious methods of locating, procuring, purifying, and storing water are included in the recorded experiences of downed aircrew members. If survivors are located in temperate, tropic, or dry climates, water may be their first and most important need. The priority of finding water over that of obtaining food must be emphasized to potential survivors. An individual may be able to live for weeks without food, depending on the temperature and amount of energy being exerted. A person who has no water can be expected to die within days. Even in cold climate areas or places where water is abundant, survivors should attempt to keep their body fluids at a level that will maintain them in the best possible state of health. Even in relatively cold climates, the body needs 2 quarts of water per day to remain efficient (Figure 15-1).

15-2. Water Requirements. Normally, with atmospheric temperature of about 68 °F, the average adult requires 2 to 3 quarts of water daily.

(a) This water is necessary to replace that lost daily in the following ways:

(1) Urine. Approximately 1.4 quart of water is lost in the urine.

(2) Sweat. About 0.1 quart of water is lost in the sweat.

(3) Feces. Approximately 0.2 quart of water is lost in the feces.

(4) Insensible Water Loss. When the individual is unaware water loss is actually occurring, it is referred to as insensible water loss. Insensible water loss occurs by the following mechanisms:

(a) Diffusion through the skin. Water loss through the skin occurs as a result of the actual diffusion of water molecules through the cells of the skin. The average loss of water in this manner is approximately 0.3 to 0.4 quart. Fortunately, loss of greater quantities of water by diffusion is prevented by the outermost layer of the skin, the epidermis, which acts as a barrier to this type of water loss.

(b) Evaporation through the lungs. Inhaled air initially contains very little water vapor. However, as soon as it enters the respiratory passages, the air is exposed to the fluids covering the respiratory surfaces. By the time this air enters the lungs, it has become totally saturated with moisture from these surfaces. When the air is exhaled, it is still saturated with moisture and water is lost from the body.

(c) Larger quantities of water are required when water loss is increased in any one of the following circumstances:

NO WALKING AT ALL

MAXIMUM DAILY TEMPERATURE (°F) IN SHADE	AVAILABLE WATER PER MAN, US QUARTS					
	0	1 Qt	2 Qts	4 Qts	10 Qts	20 Qts
	DAYS OF EXPECTED SURVIVAL					
120°	2	2	2	2.5	3	4.5
110	3	3	3.5	4	5	7
100	5	5.5	6	7	9.5	13.5
90	7	8	9	10.5	15	23
80	9	10	11	13	19	29
70	10	11	12	14	20.5	32
60	10	11	12	14	21	32
50	10	11	12	14.5	21	32

WALKING AT NIGHT UNTIL EXHAUSTED AND RESTING THEREAFTER

MAXIMUM DAILY TEMPERATURE (°F) IN SHADE	AVAILABLE WATER PER MAN, US QUARTS					
	0	1 Qt	2 Qts	4 Qts	10 Qts	20 Qts
	DAYS OF EXPECTED SURVIVAL					
120°	1	2	2	2.5	3	
110	2	2	2.5	3	3.5	
100	3	3.5	3.5	4.5	5.5	
90	5	5.5	5.5	6.5	8	
80	7	7.5	8	9.5	11.5	
70	7.5	8	9	10.5	13.5	
60	8	8.5	9	11	14	
50	8	8.5	9	11	14	

Figure 15-1. Water Requirements.

(1) Heat Exposure. When an individual is exposed to very high temperatures, water lost in the sweat can be increased to as much as 3.5 quarts an hour. Water loss at this increased rate can deplete the body fluids in short time.

(2) Exercise. Physical activity increases the loss of water in two ways as follows:

(a) The increased respiration rate causes increased water loss by evaporation through the lungs.

(b) The increased body heat causes excessive sweating.

(3) Cold Exposure. As the temperature decreases, the amount of water vapor in the air also decreases. Therefore, breathing cold air results in increased water loss by evaporation from the lungs.

(4) High Altitude. At high altitudes, increased water loss by evaporation through the lungs occurs not only as a result of breathing cooler air but also as a result of the increased respiratory efforts required.

(5) Burns. After extensive burns, the outermost layer of the skin is destroyed. When this layer is gone, there is no longer a barrier to water loss by diffusion, and the rate of water loss in this manner can increase up to 5 quarts each day.

(6) Illness. Severe vomiting or prolonged diarrhea can lead to serious water depletion.

(d) Dehydration (body fluid depletion) can occur when required body fluids are not replaced.

(1) Dehydration is accompanied by the following symptoms:

(a) Thirst.

(b) Weakness.

(c) Fatigue.

(d) Dizziness.

(e) Headache.

(f) Fever.

(g) Inelastic abdominal skin.

(h) Dry mucous membranes, that is, dry mouth and nasal passages.

(i) Infrequent urination and reduced volume. The urine is concentrated so that it is very dark in color. In severe cases, urination may be quite painful.

(2) Companions will observe the following behavioral changes in individuals suffering from dehydration:

(a) Loss of appetite.

(b) Lagging pace.

(c) Impatience.

(d) Sleepiness.

(e) Apathy.

(f) Emotional instability.

(g) Indistinct speech.

(h) Mental confusion.

(3) Dehydration is a complication which causes decreased efficiency in the performance of even the simplest task. It also predisposes survivors to the development of severe shock following minor injuries. Constriction of blood vessels in the skin as a result of dehydration increases the danger of cold injury during cold exposure. Failure to replace body fluids ultimately results in death.

(a) Proper treatment for dehydration is to replace lost body fluids. The oral intake of water is the most readily available means of correcting this deficiency. A severely dehydrated person will have little appetite. This person must be encouraged to drink small quantities of water at frequent intervals to replenish the bowdy's fluid volume. Cold water should be warmed so the system will accept it easier.

(b) To prevent dehydration, water loss must be replaced by periodic intake of small quantities of water throughout the day. As activities or conditions intensify, the water intake should be increased accordingly. Water intake should be sufficient to maintain a minimum urinary output of 1 pint every 24 hours. Thirst is not an adequate stimulus for water intake, and a person often dehydrates when water is available. Therefore, water intake should be encouraged when the person is not thirsty. Humans cannot adjust to decreased water intake for prolonged periods of time. When water is in short supply, any available water should be consumed sensibly. If sugar is available, it should be mixed with the water, and efforts should be made to find a local water source. Until a suitable water source is located, individual water losses should be limited in the following ways:

-1. Physical activity should be limited to the absolute minimum required for survival activities. All tasks should be performed slowly and deliberately with minimal expenditure of energy. Frequent rest periods should be included in the daily schedule.

-2. In hot climates, essential activity should be conducted at night or during the cooler part of the day.

-3. In hot climates, clothing should be worn at all times because it reduces the quantity of water loss by sweating. Sweat is absorbed into the clothing evaporated from its surface in the same manner it evaporates from the body. This evaporation cools the air trapped between the clothing and the skin, causing a decrease in the activity of the sweat glands and a subsequent reduction in water loss.

-4. In hot weather, light-colored clothing should be worn rather than dark-colored clothing. Dark-colored clothing absorbs the sun's light rays and converts them into heat. This heat causes an increase in body temperatures which activates the sweat glands and increases water loss through sweating. Light-colored clothing, however, reflects the sun's light rays, minimizing the increase in body temperature and subsequent water loss.

15-3. Water Sources. Survivors should be aware of both the water sources available to them and the resources at their disposal for producing water.

(a) Survivors may obtain water from solar stills, desalter kits, or canned water packed in various survival kits. It would be wise for personnel, who may one day have to use these methods of procuring water, to be knowledgeable of their operating instructions and the amount of water they produce.

(1) Canned water provides 10 ounces per can.

(2) Desalter kits are limited to 1 pint per chemical bar—kits contain eight chemical bars.

(3) A "sea solar still" can produce as much as 2½ pints per day.

(4) "Land solar stills" produce varied amounts of water. This amount is directly proportionate to the amount of water available in the soil or placed into the still (vegetation, entrails, contaminated water, etc.), and the ambient temperature.

(b) Aircrew members would be wise to carry water during their missions. Besides the fact that the initial shock of the survival experience sometimes produces feelings of thirst, having an additional water container can benefit survivors. The issued items (canned water, desalter kits, and solar stills) should be kept by survivors for times when no natural sources of freshwater are available.

(c) Naturally occurring indicators of water are:

(1) Surface water, including streams, lakes, springs, ice, and snow.

(2) Precipitation, such as rain, snow, dew, sleet, etc.

(3) Subsurface water, which may not be as readily accessible as wells, cisterns, and underground springs and streams, can be difficult for survivors to locate and use.

(d) Several indicators of possible water are:

(1) Presence of abundant vegetation of a different variety, such as deciduous growth in a coniferous area.

(2) Drainages and low-lying areas.

(3) Large clumps of plush grass.

Animal trails which may lead to water. The "V" formed by intersecting trails often point toward water sources.

(e) Survivors may locate and procure water as follows:

(1) Precipitation may be procured by laying a piece of nonporous material such as a poncho, piece of canvas, plastic, or metal material on the ground. If rain or snow is being collected, it may be more efficient to create a bag or funnel shape with the material so the water can be easily gathered. Dew can be collected by wiping it up with a sponge or cloth first, and then wringing it into a container (Figure 15-2). Consideration should be given to the possibility of contaminating the water with dyes, preservatives, or oils on the surfaces of the objects used to collect the precipitation. Ice will yield more water per given volume than snow and requires less heat to do so. If the sun is shining, snow or ice may be placed on a dark surface to melt (dark surfaces absorb heat, whereas light surfaces reflect heat). Ice can be found in the form of icicles on plants and trees, sheet ice on rivers, ponds, and lakes, or sea ice. If snow must be used, survivors should use snow closest to the ground. This snow is packed and will provide more water for the amount of snow than will the upper layers. When snow is to be melted for water, place a small amount of snow in the bottom of the container being used and place it over or near a fire. Snow can be added a little at a time. Survivors should allow water in the container bottom to become warm so that when more snow is added, the mixture remains slushy. This will prevent burning the bottom out of the container. Snow absorbs water, and if packed, forms an insulating airspace at the bottom of the container. When this happens, the bottom may burn out.

(2) Several things may help survivors locate ground water, such as rivers, lakes, and streams.

(a) The presence of swarming insects indicates water is near. In some places, survivors should look for signs of animal presence. For example, in damp places, animals may have scratched

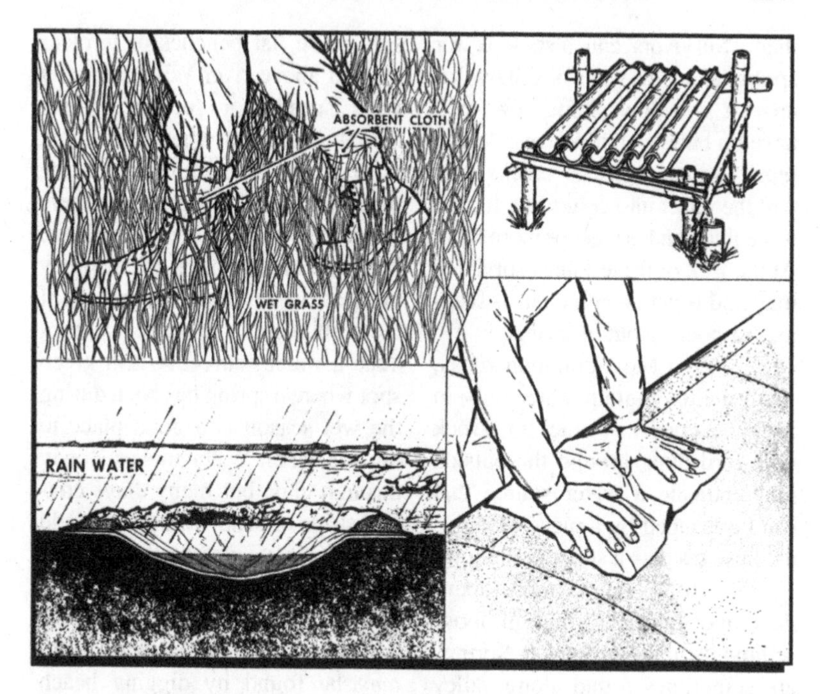

Figure 15-2. Methods of Procuring Water.

depressions into the ground to obtain water; insects may also hover over these areas.

(b) In the Libyan Sahara, donut-shaped mounds of camel dung often surround wells or other water sources. Bird flights can indicate direction to or from water. Pigeons and doves make their way to water regularly. They fly from water in the morning and to it in the evening. Large flocks of birds may also congregate around or at areas of water.

(c) The presence of people will indicate water. The location of this water can take many forms—stored water in containers that are carried with people who are traveling, wells, irrigation systems, pools, etc. Survivors who are evaders should be extremely cautious when approaching any water source, especially if they are in dry areas; these places may be guarded or inhabited.

(3) When no surface water is available, survivors may have to tap the earth's supply of ground water. Access to this depends upon the type of ground—rock or loose material, clay, gravel, or sand.

(a) In rocky ground, survivors should look for springs and seepages. Limestone and lava rocks will have more and larger springs than any other rocks. Most lava rocks contain millions of bubble holes; ground water may seep through them. Survivors can also look for springs along the walls of valleys that cross a lava flow. Some flows will have no bubbles but do have "organ pipe" joints—vertical cracks that part the rocks into columns a foot or more thick and 20 feet or more high. At the foot of these joints, survivors may find water creeping out as seepage, or pouring out in springs.

(b) Most common rocks, like granite, contain water only in irregular cracks. A crack in a rock with bird dung around the outside may indicate a water source that can be reached by a piece of surgical hose used as a straw or siphon.

(c) Water is more abundant and easier to find in loose sediments than in rocks. Springs are sometimes found along valley floors or down along their sloping sides. The flat benches or terraces of land above river valleys usually yield springs or seepages along their bases, even when the stream is dry. Survivors shouldn't waste time digging for water unless there are signs that water is available. Digging in the floor of a valley under a steep slope, especially if the bluff is cut in a terrace, can produce a water source. A lush green spot where a spring has been during the wet season is a good place to dig for water. Water moves slowly through clay, but many clays contain strips of sand which may yield springs. Survivors should look for a wet place on the surface of clay bluffs and try digging it out.

(d) Along coasts, water may be found by digging beach wells (Figure 15-3). Locate the

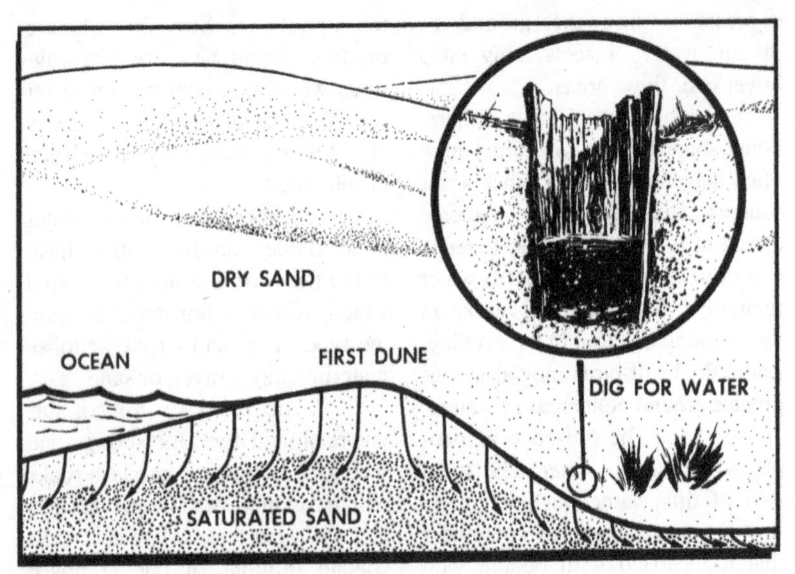

Figure 15-3. Beach Well.

wells behind the first or second pressure ridge. Wells can be dug 3 to 5 feet deep and should be lined with driftwood to prevent sand from refilling the hole. Rocks should be used to line the bottom of the well to prevent stirring up sand when procuring the water. The average well may take as long as 2 hours to produce 4 to 5 gallons of water. (Do not be discouraged if the first try is unsuccessful—dig another.)

15-4. Water in Snow and Ice Areas. Due to the extreme cold of arctic areas, water requirements are greatly increased. Increased body metabolism, respiration of cold air, and extremely low humidity play important roles in reducing the body's water content. The processes of heat production and digestion in the body also increase the need for water in colder climatic zones. The constructing of shelters and signals and the obtaining of firewood are extremely demanding tasks for survivors. Physical exertion and heat production in extreme cold place the water requirements of a survivor close to 5 or 6 quarts per day to maintain proper hydration levels. The diet of survivors will often be dehydrated rations and high protein food sources. For the body to digest and use these food sources effectively, increased water intake is essential.

(a) Obtaining water need not be a serious problem in the arctic because an abundant supply of water is available from streams, lakes, ponds, snow, and ice. All surface water should be purified by some means. In the summer, surface water may be discolored but is drinkable when purified. Water obtained from glacier-fed rivers and streams may contain high concentrations of dirt or silt. By letting the water stand for a period of time, most silt will settle to the bottom; the remaining water can be strained through porous material for further filtration.

(b) A "water machine" can be constructed which will produce water while the survivors are doing other tasks. It can be made by placing snow on any porous material (such as parachute or cotton), gathering up the edges, and suspending the "bag" of snow from any support near the fire. Radiant heat will melt the snow and the water will drip from the lowest point on the bag. A container should be placed below this point to catch the water (Figure 15-4).

(c) In some arctic areas, there may be little or no fuel supply with which to melt ice and snow for water. In this case, body heat can be used to do the job. The ice or snow can be placed in a waterproof container like a waterbag and placed between clothing layers next to the body. This cold substance should not be placed directly next to the skin; it causes chilling and lowering of the body temperature.

(d) Since icebergs are composed of freshwater, they can be a readily

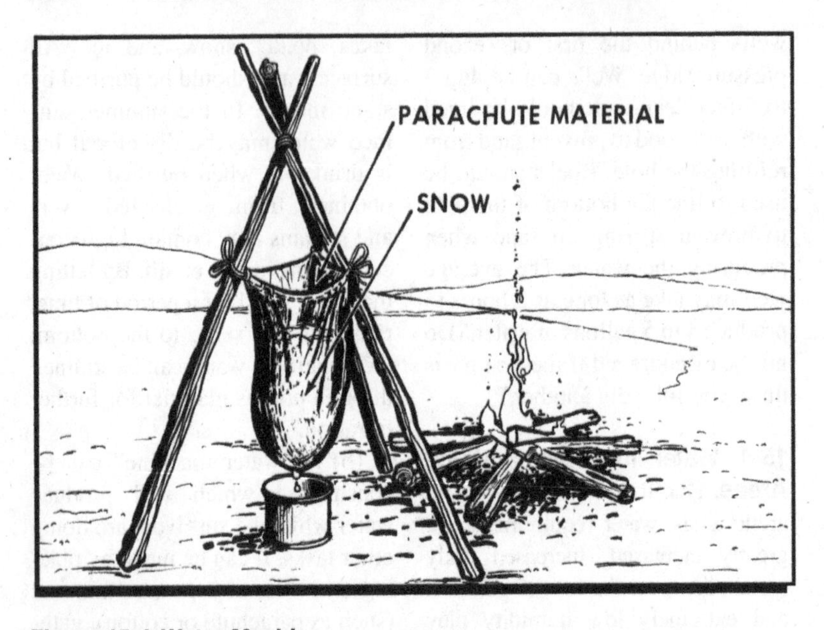

Figure 15-4. Water Machine.

available source of drinking water. Survivors should use extreme caution when trying to obtain water from this source. Even large icebergs can suddenly roll over and dump survivors into the frigid sea water. If sea ice is the primary source of water, survivors should recall that like seawater itself, saltwater ice should never be ingested. To obtain water in polar regions or sea ice areas, survivors should select old sea ice, a bluish or blackish ice which shatters easily and generally has rounded corners. This ice will be almost salt- free. New sea ice is milky or gray colored with sharp edges and angles. This type of ice will not shatter or break easily. Snow and ice may be saturated with salt from blowing spray; if it tastes salty, survivors should select different snow or ice sources.

(e) The ingesting of unmelted snow or ice is not recommended. Eating snow or ice lowers the body's temperature, induces dehydration, and causes minor cold injury to lips and mouth membranes. Water consumed in cold areas should be in the form of warm or hot fluids. The ingestion of cold fluids or foods increases the body's need for water and requires more body heat to warm the substance.

15-5. Water on the Open Seas. The lack of drinkable water could be a major problem on the open seas. Seawater should never be ingested in its natural state. It will cause an individual to become violently ill in a very short period of time. When water is limited and cannot be replaced by chemical or mechanical means, it must be used efficiently. As in the desert, conserving sweat not water, is the rule. Survivors should keep in the shade as much as possible and dampen clothing with seawater to keep cool. They should not over exert but relax and sleep as much as possible.

a. If it rains, survivors can collect rainwater in available containers and store it for later use. Storage containers could be cans, plastic bags, or the bladder of a life preserver. Drinking as much rainwater as possible while it is raining is advisable. If the freshwater should become contaminated with small amounts of seawater or salt spray, it will remain safe for drinking (Figure 15-5). At night and on

Figure 15-5. Collecting Water from Spray Shield.

foggy days, survivors should try to collect dew for drinking water by using a sponge, chamois, handkerchief, etc.

b. Solar stills will provide a drinkable source of water. Survivors should read the instructions immediately and set them up, using as many stills as available. (Be sure to attach them to the raft.) Desalter kits, if available, should probably be saved for the time when no other means of procuring drinking water is available. Instructions on how to use the desalter kit are on the container.

c. Only water in its conventional sense should be consumed. The so-called "water substitutes" do little for the survivor, and may do much more harm than not consuming any water at all. There is no substitute for water. Fish juices and other animal fluids are of doubtful value in preventing dehyration. Fish juices contain protein which requires large amounts of water to be digested and the waste products must be excreted in the urine which increases water loss. Survivors should never drink urine—urine is body waste material and only serves to concentrate waste materials in the body and require more water to eliminate the additional waste.

15-6. Water in Tropical Area. Depending on the time of the year and type of jungle, water in the tropical climates can be plentiful; however, it is necessary to know where to look and procure it. Surface water is normally available in the form of streams, ponds, rivers, and swamps. In the savannas during the dry season, it may be necessary for the survivor to resort to digging for water in the places previously mentioned. Water obtained from these sources may need filtration and should be purified. Jungle plants can also provide survivors with water.

a. Many plants have hollow portions which can collect rainfall, dew, etc. (Figure 19-6). Since there

Figure 15-6. Water Collectors.

is no absolute way to tell whether this water is pure, it should be purified. The stems or the leaves of some plants have a hollow section where the stem meets the trunk. Look for water collected here. This includes any Y-shaped plants (palms or air plants). The branches of large trees often support air plants (relatives of the pineapple) whose overlapping, thickly growing leaves may hold a considerable amount of rainwater. Trees may also catch and store rainwater in natural receptacles such as cracks or hollows.

b. Pure freshwater needing no purification can be obtained from numerous plant sources. There are many varieties of vines which are potential water sources. The vines are from 50 feet to several hundred feet in length and 1 to 6 inches in diameter. They also grow like a hose along the ground and up into the trees. The leaf structure of the vine is generally high in the trees. Water vines are usually soft and easily cut. The smaller species may be twisted or bent easily and are usually heavy because of the water content. The water from these vines should be tested for potability. The first step in testing the water from vines is for survivors to nick the vine and watch for sap running from the cut. If milky sap is seen, the vine should be discarded; if no milky sap is observed, the vine may be a safe water vine. Survivors should cut out a section of the vine, hold that piece vertically, and observe the liquid as it flows out. If it is clear and colorless, it may be a drinkable source. If it is cloudy or milky-colored, they should discard the vine. They should let some of the liquid flow into the palm of the hand and observe it. If the liquid does not change color, they can now taste it. If it tastes like water or has a woody or sweet taste, it should be safe for drinking. Liquid with a sour or bitter taste should be avoided. Water trapped within a vine is easily obtained by cutting out a section of the vine. The vine should first be cut high above the ground and then near the ground. This will provide a long length of vine and, in addition, will tend to hide evidence of the cuts if the survivors are in an evasion situation. When drinking from the vine, it should not touch the mouth as the bark may contain irritants which could affect the lips and mouth (Figure 15-7). The pores in the upper end of the section of vine may reclose, stopping the flow of water. If this occurs, survivors should cut off the end of the vine opposite the drinking end. This will reopen the pores allowing the water to flow.

c. Water from the rattan palm and spiny bamboo may be obtained in the same manner as from vines. It is not necessary to test the water if positive identification of the plant can be made. The slender stem (runner) of the rattan palm is an excellent water source. The joints are overlapping in appearance, as if one section is fitted inside the next.

d. Water may be trapped within sections of green bamboo. To determine if water is trapped within a section of bamboo, it should be shaken. If it contains water, a sloshing sound can be heard. An opening may be made in the section by making two 45-degree angle cuts, both on the same side of the section, and prying loose a piece of the section wall. The end of the section may be cut off and the water drunk or poured from the open end. The inside of the bamboo should be examined before

Figure 15-7. Water Vines and Bamboo.

consuming the water. If the inside walls are clean and white, the water will be safe to drink. If there are brown or black spots, fungus growth, or any discoloration, the water should be purified before consumption. Sometimes water can also be obtained by cutting the top off certain types of green bamboo, bending it over, and staking it to the ground (Figure 15-7). A water container should be placed under it to catch the dripping water. This method has also proven effective on some vines and the rattan palm.

e. Water can also be obtained from banana plants in a couple of different ways, neither of which is satisfactory in a tactical situation. First, survivors should cut a banana plant down, then a long section should be cut off which can be easily handled. The section is taken apart by slitting from one end to the other and pulling off the layers one at a time. A strip 3 inches wide, the length of the section, and just deep enough to expose the cells should be removed from the convex side. This section is folded toward the convex side to force the water from the cells of the plant. The layer must be squeezed gently to avoid forcing out any tannin into the water. Another technique for obtaining water from the banana plant is by making a "banana-well." This is done by making a bowl out of the plant stump, fairly close to the ground, by cutting out and removing the inner section of the stump (Figure 15-8). Water which first enters the bowl may contain a

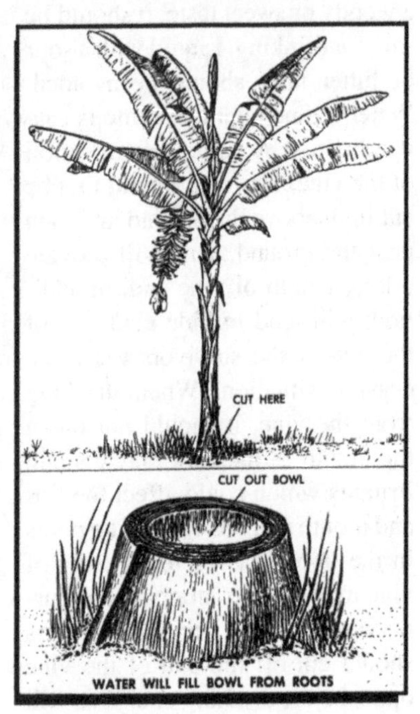

Figure 15-8. Water from Banana Plant.

concentration of tannin (an astringent which has the same effect as alum). A leaf from the banana plant or other plant should be placed over the bowl while it is filling to prevent contamination by insects, etc.

f. Water trees can also be a valuable source of water in some jungles. They can be identified by their blotched bark which is fairly thin and smooth. The leaves are large, leathery, fuzzy, and evergreen, and may grow as large as 8 or 9 inches. The trunks may have short outgrowths with fig-like fruit on them or long tendrils with round fruit comprised of corn kernel-shaped nuggets. In a nontactical situation, the tree can be tapped in the same manner as a rubber tree, with either a diagonal slash or a "V." When the bark is cut into, it will exude a white sap which if ingested causes temporary irritation of the urinary tract. This sap dries up quite rapidly and can easily be removed. The cut should be continued into the tree with a spigot (bamboo, knife, etc.) at the bottom of the tap to direct the water into a container. The water flows from the leaves back into the roots after sundown, so water can be procured from this source only after sundown or on overcast (cloudy) days. If survivors are in a tactical situation, they can obtain water from the tree and still conceal the procurement location. If the long tendrils are growing thickly, they can be separated and a hole bored into the tree. The white sap should be scraped off and a spigot placed below the tap with a water container to catch the water. Moving the tendrils back into place will conceal the container. Instead of boring into the tree, a couple of tendrils can be cut off or snapped off if no knife is available. The white sap should be allowed to dry and then be removed. The ends of the tendrils should be placed in a water container and the container concealed.

g. Coconuts contain a refreshing fluid. Where coconuts are available, they may be used as a water source. The fluid from a mature coconut contains oil, which when consumed in excess can cause diarrhea. There is little problem if used in moderation or with a meal and not on an empty stomach. Green unripe coconuts about the size of a grapefruit are the best for use because the fluid can be taken in large quantities without harmful effects. There is more fluid and less oils so there is less possibility of diarrhea.

h. Water can also be obtained from liquid mud. Mud can be filtered through a piece of cloth. Water taken by this method must be purified. Rainwater can be collected from a tree by wrapping a cloth around a slanted tree and arranging the bottom end of the cloth to drip into a container (Figure 15-9).

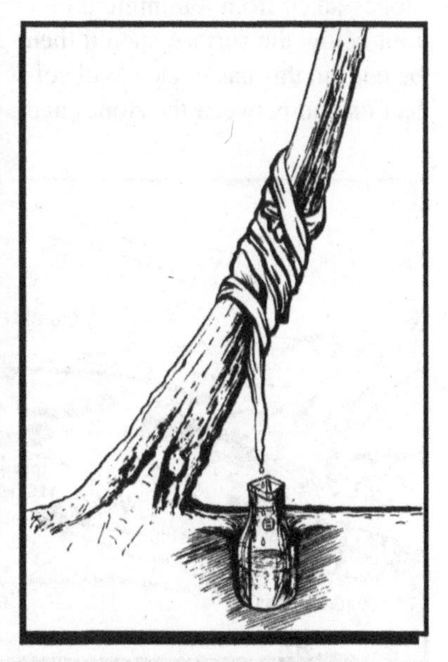

Figure 15-9. Collecting Water from Slanted Tree.

15-7. Water in Dry Areas. Locating and procuring water in a dry environment can be a formidable task. Some of the ways to find water in this environment have been explored, such as locating a concave bend in a dry riverbed and digging for water (Figure 19-10). If there is any water within a few feet of the surface, the sand will become slightly damp. Dig until water is obtained.

a. Some deserts become humid at night. The humidity may be collected in the form of dew. This dew can be collected by digging a shallow basin in the ground about 3 feet in diameter and lining it with a piece of canvas, plastic, or other suitable material. A pyramid of stones taken from a minimum of 1 foot below the surface should then be built in this basin. Dew will collect on and between the stones and trickle down onto the lining material where it can be collected and placed in a container.

b. Plants and trees having roots near the surface may be a source of water in dry areas. Water trees of dry Australia are a source of water, as their roots run out 40 to 80 feet at a depth of 2 to 9 inches under the surface. Survivors may obtain water from these roots by locating a root 4 to 5 feet from the trunk and cutting the root into 2- or 3-foot lengths. The bark can then be peeled off and the liquid from each section of root drained into a container. The liquid can also be sucked out. The trees growing in hollows or depressions will have the most water in their roots. Roots that are 1 to 2 inches thick are an ideal size. Water can be carried in these roots by plugging one end with clay.

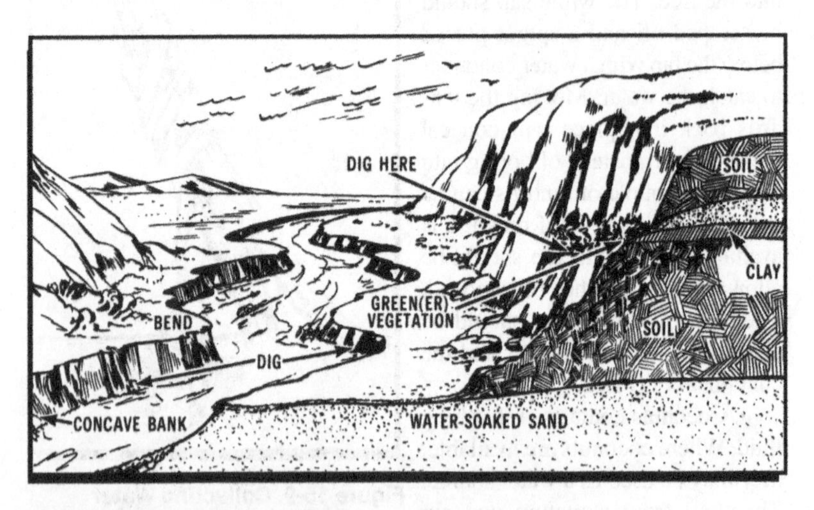

Figure 15-10. Dry Stream Bed.

c. Cactus-like or succulent plants may be sources of water for survivors, but they should recall that no plants should be used for water procurement which have a milky sap. The barrel cactus of the United States provides a water source. To obtain it, survivors should first cut off the top of the plant. The pulpy inside portions of the plant should then be mashed to form a watery pulp. Water may ooze out and collect in the bowl; if not, the pulp may be squeezed through a cloth directly into the mouth.

d. The solar still is a method of obtaining water that uses both vegetation and ground moisture to produce water (Figure 15-11). A solar still can be made from a sheet of clear plastic stretched over a hole in the ground. The moisture in the soil and from plant parts (fleshy stems and leaves) will be extracted and collected by this emergency device. Obviously, where the soil is extremely dry and no fleshy plants are available, little, if any, water can be obtained from the still. The still may also be used to purify polluted water.

(1) The parts for the still are a piece of plastic about 6 feet square, a water collector-container or any waterproof material from which a collector-container can be fashioned, and a piece of plastic tubing about one-fourth inch in diameter and 4 to 6 feet long. The tubing is not absolutely essential but makes

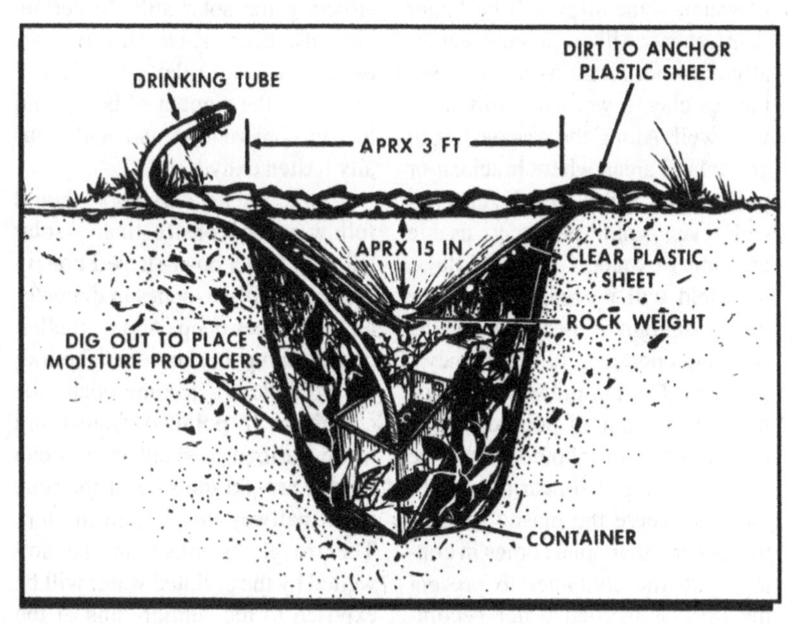

Figure 15-11. Solar Still.

the still easier to use. A container can be made from such materials as plastic, aluminum foil, poncho, emergency ration tins, or a flight helmet. The tubing, when available, is fastened to the bottom of the inside of the container and used to remove drinking water from the container without disturbing the plastic. Some plastics work better than others, although any clear plastic should work if it is strong.

(2) If plants are available or if polluted water is to be purified, the still can be constructed in any convenient spot where it will receive direct sunlight throughout the day. Ease of digging will be the main consideration. If soil moisture is to be the only source of water, some sites will be better than others. Although sand generally does not retain as much moisture as clay, a wet sand will work very well. Along the seacoast or in any inland areas where brackish or polluted water is available, any wet soil, even sand, produces usable amounts of water. On cloudy days, the yield will be reduced because direct sunlight is necessary if the still is to operate at full efficiency.

(3) Certain precautions must be kept in mind. If polluted water is used, survivors should make sure that none is spilled near the rim of the hole where the plastic touches the soil and that none comes in contact with the container to prevent the freshly distilled water becoming contaminated. Survivors should not disturb the plastic sheet during daylight "working hours" unless it is absolutely necessary. If a plastic drinking tube is not available, raise the plastic sheet and remove the container as few times as possible during daylight hours. It takes one-half hour for the air in the still to become resaturated and the collection of water to begin after the plastic has been disturbed. Even when placed on fairly damp soil and in an area where 8 hours of light per day is directed on the solar still, the average yield is only about 1 cup per day per still. Due to the low yields obtained from this device, survivors must give consideration to the possible danger of excessive dehydration brought about by constructing the solar still. In certain circumstances, solar still returns, even over 2- or 3-day periods, will not equal the amount of body fluid lost in construction and will actually hasten dehydration.

(4) Steps survivors should follow when constructing a solar still are: Dig a bowl-shaped hole in the soil about 40 inches in diameter and 20 inches deep. Add a smaller, deeper sump in the center bottom of the hole to accommodate the container. If polluted waters are to be purified, a small trough can be dug around the side of the hole about halfway down from the top. The trough ensures that the soil wetted by the polluted water will be exposed to the sunlight and at the same time that the polluted water

is prevented from running into the container. If plant material is used, line the sides of the hole with pieces of plant or its fleshy stems and leaves. Place the plastic over the hole and put soil on the edges to hold it in place. Place a rock no larger than a plum in the center of the plastic until it is about 15 inches below ground level. The plastic will now have the shape of a cone. Put more soil on the plastic around the rim of the hole to hold the cone securely in place and to prevent water-vapor loss. Straighten the plastic to form a neat cone with an angle of about 30 degrees so the water drops will run down and fall into the container. It takes about 1 hour for the air to become saturated and start condensing on the underside of the plastic cone.

e. The vegetation bag is a simpler method of water procurement. This method involves cutting foliage from trees or herbaceous plants, sealing it in a large clear plastic bag, and allowing the heat of the sun to extract the fluids contained within. A large, heavy-duty clear plastic bag should be used. The bag should be filled with about 1 cubic yard of foliage, sealed, and exposed to the sun. The average yield for one bag tested was 320 ml/bag 5-hour day. This method is simple to set up. The vegetation bag method of water procurement does have one primary drawback. The water produced is normally bitter to taste, caused by biological breakdown of the leaves as they lay in the water produced and super heated in the moist "hothouse" environment. This method can be readily used in a survival situation, but before the water produced by certain vegetation is consumed, it should undergo the taste test. This is to guard against ingestion of cyanide-producing substances and other harmful toxins, such as plant alkaloids. (Figure 15-12.)

f. One more method of water procurement is the water transpiration bag, a method that is simple to use and has great potential for enhancing survival. This method is the vegetation bag process taken one step further. A large plastic bag is placed over a living limb of a medium-size tree or large shrub. The bag opening is sealed at the branch, and the limb is then tied down to allow collected water to flow to the comer of the bag. For a diagram of the water transpiration method, Figure 15-13.

(1) The amount of water yielded by this method will depend on the species of trees and shrubs available. During one test of this method, a transpiration bag produced approximately a gallon per day for 3 days with a plastic bag on the same limb, and with no major deterioration of the branch. Other branches yielded the same amount. Transpired water has a variety of tastes depending on whether or not the vegetation species is allowed to contact the water.

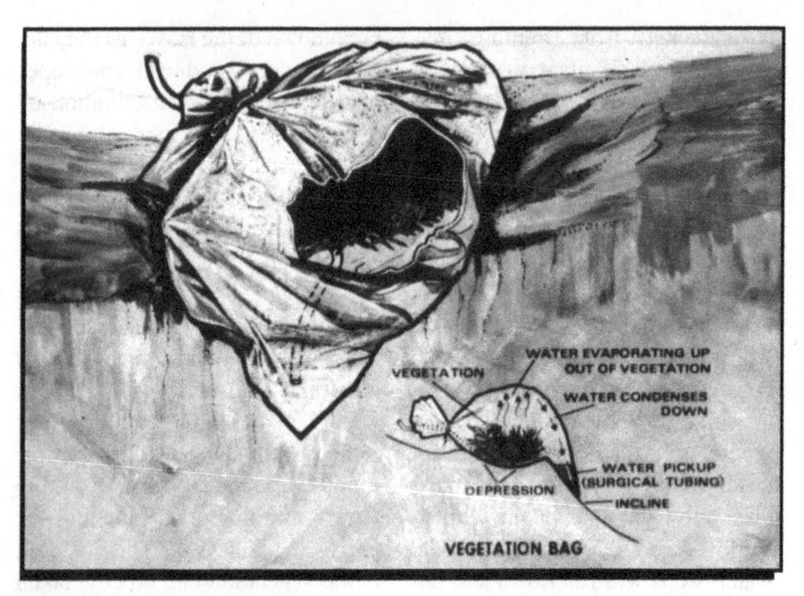

Figure 15-12. Vegetation Bag.

(2) The effort expended in setting up water transpiration collectors is minimal. It takes about 5 minutes' work and requires no special skills once the method has been described or demonstrated. Collecting the water in a survival situation would necessitate survivors dismantling the plastic bag at the end of the day, draining the contents and setting it up again the following day. The same branch may be reused (in some cases with almost similar yields); however, as a general rule, when vegetation abounds, a new branch should be used each day.

(3) Without a doubt, the water transpiration bag method surpasses other methods (solar stills, vegetation bag, cutting roots, barrel cactus) in yield, ease of assembly, and in most cases, taste. The benefits of having a simple plastic bag can't be over-emphasized. As a water procurer, in dry, semi-dry, or desert environments where low woodlands predominate, it can be used as a water transpirator; in scrubland, steppes, or treeless plains, as a vegetation bag; in sandy areas without vegetation, it can be cut up and improvised into solar stills. Up to three large, heavy-duty bags may be needed to sustain one survivor in certain situations.

15-8. Preparation of Water for Consumption:

a. The following are ways survivors can possibly determine the presence of harmful agents in the water:

(1) Strong odors, foam, or bubbles in the water.

Figure 15-13. Transpiration Bag.

(2) Discoloration or turbid (muddy with sediment).

(3) Water from lakes found in desert areas are sometimes salty because they have been without an outlet for extended periods of time. Magnesium or alkali salts may produce a laxative effect; if not too strong, it is drinkable.

(4) If the water gags survivors or causes gastric disturbances, drinking should be discontinued.

(5) The lack of healthy green plants growing around any water source.

b. Because of survivors' potential aversion to water from natural sources, it should be rendered as potable as possible through filtration. Filtration only removes the solid particles from water—it does not purify it. One simple and quick way of filtering is to dig a sediment hole or seepage basin along a water source and allow the soil to filter the water (figure 15-14). The seepage hole should be covered while not in use. Another way is to construct a filter—layers of parachute material stretched across a tripod (figure 15-15). Charcoal is used

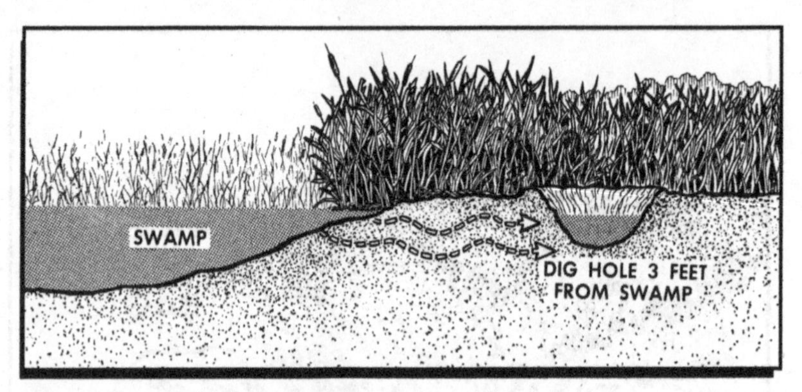

Figure 15-14. Sediment Hole.

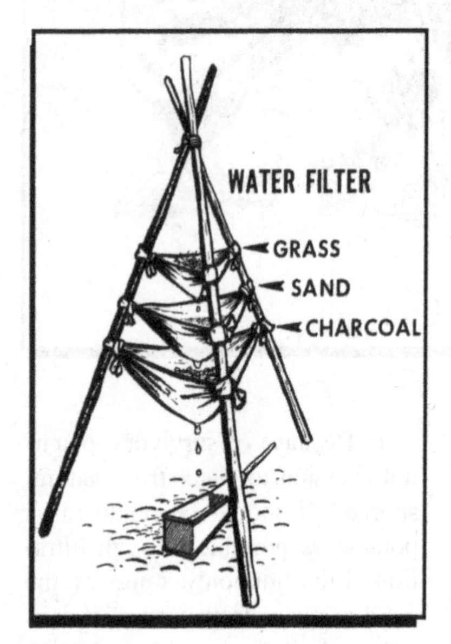

Figure 15-15. Water Filter.

to eliminate bad odors and foreign materials from the water. Activated charcoal (obtained from freshly burned wood is used to filter the water). If a solid container is available for making a filter, use layers of fine-to-coarse sand and gravel along with charcoal and grass.

c. Purification of water may be done a variety of ways. The method used will be dictated by the situation (such as tactical or nontactical).

(1) Boil the water for at least 10 minutes.

(2) To use purification tablets survivors should follow instructions on the bottle. One tablet per quart of clear water; two tablets if water is cloudy. Let water stand for 5 minutes (allowing the tablet time to dissolve), then shake and allow to stand for 15 minutes. Survivors should remember to turn the canteen over and allow a small amount of water to seep out and cover the neck part of the canteen. In an evasion situation, water purification tablets should be used for purifying water. If these are not available, plant sources or non- stagnant, running water obtained from a location upstream from habitatio should be consumed.

(3) Eight drops of 2*/2-percent iodine per quart—stir or shake and let stand for at least 10 minutes.

d. After water is found and purified, survivors may wish to store it for later consumption. The following make good containers:

(1) Waterbag.

(2) Canteen.

(3) Prophylactic inside a sock for protection of bladder.

(4) Segment of bamboo.

(5) Birch bark and pitch canteen.

(6) LPU bladder.

(7) Hood from antiexposure suit.

Part Six

EVASION AND CAMOUFLAGE

Chapter 16

FACTORS OF SUCCESSFUL EVASION

16-1. Basic Principles. All potential evaders must have three things in their favor. These are the same three things needed by a potential escapee. The three factors which increase chances of successful evasion are preparation, opportunity, and motivation.

16-2. Preparation:

a. Preparation is one of the most important factors for successful evasion. The actions crewmembers take before the evasion episode can make the difference between being able to evade or being captured. In a hostile area the survivors should remember evasion is an integral part of their mission and plan accordingly. The enemy may make mistakes of every conceivable form and not suffer more than indignation, anger, and fatigue. The evader, on the other hand, must constantly guard against mistakes of any sort. Being seen is the greatest mistake an evader can make. The evader must prepare for this task.

b. Three basic problems during evasion are:

(1) Evading the enemy.

(2) Surviving.

(3) Returning to friendly control.

c. Chances for a successful evasion are improved if evaders:

(1) Observe the elementary rules of movement, camouflage, and concealment.

(2) Have a definite plan of action.

(3) Be patient, especially while traveling. Hurrying increases fatigue and decreases alertness. Patience, preparation, and determination are key words in evasion.

(4) Conserve food.

(5) Conserve as much strength as possible for critical periods.

(6) Rest and sleep as much as possible.

(7) Maintain a highly developed "will to survive" and "can do" attitude. Evasion may require living off the land for extended periods of time and traveling on foot over difficult terrain, often during inclement weather.

(8) Study the physical features of the land. They should note the location of mountains, swamps, plains, deserts, or forests, type of vegetation, and availability of water.

(9) Consider the climate. Aircrew members should know the climatic characteristics and typical weather conditions

of the area which may be flown over.

(10) Study ethnic briefs and survival, evasion, resistance, and escape (SERE) contingency guides before a mission and learn some of the customs and habits of the local people. Such knowledge will aid in planning missions and evasion plans of action. For example, it may give the evader the ability to avoid hostile people or groups or to identify and deal with "friendlies." This knowledge may also allow for blending into the local populace.

(11) Know the equipment well! One must know the location of each item in the kit, its operation, and its value. An evader must preplan which equipment should be retained and which should be left behind.

d. Once in the evasion situation, planning for travel will be a consideration for evaders. They must have a definite objective and be confident in their approach and ability to achieve it. They will normally have several options with variations to choose from in selecting a plan of action or destination. The enemy force deployment, search procedures, terrain, population distribution, climate, distance, and environment (that is, NBC) will influence destination selection. Examples of options and destinations:

(1) Await SAR forces.

(2) Evade to a SAFE area.

(3) Evade to a neutral country. (NOTE: Border areas not disrupted by combat may have a security system intact.)

(4) If evaders are in the forward edge of the battle area (FEBA) and feel sure that friendly forces are moving in their direction, they should seek concealment and allow the FEBA to overrun their position. Evaders' attempts at penetrating the FEBA should be avoided. Evaders face stiff opposition from both sides.

e. The chances that one of these destinations may be close by will be determined by many things including the time and location of the bailout. Other determining factors are: the location and direction of movement of the FEBA, the presence or absence of willing assisters, and the knowledge of the evader's whereabouts possessed by rescue personnel. If the survivor does not land close to one of the above areas or if the previously mentioned factors do not favor immediate air pickup, the survivor may have to travel some distance to reach one of the destinations.

f. One consideration in choosing destination and direction of travel after bailout is whether one of these suitable areas for a pickup or contact with friendly forces exists and, if so, its location. Some preplanning should have been done before the mission. Information upon which to base a decision is derived from command area briefings, area studies, SERE contingency guides, and premission intelligence briefings.

g. Another consideration is physical condition. One's physical condition is the responsibility of the potential evader and has a great effect on the evader's ability to survive. Once on the ground, it is too late to get in shape. One more aspect is an aircrew member's personal habits. Upon first consideration, personal grooming habits might not be considered an important premission briefing item. However, using aftershave lotion, hair dressing, or cologne could add to the problems of an evader. The odor can carry for great distances and give away the evader's presence.

16-3. Opportunity. Potential evaders must take advantage of any and all opportunities to evade. This starts in the aircraft when an emergency is declared. Following current, approved emergency in-flight procedures for the theater of operations (when ejection, bailout, or ditching appears imminent), the aircraft commander will attempt to establish radio contact by first calling on the secure frequency of last contact; second, on an established common secure frequency; and third, on the international emergency frequency. When communication is established, the tactical call sign, type of aircraft, position, course, speed, altitude, nature of difficulties, and intentions will be transmitted. The identification friend or foe (IFF) should be set to the emergency position. When possible, ejection or bailout should be attempted over or near a SAFE area, lifeguard station, or submarine pickup point. This minimizes threat involvement for evaders and SAR forces alike. After ejection or bailout and during descent, the aircrew member must remain alert and steer the parachute away from potential threats (populated areas, gun emplacements, troop concentrations, etc.) or out to sea (feet wet). Once on the ground, the evader must be proficient in the use of the survival/evasion equipment to facilitate evasion (for example, use of the compass in conjunction with the survival radio to call in airstrikes on enemy forces threatening the evader). In addition to the opportunity to evade, motivation is essential to the evader's success.

16-4. Motivation. A strong, central drive will give the evader the necessary push to succeed. It may be personal, ranging in nature from a frame of reference gained through training to a desire to return to a family or loved ones. Motivation may be strictly military, involving one or all of the following reasons applicable to all military men:

a. To return and fight again.

b. To deny the enemy a source of military information.

c. To deny the enemy a source of propaganda.

d. To deny the enemy a source of forced labor.

e. To tie up enemy forces, transportation, and communications that otherwise might be committed to the war effort.

f. Return with intelligence information.

g. In addition, the Code of Conduct calls upon the military members not to surrender of their own free will.

h. Additionally, it is suggested that other personal reasons for being motivated to evade include: fear of death, pain, suffering, humiliation, degradation, disease, illness, torture, uncertainty, and fear of the unknown. From the evader's point of view, ***evasion is far more desirable than captivity or death.***

16-5. Evasion Principles:

a. Besides the preparation, opportunity, and motivation factors important to evasion, there are other important principles. The evader should try to recall any previous briefings, standard operating procedures, or training. A course of action should then be chosen which has the greatest likelihood of resulting in the return to friendly forces.

b. Evader actions should be flexible. Flexibility is one of the most important keys to successful evasion. The evader, basically, must never be so firmly set in a course of action that a change is out of the question. The best thing an evader can do is to stay open to new ideas, suggestions, and changes of events. Having several backup plans of action can give the evader organized flexibility. If one plan of action is upset by enemy activity, the evader could rapidly switch to a backup plan without panic.

c. The evader is primarily interested in avoiding detection. Each evader should remember that people catch people. If the evader avoids detection, success is almost assured. Evaders should:

(1) Observe and listen for sounds of enemy fire and vehicle activity during parachute descent and moving away from those enemy positions once on the ground. Flyers downed during daylight should assume they were seen during descent and expect a search to center on their likely point of landing.

(2) Be patient and determined while traveling.

(3) Use poor weather conditions as an aid in evading.

(4) Circumstances permitting, select times, routes, and methods of travel to avoid detection.

(5) Avoid lines of communication (waterways, roads, etc.

d. The evaders' main objective is immediate recovery. In hostile areas or situations, survivors must sanitize all evidence of presence and direction of travel.

Survivors may never be certain that rescue is imminent.

e. Although evaders would not normally move too far if rescue is imminent, in many situations they will have to leave the landing area quickly and travel as far as practical before selecting a hiding place. They should leave no sign which indicates the direction or presence of travel. All hiding places should be chosen with extreme care. The time evaders will remain in the first location is governed by enemy activity in the area, their physical condition, availability of water and food, rescue capabilities, and patience. It is in this place of initial concealment that the evaders should regain strength, examine the current situation, and plan for the evasion problems ahead.

f. Once in a place of concealment, evaders should make use of all available navigation aids to orient themselves. After finding their location, evaders should also select an ultimate destination and any necessary alternate destinations. The best possible route of travel should then be decided upon. When the time comes to move, they should have a primary plan and alternate plans for travel that cover eventualities they may encounter.

(1) Evasion in a forward area has one great advantage which is not present further to the rear. Assistance may be close at hand. This assistance may come from serveral sources, each of which, under particular circumstances, may prove to be the most effective. These sources may be air cover by tactical fighter flights, helicopter recovery, and rescue by ground forces. Contact with friendly forces in forward areas requires extreme caution. Do not surprise them or move suddenly. They may mistake the evader as the enemy.

(2) The situation at the time of the emergency will determine the evaders' best course of action. High ground is normally the best position from which to await rescue; evaders may expect the best results from signaling devices, may observe the surrounding terrain, and may be kept under observation by friendly air cover. Whatever position is chosen, it must be clear of obstacles that would prevent a successful rescue.

(3) If not rescued immediately, the situation may compel evaders to move. Evaders must plan a course of action before leaving their position. When the evaders are certain their position is known to friendly elements, they might expect ground forces to attempt a rescue. They should remember their position may be detected as the enemy search parties approach. They must be prepared to evade to a new position.

(4) Evaders should remember that when traveling they are probably more vulnerable to capture. Once past the danger of an immediate search, evaders

must avoid people. Inhabited areas should be bypassed rather than penetrated, even if it means miles of added travel. Many evaders have been captured because they followed the easiest and shortest route, or failed to employ simple techniques such as scouting, patrolling, camouflage, and concealment. As a rule, the safest route avoids major roads and populated areas, even if it takes more time and energy. Unaccompanied evasion requires self-reliance and independent action.

Chapter 17

CAMOUFLAGE

17-1. Introduction. Presence of evaders in an area controlled by the enemy may require the evaders to adopt and maintain camouflage to avoid observation. Camouflage consists of those measures evaders use to conceal their presence from the enemy. Camouflage is a French word meaning disguise, and it is used to describe action taken to mislead the enemy by misrepresenting the true identity of an installation, an activity, an item of equipment, or an evader. As a tool for evasion, it enables evaders to carry out life supporting activities and to travel unseen, undetected, and free to return to friendly control. Camouflage allows them to see without being seen. They should try to blend in with the surrounding environment. Effective individual concealment often depends primarily on the choice of background and its proper use. Background is that portion of the surroundings against which an evader will be seen from the ground and the air. It may consist of a barren rocky desert, a farm yard, or a city street. It is the controlling element in individual camouflage and governs every concealment measure. At all times, camouflage is the responsibility of the individual evader. In the event of group evasion, the group leader and each individual are responsible for the camouflage of the group. Evaders should remember that camouflage is a continuous, never-ending process if they want to protect themselves from enemy observation and capture (Figure 17-1).

17-2. Types of Observation:

a. Of the five senses, sight is by far the most useful to the enemy, hearing is second, while smell is of only occasional importance. But these same senses can be of equal value to the evader and observer.

b. How useful these senses are depends primarily on range. For this reason, basic camouflage stresses visual concealment which is relatively long range. Most people are accustomed to looking from one position on the ground to another position on the ground.

c. Before evaders can conceal themselves from aerial observation,

Figure 17-1. Camouflaging.

they should become familiar with what their activities look like from the air, both in an aerial photograph and from direct observation. The evaders must also have an understanding of the types of observation used by the enemy. There are two categories of observation—direct and indirect.

(1) Direct Observation:

(d) Direct observation refers to the process whereby the observer looks directly at the object itself without the use of telescopes, field glasses, or sniper- scopes. Direct observation may be made from the ground or from the air. Direct aerial observation becomes more and more important because of the rapid changes in weapons and in tactical situations due to greater mobility of troops. Reconnaissance aircraft over enemy lines report locations of troops, vehicles, and installations (or shelter areas) as seen from the air-to-ground control stations. Reported targets can be immediately fired upon or troops can be sent in to investigate shelter areas or other suspicious areas.

(e) The enemy may also use dogs, foot patrols, and mechanized units to patrol a given area. Such teams could physically search an area for signs of the passage of strangers, such as footprints, old campfires, discarded or lost equipment, and other "telltale" signs which would indicate that someone had been in the area.

(f) Observation by the local populace is also a possibility. Upon seeing an evader or "telltale" signs an evader left behind, they may contact the local authorities, who initiate organized searches.

(2) Indirect Observation:

a. Indirect observation refers to the study of a photograph or an image of the subject via photography, radar, or television. This form of observation is becoming increasingly more varied and widespread, and may be used from either manned or unmanned positions.

b. Views from the ground are familiar, but views from the air are usually quite unfamiliar. In modern warfare, the enemy may put emphasis on aerial photographs for information. It is important to become familiar with the "bird's-eye view" of the terrain as well as the ground view in order to learn how to guard against both kinds of observation.

17-3. Comparison of Direct and Indirect Observation:

a. The main advantage of direct observation is that observers see movement of an evader without camouflage. An observation can be maintained over relatively long periods of time. The main disadvantage lies in human frailty. For example, the observer's attention may be diverted to another area, or the observer may be fatigued and unable to concentrate.

b. Indirect observation has many advantages. Indirect observation can be far-reaching, cover large areas, and be very accurate. It also produces a record of the area observed so that the recorded picture can be studied in detail, compared, and evaluated. The principal disadvantage is a photograph covers a very short period of time, making detection of movement difficult. This disadvantage can be partially overcome by taking pictures of the same area at different intervals and comparing them for changes (Figure 17-2).

Figure 17-2. Indirect Observations.

17-4. Preventing Recognition:

a. Recognition is the determination (through appearance, behavior, or movement of the hostile or friendly nature) of objects or persons. One objective of camouflage concealment is to prevent recognition. Another objective is to deceive or induce false recognition. This implies that camouflage is not always designed to be a "cloak of invisibility." In some instances, camouflage is used to allow deception. The camouflaged object or person is then seen as a natural feature of the landscape.

b. Recognition through appearance is the result of conclusions drawn by the observer from the position, shape, shadow, texture, or color of the objects or persons. Recognition through behavior or movement includes deductions made from the actual movements themselves or from the record left by tracks of persons or vehicles or by other violations of camouflage discipline. Camouflage disciplines are those actions which contribute to an evader's ability to remain undetected. Proper use of camouflage discipline avoids any activity that changes an area or reveals objects to an enemy. Examples of common breaches of camouflage discipline include reflections from brightly shining objects (watches, glasses, rings, etc.) (Figure 17-3), overcamouflaging, or using camouflage materials which are foreign to the area presently occupied by an evader. Evaders must also watch for signs that may reveal enemy camouflage efforts. Inadvertently walking into a camouflaged enemy position may result in capture.

Figure 17-3. Reflections.

Figure 17-4. Position.

17-5. Factors of Recognition.

Regardless of the type of observation, there are certain factors which help to identify an object. They are called the factors of recognition and are the elements which determine how quickly an object will be seen or how long it will remain unobserved. The eight factors of recognition are position, shape, shadow, texture, color, tone, movement, and shine. These factors must be considered in camouflage to ensure that one or more of these factors do not reveal the location of the evaders.

a. Position. Position is the relation of an object or person to its background. When choosing a position for concealment, a background should be chosen which will virtually absorb the evader (Figure 17-4).

b. Shape. Shape is the outward or visible form of an object or person as distinguished from its surface characteristics and color. Shape refers to outline or form. Color or texture is not considered. At a distance, the forms or outlines of objects can be recognized before the observer can make out details in their appearance. For this reason, camouflage should disrupt the normal shape of an object or person (Figure 17-5).

c. Shadow. A shadow may be more revealing than the object itself, especially when seen from the air. Objects such as factory chimneys, utility poles, vehicles, and tents (or people) have distinctive shadows. Conversely, shadows may sometimes assist in concealment. Objects in the shadow of another object are more likely to be overlooked. As with shape, it is more important to disrupt the

Figure 17-5. Shape.

shadow pattern than to totally conceal the object or person. The identifiable shadows can be broken up by the addition of natural vegetation at various points on the body. Wearing "shapeless" garments will also disrupt the outline. For example, a soft and shapeless field cap can be used instead of a helmet or flight cap (Figures 17-6 and 17-7).

d. Texture. Texture is a term used to describe the relative characteristics of a surface, whether that surface is a part of an object or an area of terrain. Texture affects the tone and apparent coloration of things because of its absorption and scattering of light. Highly textured surfaces tend to appear dark and remain constant in tone regardless of the direction of view and lighting, whereas relatively smooth surfaces change from dark to light with a change in direction of viewing or lighting. The application of texture to an object often has the added quality of disrupting its shape and the shape of its shadow, making it more difficult to detect and identify as something foreign to the surroundings in which it exists. As an example, a surface having the same color but with heavy "nap" or texture is tall grass. Each separate blade is capable of casting a shadow upon itself and its surroundings. The light reflecting properties have been cut to a minimum. It will look and photograph dark gray. Looking straight down, the aerial observer sees all of the shadows, whereas a person on the ground may not. The textured surface may look light at ground level, but to the aerial observer the same surface produces an effect of rela-

Figure 17-6. Shadow.

Figure 17-7. Shadow Breakup

tive darkness. The material used to conceal a person or an object must approximate the texture of the terrain in order to blend in with the terrain. Personnel walking or vehicles moving across the terrain will change the texture by mashing down the growth. Therefore, this will show up clearly from the air as vehicle tracks or foot paths.

e. Color:

(1) Pronounced color differences at close range distinguish one object from another. The contrast between the color of the object and the color of its background can be an aid to enemy observers. The greater the contrast in color, the more visible the object appears (Figure 17-8).

(2) Color differences or differences in hue, such as red and green-yellow, become increasingly difficult to distinguish as the viewing range is increased. This happens because of atmospheric effects. Colors in nature, except for certain floral and tropical animal life, are not brilliant. The impression of the vividness of nature's colors results from the large areas of like colors involved and contrast of these areas with each other. The principal contrast is in their dark and light qualities. However, the dark and

Figure 17-8. Contrast.

light color contrast does not fade out quickly and is distinguishable at greater distances. Therefore, as a first general principle, the camouflage should match the darker and lighter qualities of the background and be increasingly concerned with the colors involved as the viewing range is decreased or the size of the object or installation becomes larger. A second general rule to follow is to avoid contrasts of hues. This is especially true in areas with heavy vegetation. Light-toned colors, such as leaf bottoms, should be avoided as they tend to attract attention.

f. Tone. Tone is the amount of contrast between variations of the same color. It is the effect achieved by the combination of light, shade, and color. In a black-and-white photograph, the shades of gray in which an object appears is known as tone (Figure 17-9). By adding texturing material to a smooth or shiny surface, the surface can be made to produce a darker tone in a photograph, because the textured surface now absorbs more light rays. Objects become identifiable as such because of contrasts between them and their background. Camouflage blending is the process of eliminating or reducing these contrasts. The principal contrast is that of tone; that is, the dark and light relationship existing between an object and its background. The two principal means available for reducing tone contrast are the application of matching or neutral coloration and the use of texturing to form disruptive patterns. Poorly chosen, disruptive patterns tend to make the object more conspicuous instead of concealing it.

Figure 17-9. Three Half-Tone Blocks.

g. Movement. Of the eight factors of recognition, movement is the quickest and easiest to detect. The eye is very quick to notice any movement in an otherwise still scene. The aerial camera can record the fact that something has moved when two photographs of the same area are taken at different times. If an object has moved, the changed position is apparent when the two photographs are compared (Figure 17-10).

h. Shine:

(1) Shine is a particularly revealing signal to an observer. In undisturbed, natural surroundings, there are comparatively few objects which cause a reflected shine. Skin, clean clothing, metallic insignia, rings, glasses, watches, buckles, identification bracelets, and similar items produce "shine." When light strikes smooth surfaces such as these, it may be reflected directly into the observer's eye or the

Figure 17-10. Movement.

Figure 17-11. Reflected Shine.

camera lens with striking emphasis (Figure 17-11).

(2) Such items must be neutralized by staining, covering, or removing to prevent their shine from revealing the location of evaders. This is especially true at night.

17-6. Principles and Methods of Camouflage:

a. No matter how applied, camouflage can be successful only by observing three fundamental principles. These basic principles of camouflage are choice of position, camouflage discipline, and camouflage construction.

b. When these factors have been considered, the evader is ready to begin application of various methods of deceiving the enemy. These methods are:

(1) Hiding. The complete concealment of a person or object by physical screening.

(2) Disguising. Changing the physical characteristics of an object or person in such a manner as to fool the enemy.

(3) Blending. The arrangement of camouflage material on or about an object in such a manner as to make the object appear to be part of the background. To properly use these methods, three simple rules should be followed:

(d) First, the background should be changed as little as possible. When choosing a position to gain concealment, a background should be chosen that will visually absorb the elements of the position. Evaders should use a "natural" position if available. They should look for an existing position which can be used almost as is, such as a cave or thicket if there are many like it in the area. Isolated landmarks such as individual trees, haystacks, or houses should be avoided. They tend to attract attention and are likely to be searched first because they are so obvious. At times, by making use of background, complete concealment against visual and photographic detection may be gained with no construction. In terrain where natural cover is plentiful, this is a simple task. Even in areas where natural cover is scarce, concealment may be achieved through use of terrain irregularities. Regardless of the activity involved, evaders must always be mentally aware of their positions (Figure 17-12).

(e) Secondly, the evader should use camouflage discipline. This means all of the factors of concealment are continuously applied.

Figure 17-12. Background.

-1. Daytime. Camouflage discipline is the avoidance of activity which changes the appearance of an area or reveals military objects to the enemy. A well-camouflaged position is only secure as long as it is well maintained. Concealment is worthless if obvious tracks point like directional arrows to the heart of the location or if signs of occupancy are permitted to appear in the vicinity. Tracks, debris, and terrain disturbances, are the most common signs of activity. Therefore, natural lines in the terrain should be used. If practical, exposed tracks should be camouflaged by brushing or beating them out. If leaving tracks is unavoidable, they should be placed where they will be least noticed and partially concealed (along logs, under bushes, in deep grass, in shadows, etc.). If tracks cannot be concealed, brushing them out will help them disintegrate quickly. Tying rags or brush to the feet will disguise boot prints and may help disguise them as refugee tracks (Figure 17-13).

Figure 17-13. Camouflaging Tracks

-2. Nighttime. Visual concealment at night is less necessary than in the daytime; however, noises at night are more noticeable. A simple act such as snoring may prove fatal. Calling to one another, talking, and even whispering should be kept to a minimum. But by far, the most important apsect of night discipline is light discipline. Lights at night not only disclose the evaders' position but also hinder the evaders' ability to detect the enemy. Even on the darkest nights, eyes grow accustomed to the lack of light in approximately 30 minutes. Everytime a match is lit or a flashlight is used, the eyes must go through the complete process of getting adjusted to the darkness again. Smoking and lights should be prohibited at night in areas in close proximity to the enemy because the light is impossible to conceal. Additionally, a cigarette light aggravates the situation by creating a reflection which completely illuminates the face. The smell of the evader's foreign tobacco would stand out even if the enemy is smoking.

-3. Evaders can lessen the effects of sound by simply taking precautions against sound production. They should avoid any sound-producing activity. Walking on hard surfaces should be avoided and full use should be made of soft ground for digging. Hand signals or signs should be used when possible during group travel. Individual equipment should be padded and fastened in such a manner as to prevent banging noises.

(f) The evader should consider the following points regarding the use of camouflage:

(1) Take advantage of all natural concealment.

(2) Don't over-camouflage. Too much is as obvious as too little.

(3) When using natural camouflage, remember that it fades and wilts, so change it regularly.

(4) If taking advantage of shadows and shade, remember they shift with the sun.

(5) Above all, avoid unnecessary movement.

(6) When moving, keep off the skyline; use the military crest (three-quarters' way up the hill).

(7) Do not expose anything that may shine.

(8) Break up outlines of manmade objects.

(9) When observing an area, do so from a prone position, while in cover.

(10) Match vegetation used as camouflage with that in the immediate locale, and when moving from position to position, change camouflage to blend with the new area's vegetation types.

17-7. Individual Camouflage:

a. At this point, with some of the general information about camouflage presented, it is time for a more detailed examination concerning individual camouflage.

b. Generally, individual camouflage is that personal concealment which evaders must use to deceive the enemy. Evaders must know how to use the terrain for effective concealment. Evaders must dress for the best concealment and carefully select their routes to provide for as much concealment as possible. All of the methods and techniques of camouflage addressed in this section have been successfully used by past evaders. If this information is learned and practiced by today's aircrew members (tomorrow's possible evaders), they will be more successful in evading and have a greater chance of returning to friendly forces.

c. Evaders should remember that in some areas they may have to engage in camouflage activities designed to deceive two types of enemy observation—ground and air. Many objects which are concealed from ground observation, may be seen from the air. This means the evader should camouflage for both types of observation.

d. Form is basic shape (body outline) and height. Three things which give an evader away in terms of form are to reveal outline of head and shoulders, to present straight lines of sides, and to allow the inverted "V" of the crotch and legs to be distinguishable. If staying in shadows, blending with background, adopting body positions other than standing erect, and other behavioral procedures are inadequate. They can be camouflaged by using "add-ons" such as branches or twigs to break up the lines. This addition of vegetation will also help an evader blend in with the background.

e. Effective concealment of evaders depends largely on the choice and proper use of background. Background varies widely in appearance, and evaders may find themselves in a jungle setting, in a barren or desert area, in a farmyard, or in a city street. Each location will require individual treatment because location governs every concealment measure taken by the individual. Clothing which blends with the predominant color of the background is desirable. There will be occasions when the uniform color must be altered to blend with a specific background. The color of the skin must receive individual attention and be toned to blend with the background.

f. There are certain general aspects of individual body and equipment camouflage techniques which apply almost anywhere. The evaders should take each of the following areas under consideration.

(1) Exposed Skin. The contrast in tone between the skin of face and hands and that of the surrounding foliage and other background must be reduced. The skin is to be made lighter or darker, as the case may be, to blend with the surrounding natural tones. The shine areas are the forehead, the cheekbones, nose, and chin.

These areas should have a dark color. The shadow areas such as around the eyes, under the nose and under the chin should have a light color. The hands, arms, and any other exposed areas of skin must also be toned down to blend with the surroundings. Burnt cork, charcoal, lampblack, mud, camouflage stick, berry stains, carbon paper, and green vegetation can all be used as toning materials.

(b) A mesh mosquito face net, properly toned down, is an effective method of breaking up the outlines of the face and ears.

(c) Two primary methods of facial camouflage have been found to be successful patterns. They are the "blotch" method for use in deciduous forests, and the "slash" method for use in coniferous forests (Figure 17-14).

(d) Application of these two patterns are simply modified appropriately to whatever environment the evader is in. In the jungle, a broader slash method would be used to cover exposed skin; in the desert, a thinner slash; in barren snow, a wide blotch; and in grass areas, a thin type slash. To further break up the outline of the facial features, a flop hat or other loosely fitting hat may help. A beard that is not neatly trimmed may also aid the evader.

(e) When toning down the skin, evaders should not neglect to pattern all of the skin; for example, the back of the neck, the insides and backs of the ears, and the eyelids. Covering these areas may help

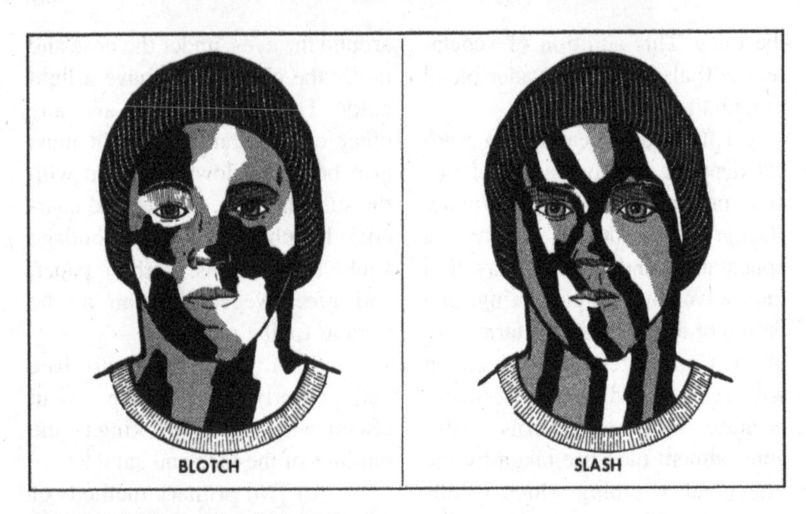

Figure 17-14. Camouflaged Faces.

somewhat, especially if there is a lack of other material to tone down the skin. Vegetation hung from the hat, collar buttoned and turned up, a scarf, or even earflaps may help. To cover the hands, evaders may use flight gloves, mittens, or loose cloth if unable to tone down wrists, backs of hands, and between fingers sufficiently. Evaders should watch for protruding white undergarments, T-shirt, long underwear sleeves, etc. They should also tone down these areas.

(f) Lack of hair or light colored hair requires some type of camouflage. This could include those applied to the skin or an appropriate hat, scarf, or mosquito netting.

(g) Odors in a natural environment stand out and may give evaders away. Americans are continually surrounded by artificial odors and are not usually aware of them. Human body odor would have to be very strong to be detected by ground troops (searchers) that have been in the field for long periods. The following odors should be of concern to the evader.

-1. Soaps and Shampoos. In combat areas, personnel should always use unscented toilet articles. Shaving cream, after shave lotion, perfume, and other cosmetics are to be avoided. The potential evaders should also realize that insect repellent is scented. They should try to use headnets, but if forced to use a repellent, the camouflage stick which has repellent in it is the least scented. Tobacco should not be used. The stain and odor should be removed from the body and clothing. Gum or candy may have strong or sweet smells—evaders should take care to rinse out their mouths after use. These odors, especially tobacco, can be detected at great distances.

-2. Smoke odors from campfires may permeate clothing, but if the potential searchers use fires for cooking and heat, they probably can't detect it on evaders.

(2) Clothing and Personal Items. These items require attention both before assignment to a combat mission and again if forced to evade. Prior preparation for the survivor may include:

(h) Ensuring that flight clothing does not smell of laundry products and is in good repair and not worn to the point where it shines or is faded.

(i) Checking zippers for shine and function.

(j) Checking rank insignia and patches for light reflection and color. Remove name tag, branch of service tag, and rank (whether they are stripes or metal insignia, bright unit patches, etc.) from the uniforms and place them either in the pack or in a secure pocket. Underwear should also be subdued for camouflage in the event that the outer layer is tom. Boots should be black but not shiny. Shiny eyelets should be repainted. Squeaky boots should be fixed or replaced. Sanitize pocket or wallet contents. Remove items which might aid in enemy exploitation attempts of the individual PW or PWs as a group; for example, credit cards, photographs, money, and addresses. Evaders should carry only those necessary pieces of identification which will prove a person is a U.S. military member.

(3) Additional Clothing. Additional clothing may be desirable and located in flight clothing, such as hat, socks, scarf, and gloves. In an evasion environment, clothing and equipment need quick camouflaging attention. Anything to be discarded should be hidden at the initial landing site.

(4) Sanitizing Clothing. Clothing should be sanitized by removing anything bright or shiny. Evaders should consider camouflaging their clothing (to include boots) just as they would their skin. This is detrimental to the insulative qualities of their clothes but not as much as bullets or prison barbed wire if they are seen.

(5) Camouflaging Clothing:

(k) One principle of camouflage is to disrupt or conceal uniform color, straight lines, and squares—things rarely found in natural features. If ground-to-air signals are designed to exploit these visual characteristics, then evaders should certainly want to eliminate them.

(l) Evaders should reduce the tone of all equipment by smearing it with camouflage stick, mud, etc., or with whatever is available in a mottled pattern. In some instances equipment will have to be lighter in tone, in others, darker.

(m) To maintain the functional capability of clothing and equipment, it must be kept clean. In some areas, however, these items may be the only natural camouflage material survivors have to work with (such as desert regions). But, in areas where they have access to vegetation and the various dyes which can be made from vegetation, the vegetation should be used. In contrast to substances which soil the material and actually break down the fibers, dyes derived from grasses or plant sap (banana trees, ash trees, etc.) will offer the evader the toning material necessary to break up the solid green of a uniform and leave the fabric grit-free and still able to "breathe." The same saps which produce stains for cloth can be used to discolor metal objects. Banana tree sap, when left on the metal blade of a knife, will produce a blue-black stain which is a permanent discoloration. Trappers still boil their traps in ash tree chips and water to produce the blue-black, rust- inhibiting coloration to the tools of their trade.

(n) All principles and techniques for care and use of clothing and equipment cannot be forgotten or ignored in an evasion situation, although some modifications may be necessary. A number of variables will influence what changes or omissions will be necessary.

(1) All cutting tools must be kept sharp. Evaders should try to coincide these noisy, yet essential, tasks with natural noise in the area (a downpour of rain for instance) or in a protected, noise-dampening area.

(2) Clothing must be kept clean if it is to protect a survivor from a harsh environment. Dirt-clogged, perspiration-soaked fibers will not give the insulating qualities of clean cloth. Clothing can be washed during the downpour of rain or possibly under the cover of darkness in a stream. Convenient, secluded puddles of water may afford the opportunity a survivor needs to clean clothing and equipment properly.

(3) Cooking and eating utensils must be kept clean on the inside to prevent dysentery and diarrhea. Simultaneously, the outsides can be toned down with soil, mud, etc., to camouflage them.

(4) Where metal pieces come into contact with one another, there should be padding between them so they will not inadvertently "clank" together. Evaders should place all items needed for environmental protection in the top of the pack where they will be most readily available. The rest of the gear can be used as padding around metal objects. In this manner, with everything stored inside the bundle (pack), it is secure from loss, damage, and enemy observation, as well as being readily available when needed. Evaders should also remove jewelry, watches, exposed pens, and glasses if possible. If

glasses are required, hat netting or mask may help reduce shine.

(5) An evader's pockets should be secured and all equipment, including dog tags, arranged so that no jingle or rattle sounds are made. This can be done with cloth, vegetation, padding, or tape.

(6) Evaders should minimize the sound of clothing brushing together when the body moves. Moving in a careful manner can decrease this sound. Evaders should remember that camouflaged clothing and equipment alone won't conceal, but it must be used intelligently in accordance with the other principles of camouflage and movement. As one example, even if evaders are perfectly camouflaged for the arctic, there could still be problems. Because snow country is not all white, shadows and dark objects appear darker than usual. A snowsuit cannot conceal the small patches of shadow caused by the human figure, but that is not necessary if the background contains numerous dark areas. If the background does not contain numerous dark areas, maximum use is to be made of snowdrifts and folds in the ground to aid in individual concealment (Figure 17-15).

(7) The concept of blending in with the background is indeed an important one for the evader to understand. One major point in blending with the background is not to show a body silhouette.

(8) Losing the body silhouette is done by making use of the shadows in the background.

Figure 17-15. Arctic Travel.

Evaders should be constantly aware of two factors—silhouette and shadow. From a concealment point of view, backgrounds consist of terrain, vegetation, artificial objects, sunlight, shadows, and color. The terrain may be flat and smooth, or it may be wrinkled with gullies, mounds, or rock outcroppings. Vegetation may be dense jungle growth or no more than small patches of desert scrub growth. The size of artificial objects may range from a signpost to a whole city block. There may be many colors in a single background, and they may vary from the almost black of a deep woods to the sand pink of some desert valleys. Blending simply means the matching with as many of these backgrounds as possible and avoiding contrast. If it is necessary for evaders to be positioned in front of a contrasting or fixed background, they must be aware of their position and take cover in the shortest possible time. The next point to which they will move for concealment must be selected in advance and reached as quickly as possible.

(9) As in the daytime, silhouette and background at night are still the vital elements in concealment (Figure 17-16). A silhouette is always black against a night sky, and care must be taken at night to keep off the skyline. On moonlit nights, the same precautions must be taken as in daylight. It should be remembered that the position of the enemy observer, and not the topographic crest, fixes the skyline. At night, sound is an amplified, revealing signal. Movement must be careful, quiet, and close to the ground.

Figure 17-16. Silhouetting.

If the pop of a flare is heard before the illuminating burst, evaders must drop to the ground instantly and remain motionless. If they are surprised by the light, they must freeze in place with their faces down.

17-8. Concealment in Various Geographic Areas:

a. When not otherwise specified, temperate zone terrain is to be assumed in this section. Desert, snow, and ice areas are mostly barren and concealment may require considerable effort. Jungle and semitropical areas usually afford excellent concealment if the evader employs proper evasion techniques.

b. First, some general observations and rules regarding the addition of vegetation to the uniform and equipment. The cycles of the seasons bring marked changes in vegetation, coloring, and terrain pattern requiring corresponding changes in camouflage. Concealment which is provided in wooded areas during the summer is lost when leaves fall in the autumn. This will create a need for additional camouflage construction. Also, vegetation must be of the variety in the evaders' immediate location. It must be changed if it wilts or the evaders move into a different vegetation zone. Evidence of discarding the old and picking the new should be hidden. The vegetation should not be cut, this will give evidence of human presence.

c. Any type of material indigenous to the locality of the evaders may be classified as natural material. Natural materials consist of foliage, grasses, debris, and earth. These materials match local colors and textures and when properly used are an aid against both direct and indirect observation. The use of natural materials provides the best type of concealment. The chief disadvantage of natural foilage is that it cannot be prepared ahead of time, is not always available in usable types and quantities, wilts after gathering, and must be replaced periodically. Foliage of coniferous trees (evergreens) retains its camouflage qualities for considerable periods, but foliage that sheds leaves will wilt in a day or less, depending on the climate and type of vegetation.

(1) The principal advantage in using live vegetation is its ability to reflect infrared waves and to blend in with surrounding terrain. When vegetation is used as garnishing or screening, it must be replaced with fresh materials before it has wilted sufficiently to change the color or the texture. If vegetation is not maintained, it is ineffective. Thorn bushes, cacti, and other varieties of desert growth retain growing characteristics for long periods after being gathered.

(2) The arrangement of foliage is important. The upper sides of leaves are dark and waxy; the undersides are lighter. In camouflage,

therefore, foliage must be placed as it appears in its natural growing state, top sides of leaves up and tips of branches toward the outside of the leaves (Figure 17-17).

(3) Foliage gathered by survivors must be matched to existing foliage. For example, foliage from trees that shed leaves must not be used in an area where only evergreens are growing. Foliage with leaves that feel leathery and tough should be chosen. Branches grow in irregular bunches and, when used for camouflage, must be placed in the same way. When branches are placed to break up the regular, straight lines of an object, only enough branches to do this should be used. The evader must adopt principles that apply and know that the enemy also applies these principles.

(4) When vegetation is applied to the body or equipment of evaders, it must be secured to the clothing or equipment in such a way that:

Figure 17-17. Dark and Light Leaves.

(a) Any inadvertent movement of material will not attract attention.

(b) It appears to be part of the natural growth of the area; that is, when the evaders stop to hide, the light undersides of the vegetation are only visible from beneath. After evaders complete their camouflage, they should inspect it from the enemy's point of view. If it does not look natural, it should be rearranged or replaced.

(c) It does not fall off at the wrong moment and leave evaders exposed, or it should not show evidence of the evaders' passage through the area.

(5) Too much vegetation can give evaders away.

(6) If cloth material is used like vegetation to break up shape and outline and to help blend in with the environment, there are some points evaders should be aware of. Cloth can be used successfully when wrapped around equipment or designed into loose, irregular shaped clothing or accessories (Figure 17-18).

(g) Some of the materials evaders may use are:

-1. The colors (green, brown, white) of parachute materials plus the harness.

-2. Excess clothing the evader may have prepacked (scarf, bandana, etc.).

-3. Burlap, when found. It is used in battle areas in the form of sandbags.

Figure 17-18. Breakup of Shape and Outline.

(h) Most artificial material is versatile, but it can have drawbacks:

-1. Parachute material, for example, tends to shine and the unraveled edges may leave fine filaments of nylon on the ground as evidence of evaders in an area. Parachute material is very lightweight. A sudden breeze might cause it to move when movement is not desirable.

-2. A white suit of parachute material is excellent for winter evasion over snow and ice environments. The time required to fabricate this suit should be considered. (NOTE: Shadows cast on the snow cannot be camouflaged; they must be masked by terrain or other shadows.)

17-9. Concealment Factor for Areas Other Than Temperate:

a. Desert. Lack of natural concealment, high visibility, and bright tone (smooth texture) all emphasize the need for careful selection of a position for a campsite. Deep shadows in the desert, strict observance of camouflage discipline, and the skillful use of deception and camouflage materials aid in concealing evaders in a desert area.

(1) Deserts are not always flat, single-toned areas. They are sometimes characterized by strong shadows with heavy broken terrain lines and sometimes by a mottled pattern. Each type of desert terrain presents its own problems. When evaders and their shelters are located in the desert, their shadows are inky black and in strong contrast to their surroundings and are extremely conspicuous. To minimize the effect of these shadows when possible, use concealment that is afforded by the shadows of deep gullies, scrub growth, and rocks (Figure 17-19).

(2) Many objects which cannot be concealed from the air can be effectively viewed from the ground. Even though these objects are observed from the air, lack of reference points in the terrain will make them difficult to locate on a map.

Figure 17-19. Concealment by Shadows.

b. Snow and Ice. From the air, snow-covered terrain is an irregular pattern of white, spotted with dark tones produced by objects projecting above the snow, their shadows, and irregularities in the snow-covered surface such as valleys, hummocks, ruts, and tracks. It is necessary, therefore, to make sure dark objects have dark backgrounds for concealment to control the making of tracks in the snow and to maintain the snow cover on camouflaged objects.

(1) Mountain Areas Above Timberline and Arctic Areas. Common characteristics include an almost complete snow cover with a minimum of opportunities for concealment. Only a few dark objects protrude above the snow except for rugged mountain peaks (Figure 17-20).

(2) Mountain Areas Below the Timberline and Subarctic Areas. Common characteristics of these areas are forests, rivers, lakes, and artificial features such as trails and buildings. The appearance of the area is irregular in pattern and variable in tone and texture (Figure 17-21).

(3) Areas Between the Subarctic Zone and the Southern Boundary of the Temperate Zone. These have the same characteristics as mountain areas below the timberline and subarctic areas.

c. Blending With Background in Snow and Ice Terrains. No practical artificial material has yet been developed which will reproduce the texture of snow sufficiently well to be a protection against recognition by aerial observers. Concealment from direct ground observation is

Figure 17-20. Above Timberline.

Figure 17-21. Below Timberline.

relatively successful with the use of white "snowsuits," white pants, and whitewash; these measures offer some protection against aerial detection. (White parachute cloth should also be considered.)

(1) People evading in snow-covered frozen areas should wear a complete white camouflage outfit. A white poncho-like cape can be made easily from parachute material (Figure 17-22).

Figure 17-22. Evading in Snow-Covered Areas.

(2) A pair of white pants will normally be sufficient in a heavily wooded area. However, following or during a heavy snowfall when the trees are well covered with snow, the wearing of a complete white camouflage suit is necessary to blend in with the background. Other equipment, such as packs, should also be covered with white material.

d. Camouflage Checklist. The following checklist can be used to remember ideas concerning camouflage and to determine the completeness of individual camouflage application:

(1) Effective concealment is protection from hostile observation from the ground as well as from the air.

(2) Natural terrain lines are to be used for help in concealing the evader when possible.

(3) Every possible feature of the terrain should be used for concealment.

(4) A silhouette against the sky should be avoided.

(5) Every effort should be made to reduce tone contrast and eliminate shine.

(6) Evaders should be especially careful at night due to infrared and low light detection equipment which may be used by the enemy. Keeping close to the ground and using terrain masking for concealment provide the best protection.

17-10. Camouflage Techniques for Shelter Areas:

a. As used in this section, the word shelter refers to the concept of personal protection synonymous with the terms refuge, haven, or retreat. Readers should not visualize the word "shelter" when mentioned in this text to mean a dwelling traditionally occupied by survivors in nontactical situations; such as, tents, cabins, or other such places of habitation. While it is true that these structures can and do provide safety and relief, it must be understood that where an evasion situation is concerned, concealment, not personal comfort or convenience, will be the primary concern for the evader who seeks "shelter" or sanctuary.

b. Besides resting or sleeping, there are a number of reasons why survivors may need to conceal themselves for varying amounts of time. They may need to take care of problems concerning personal hygiene, adjustment of clothing, maintenance or alteration of camouflage, triangulation for position determination, food and water procurement, etc. Concealment cannot be overstressed in respect to the areas survivors (evaders) may select for shelter. One important reason evaders must be able to select secure areas for refuge is to avoid and prevent detection by the enemy. This is especially important if the haven selected is to be used for resting or sleeping. Anyone who is resting or sleeping will not be totally alert, and added precautions may be necessary to maintain security. Another factor to consider, since evaders are also survivors, is to protect themselves from the elements as much as possible.

c. At no time will evaders be able to safely assume they are free from the threat of either ground or aerial observation. Therefore, not only is the shelter area and type determined by the needs of the moment (enemy presence, etc.), but consideration must also be given to the terrain and climatic conditions of the area. Evaders must constantly be aware of how long they may have to remain in the area and, most importantly, what type of enemy observation may be employed.

d. The shelters may be naturally present, or they may be those which are "assembled" and camouflaged by the evaders. Full use must be made of concealment and camouflage no matter what types of shelter areas are selected. The use of the natural concealment afforded by darkness, wooded areas, trees, bushes, and terrain features are recommended; however, any method used for disguise or hiding from view will increase the chances for success. There is much for evaders to consider concerning the many facets of evasion shelter site selection if they expect to establish and maintain the security of their area (Figure 17-23).

e. Evaders should locate their shelter areas carefully. They should choose areas which are least likely to be searched. They must be in the least obvious locations. The

Figure 17-23. Natural Shelter.

chosen areas should look typical of the whole environment at a distance. They should not be near prominent landmarks. Areas that look bland get a cursory glance. The areas should also be those least likely to be searched; for example, rough terrain and thickly vegetated areas. The shelter sites should also be situated so that in the event of impending discovery, the evaders will be able to depart the area via at least one concealed escape route. The shelter areas should never be in areas which may trap them if the enemy discovers the places of concealment.

f. Evaders should choose natural concealment areas— a "natural shelter." Examples include small, concealed caves, hollow logs, holes or depressions, clumps of trees, or other thick vegetation (tall grass, bamboo, etc.). The site should have as much natural camouflage as possible. There should be cover on all sides; this includes natural formations or vegetation which can also protect evaders from aerial observations. The site should be as concealed as possible with a minimum of work. Sites chosen this way will make concealment easier and require less activity and movement. This is most important if the evaders are close to population centers or if the enemy is present.

g. The evaders should attempt to stay as high as possible and to select concealment sites near the military crest of a hill if cover is available. Noises from ridge to ridge tend to dissipate. Whispers or other sounds made in a valley tend to magnify as listeners get further up a hill. Shelter areas located on a slope are subject to higher daytime and lower nighttime winds, thereby minimizing the chance of detection through the sense of smell.

h. If possible, evaders should be in such a position in the shelter area that shadows will fall over the side of the area throughout the day. This can best be done in heavy brush and timber.

i. Evaders should try to locate alternate entrance and exit routes along small ridges or bumps, ditches, and rocks to keep the ground around the shelter area from becoming worn and forming "paths" to the site. They should avoid staying in one area so long that it has the appearance of being "lived in." Evaders should try to stay away from and out of sight of any open areas; examples are roads and meadows. Several miles distance from those may be desirable.

j. Waterways such as lakes, large rivers, and streams, especially at the junctions, are dangerous places. Power and fence lines or any prominent landmarks may indicate places where people may be. Evaders will want to stay clear of these areas. The enemy may patrol bridges frequently. Evaders should avoid any areas close to population centers. The evaders should be able to observe the enemy and their movements and the surrounding country from this hiding area if at all possible. If any assemblage of camouflage is necessary at the shelter site, evaders should keep in mind that they should always "construct" to blend. They should match the shelter area with natural cover and foliage, remembering that over-camouflage is as bad as no camouflage. Natural materials should be taken from areas of thick growth. Any place from which materials have been taken should be camouflaged. The following is an easy to remember acronym (BLISS) for evasion shelter principles:

B - Blend.
L - Low silhouette.
I - Irregular shape (outline).
S - Small.
S - Secluded location.

k. Other facilities evaders may use, such as latrines, caches, garbage pits, etc., must be located and camouflaged in the same manner as the shelter sites. Evaders should avoid forming a line of installations which lead from point to point to their location. They should dog-leg through ground cover to use concealment to its best advantage. A dry, level sleeping spot is ideal, but the ideal spot to provide non-visibility and comfort may be difficult to find. Evaders must have patience and perseverance to stay hidden until danger has passed or until they are prepared and rested enough to safely move on. They must be constantly on the watch for shelter areas which need little or no improvement for camouflage, protection from the elements, or security.

17-11. Firecraft Under Evasion Conditions:

a. Whether or not to build a fire under evasion conditions is indeed a difficult decision evaders must make at times. Basically, fire should only be used when it is absolutely necessary in a life or death situation. Potential evaders must understand that the use of fire can greatly increase the probability of discovery and subsequent capture.

b. If a fire is required, location, time, selection of tinder, kindling, and fuel, and construction should be major considerations.

(1) Evaders should keep the fire as inconspicuous as possible. The location of an evasion fire is of primary importance. All of the small evasion-type fires must be built in an area where the enemy is least likely to see them. If possible, in hilly terrain with cover, the fire should be built on the side of a ridge (military crest). No matter where the fire is built, it should be as small and smokeless as possible (Figure 17-24).

(2) Fires are easier to disguise and will blend in better during the times of dawn and dusk and during times of bad weather. At these times, there is a haze or vapor trap that hinges in and around hills and depressions and is prevalent on the horizon. Any smoke from the fire will be masked by this haze in the early morning and at sunset. This is the time when the local populace is most likely to have their cooking or heating fires lit. Another method of disguising the smoke from a fire is

Figure 17-24. Fire.

to build it under a tree. The smoke will tend to dissipate as it rises up through the branches, especially if there is thick growth or the boughs are low hanging. If this is not possible, it will be helpful to camouflage the fire with earthen walls, stone fences, bark, brush, or snow mounds to block the light rays and help disperse the smoke.

(3) The best wood to use on an evasion fire is dry, dead hardwood no larger than a pencil with all the bark removed. This wood will produce more heat and burn cleaner with less smoke. Wood that is wet, heavy with pitch, or green will produce large amounts of smoke. When the wood has been gathered, evaders should select small pieces or make small pieces out of the wood collected. Small pieces of wood will burn more rapidly and cleanly thus reducing the chance of smoldering and creating smoke. The wood selected should be stacked so the fire gets plenty of air as ventilation will make the fire burn faster with less smoke.

c. One type of evasion fire which has the capability of being inconspicuous is called the Dakota Hole Fire (Figure 17-25). After selecting a site for the fire hole, a "fireplace" must be prepared. This is done by digging two holes in the ground, one for air or ventilation and the other to actually lay the fire in. These holes should be roughly 8 to 12 inches deep and about 12 inches apart with a wide tunnel dug to connect the bottoms of the holes. The depth of the holes depends on the intended use of the fire. Place

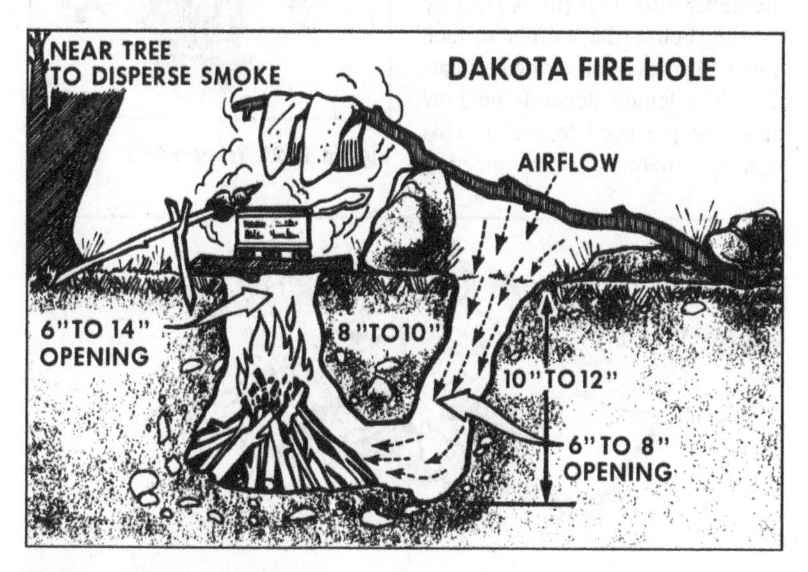

Figure 17-25. Dakota Hole.

dirt on a piece of cloth so it can be used to rapidly extinguish the fire and conceal the fire site. Evasion fires should be small. Evaders who build these fires should strive to keep the flame under the surface of the ground. Initially the fire may appear to be smoking due to the moisture in the ground. At night the area may glow, but there should be no visible flame. A fire built in a hole this way will burn fast as all the heat is concentrated in a small area. This is a good type of fire for a single evader as opposed to many persons taking turns cooking over this one hole. Everything about the evasion situation will have to be examined before deciding which fire configuration would be most useful if a fire must be built.

d. Another good evasion fire is the trench fire. This fire is built by digging below the earth's surface 8 to 12 inches in an elongated pattern. The length depends on how many people need to use it. This is a fire more suited to meeting the needs of a small group. The fire should not crowd either end of the excavation, as it must be able to "draw" an adequate amount of air to help it burn hot and eliminate smoke (Figure 17-26).

e. An evasion fire can also be built just below the ground cover (Figure 17-27). Here the emphasis is on quick concealment of the area. It should be kept in mind, however, that some type of screen

Figure 17-26. Trench Fire.

Figure 17-27. Fire Concealment.

should be built to hide the flames. The small fire is built on the bare ground after a layer of sod or earth has been sliced and rolled back. After use, evaders simply scatter the fire remnants over the bare area and roll or fold the piece of ground back into place.

f. If evaders are in areas where holes can't be dug or sod can't be lifted, they will have to settle for some type of screen around the fire. They should also keep the fire small and finish using it quickly.

g. All traces of the fire should be removed. Unburned firewood should be buried. Holes should be totally filled in. Placing the soil on a holder, such as a map or piece of equipment, will aid in replacing it. This way there will be no leftover dirt patches on the ground after the holes were filled in. Once evaders feel all available measures have been taken to obliterate any leftover evidence, they should move out of the area if possible. Since evaders can never be positive the fire wasn't detected, they must assume it was spotted and take all necessary precautions.

17-12. Sustenance for Evasion:

a. As previously stated in this section, not only do evaders face the problem of remaining undetected by the enemy, they must also have the knowledge which will enable them to "live off the land" as they evade. They must be prepared to use a wide variety of both "wild" and domestic food sources, obtain water from different sources, and use many methods for preparing and preserving this food and water.

b. No matter what the circumstances of the evasion situation are, evaders should never miss an opportunity to obtain food. Ordinarily food will be obtained from wild plant and animal sources. If possible, evaders should stay away from domestic plants (crops) and animals. Using wild animals and plant sources for food will reduce the probability of capture.

c. Animal foods are a prime source of sustenance for evaders, having more nutritional value for their weight than do plant foods. Evaders may obtain enough animal food in one place to last for several days while they travel.

d. There are several ways which evaders may procure animal foods (trapping or snaring, fishing, hunting, and poisoning). A few modifications should be considered regarding the use of traps and snares in a tactical situation even though the same basic principles apply in both tactical and nontactical environments.

(1) Because small game is more abundant than is large game in most areas, evaders should confine traps and snares to the pursuit of small game. There are other advantages to restricting trap and snare size. Evaders will find it is easier to conceal a small trap from

the enemy. Small animals make less noise and create less disturbance of the area when caught.

(2) Conversely, there are also a few disadvantages pertaining to the use of snares or traps during evasion episodes. Two disadvantages are: Evaders must remain in one place while the snares are working, and there may be some disturbance of the area where materials have been removed.

e. Fishing is another effective means of procuring animal food. Fish are normally easy for evaders to catch, and they are easy to cool.

(1) There are several methods which evaders may use to catch fish. A simple hook and line is one of these methods; another is a "trotline." Evaders may construct a trotline by fixing numerous hooks to a pole and by sliding it into the water from a place of concealment (Figure 17-28).

(2) Nets and traps may also be used; however, they should be set below the water line to avoid detection. Spearing is another option. Here again, exposure in open waterways can be very dangerous to the evader.

(3) "Tickling" the fish (Figures 17-29, 17-30, and 17-31) is also effective if evaders can remain concealed. This method requires no equipment to be successful. The main disadvantage to fishing is people live by water bodies and travel on them. This greatly increases the chances of being detected. (NOTE: Caution should be

Figure 17-28. Fishing.

Figure 17-29. Fish Tickling - A.

Figure 17-30. Fish Tickling - B.

Figure 17-31. Fish Tickling - C.

used when tickling fish in areas with carnivorous fish or reptiles.)

f. Weapons may be used by evaders to procure animals. The best weapons are those which can be operated silently, such as a blowgun, slingshot, bow and arrow, rock, club or spear. These should be used primarily against small game. One major advantage of using weapons is game can be taken while evaders travel. Because of noise, firearms should never be used in an evasion situation.

g. Plant foods are very abundant in many areas of the world. Evaders may be able to procure plant food types that require no cooking. One advantage of procuring plant foods during evasion is that by collecting natural fruits and nuts, evaders can remain deep in unpopulated areas. In some areas, it is possible to find old garden plots where vegetables may be obtained. When possible, select foods which can be eaten raw. (Refer to Chapter 14-Food.)

(1) The disadvantage of plant food procurement is that evaders may not be the only ones looking for food. The natives of the country could also be out looking for food. If natives know of a good area, they may visit that place many times. If evaders have been in the area, their presence could be discovered.

(2) Some other considerations and methods concerning plant food procurement are as follows. Evaders should:

(a) Never take all the plants or fruits from one area.

(b) Pick only a few berries off of any one bush.

(c) When digging plants, take only one plant, then move on some distance before digging up another.

(d) When digging plants from old garden plots, make sure the plot is old. In many countries the people plant their crops and do not return to the plot until harvest time.

(e) Camouflage all signs of presence.

(a) Most domestic foods must be procured by theft which is very dangerous. However, if proper methods are used and the opportunity presents itself, plants and animals may be stolen. The main reason thieves are captured is the boldness they display after several successful thefts. The basic rules of theft are:

(a) If at all possible, the theft should take place at night.

(b) Evaders should thoroughly observe the area of intended theft from a safe vantage point.

(c) Evaders should find the vantage points just before dusk and look the place over to make sure everything is the same as it was the last time a check was made.

(d) Evaders should check for dogs, which could be a big hazard. Barking draws attention; also, some dogs are vicious and can harm evaders. Besides dogs, other animals or fowl can alert the enemy to the evaders' position.

(e) Evaders should never return twice to the scene of a theft.

(f) Every theft should be planned, and after its accomplishment, evaders should leave no evidence of either their presence or the theft itself (Figure 17-32).

(g) Only small amounts should be taken (Figure 17-33).

(4) When evaders find it necessary to take cultivated plant foods, they should never take the complete plant. Taking plants from the inside of the field, not the edge, and leaving the top of plants in place may help conceal the theft.

Figure 17-32. Stealing Vegetables.

Figure 17-33. Stealing Small Amounts of Food.

(h) The rules of theft also apply to taking domesticated animals. The evader should concentrate on animals that don't make much noise. If a choice has to be made as to which animal to steal, evaders should take the smallest one.

(i) Water is very essential, but it can be difficult to acquire.

(1) When procuring water, evaders should try to find small springs or streams well away from populated areas. The enemy knows evaders need water and may check all known water sources. No matter where water is procured, evaders should try to remain completely concealed when doing so. The area around the water source should be observed to make sure it isn't patrolled or watched. While obtaining water and when leaving the area, evaders should conceal any evidence of their presence. Good sources of water include trapped rainwater in holes or depressions and plants which contain water.

(2) The preparation and purification of food and water by cooking is a precaution which should be weighed against the possibility of capture. It might be necessary to eat raw plant and animal foods at times. Some plant foods will require cutting off thorns, peeling off the outer layer, or scraping off fuzz before consumption. Raw animal foods may contain parasites and microorganisms which may not effect evaders for days, weeks, or months, at which time, hopefully, they will be under competent medical care. If the environment in which the parasites live is altered by cooking, cooling, or drying, the organisms

may be killed. Meat can be dried and cooled by cutting into thin strips and air-drying. Salting makes the meat more palatable. If the meat must be cooked, small pieces should be cooked over a small hot fire built in an unpopulated area. The best methods of preserving food during evasion are drying or freezing.

(3) In an evasion situation, boiling water for purification should not be used as a method except as a last resort. Iodine tablets are the best method of purification. If evaders do not have purification tablets, and the danger of the enemy detecting their fires is too great, evaders may have to forego purification. The only problem with this is if water is not purified, it may cause vomiting and(or) diarrhea. These ailments will slow down the evaders and make them susceptible to dehydration. Aeration and filtration may help to some degree and are better than nothing. If water cannot be purified, evaders should at least try to use water sources which are clean, cold, and clear. Rain, snow, or ice should be used if available.

17-13. Encampment Security Systems:

a. When evading for extended periods in enemy-held territories, it becomes essential for evaders to rest. To rest safely, especially if in a group, it is essential to devise and use some sort of early warning system to prevent detection and unexpected enemy infiltration. When establishing an evasion "shelter" area, there are certain things which should be done (day or night) for security purposes (Figure 17-34).

Figure 17-34. Rest Safely.

b. Evaders should scout the area around their encampment for signs of people. They should pay particular attention to crushed grass, broken branches, footprints, cigarette butts, and other discarded trash. These signs may reveal identity, size, direction of travel, and time of passage of an enemy force. If large numbers of these signs are present, the evader should consider moving to a more secure area.

c. Once the camp area has been determined to be fairly secure, some type of alarm system must be devised. For a lone evader, this may consist of actually constructing wire or line with sound-producing devices attached. However, this system works for the enemy as well and may prove to alert enemy forces to the evader's presence. A lone evader should use the natural alarm system available. Disturbances in animal life around an evader may indicate enemy activity in the area. Group situations may allow for more security. Two or more evaders may use lookout(s) or "scout(s)" at observation posts strategically located around the encampment.

d. Readiness is another aspect of security. The evader should be aware that, at any time, the shelter area may be overrun, ambushed, or security compromised, making it necessary to vacate the area. If evaders are in a group and future group travel is desired, it is essential that everyone in the group knows and memorizes certain things; such as, compass headings or direction of travel, routes of travel, destination descriptions, and rally points (locations where evaders regroup after separation). Alternate points must be designated in case the original cannot be reached or if it is compromised by enemy activity.

(1) Once everyone in the evader's group has reached the final destination, alternate point, or rally point, a new emergency evacuation and rendezvous plan must be established.

(2) Evaders should always be aware of the next rally point, its location, and direction. These places, which provide concealment and cover, should be designated along the route in case an enemy raid or ambush scatters the group. There should be a rally point for every stage of the journey. Even when approaching the supposed "final destination" of the day, evaders should have an evacuation plan ready.

e. Maintaining silence is a very important aspect of security. It is essential to be able to communicate with individual group members and scouts so that everyone is aware of what is going on at all times during evasion. Hand signals are the best method of communication during evasion as they are silent and easily understood. Instructions and commands which must be conveyed throughout the entire group are: (Figure 17-35 for examples of hand signals.)

Figure 17-35. Hand Signals.

(1) Freeze.
(2) Listen.
(3) Take cover.
(4) Enemy in sight.
(5) Rally.
(6) All clear.
(7) Right.
(8) Left.

17-14. Evasion Movement:

a. During evasion travel, the evader is probably most susceptible to capture. Many evaders have been captured as a direct result of their failure to use proper evasion movement techniques. Evasion movement is the action of a person who, through training, preparation, and application of natural intelligence, avoids capture and contact with hostiles, both military and civilian. Not only is total avoidance of the enemy desirable for evaders, it is equally important for evaders not to have their presence in any enemy controlled area even suspected. A fleeting shadow, an inopportune movement or sound, and an improper route selection are among a number of things which can compromise security, reveal the presence of evaders, and lead to capture.

b. One evasion situation will not be identical to another. There are, however, general rules which apply to most circumstances. These rules, carefully observed, will enhance the evaders' chances of returning to friendly control.

(1) Evasion begins even before a crewmember leaves an aircraft over enemy territory. Two factors which are essential to successful evasion and return are opportuntiy and motivation. Pre-mission preparation and knowledge of areas of concern are very important to a crewmember. Pre-mission knowledge gained must be based on the most current information available through command area briefings, area studies, and intelligence briefings.

(2) Some areas of interest to the prospective evader are:

(a) Topography and Terrain. An aircrew member should know the physical features and characteristics, possible barriers, best areas for travel, availability of rescue, and the type of air or ground recovery possible. A future evader should also know the requirements for long-term unassisted survival in the area of operation.

(b) Climate. The typical weather conditions and variations should be known to aid in evasion efforts.

(c) People. A very critical consideration may be knowing the people in the area. From ethnic and cultural briefs read before the mission, crewmembers should familiarize themselves with the behavior, character, customs, and habits of the people. It may, at times during the evasion episode, be necessary to emulate the natives in these respects.

(d) Equipment. Aircrew members who would hope to be successful evaders should be thoroughly familiar with all of their equipment and know where it is located. They should also preplan what equipment should be retained and what should be left behind under certain conditions of an evasion travel situation.

(3) Before addressing evasion movement techniques, it is appropriate to go over some of the factors which influence an evader's decision to travel.

(a) The first few minutes after entering enemy or unfriendly territory is usually the most critical period for the evader. The evader must avoid panic and not take any action without thinking. In these circumstances, the evader must try to recall any previous briefings, standard operating procedures, or training and choose a course of action which will most likely result in return to friendly territory.

(b) In those first few minutes, there is a great deal for the downed crewmember to think about. Quick consideration must be given to landmarks, bearings, and distance to friendly forces and from enemy forces, likely location for helicopter landing or pickup, and the initial direction to take for evasion. Knowledge of what to expect is important because when circumstances arise which have been considered in advance, they can be carried out more quickly and easily. Evaders should try to adapt this knowledge and any skills they have to their particular situation. Flexibility is most important, as there are no hard and fast rules governing what may happen in an evasion experience.

(c) In most evasion situations, evaders will be required to move if for no other reason than to leave the immediate landing area when pickup is not imminent. Because any movement has the inherent risk of exposure, some specific principles and practices must be observed. Periods of travel are the phases of evasion when evaders are most vulnerable. Many evaders have been captured because they followed the easier or shorter route and failed to employ simple techniques such as watching and listening frequently and seeking concealment sites.

Figure 17-36. Viewing Terrain - A.

17-15. Searching Terrain:

a. Evaders should visually survey the surrounding terrain from an area of concealment to determine if the route of travel is a safe one. Evaders should first make a quick overall search for obvious signs of any presence such as unnatural colors, outlines, or movement. This can be done by first looking straight down the center of the area they are observing, starting just in front of their position, and then raising their eyes quickly to the maximum distance they wish to observe. If the area is a wide one, evaders may wish to subdivide it as shown in Figure 17-36. Now all areas may be covered as follows: First, by searching the ground next to them. A strip about 6 feet deep should be looked at first. They may search it by looking from right to left parallel to their front. Secondly, by searching from left to right over a second strip farther out, but overlapping the first strip. Searching the terrain in this manner should continue until the entire area has been studied. When a suspicious spot has been located, evaders should stop and search it thoroughly (Figure 17-37).

Figure 17-37. Viewing Terrain - B.

b. The evader must question the movement:

(1) Is the enemy searching for the evader?

(2) What is the evader's present location?

(3) Are chances for rescue better in some other place?

(4) What type of concealment can be afforded in the present location?

(5) Where is the enemy located relative to the evader's position?

c. Having considered the necessity and risks of travel, evaders must:

(1) Orient themselves.

(2) Select a destination, alternates, and the best route.

(3) Have an alternate plan to cover all foreseeable events.

d. Cautious execution of plans cannot be overemphasized since capture of evaders has generally been due to one or more of the following reasons:

(1) Unfamiliarity with emergency equipment.

(2) Walking on roads or paths.

(3) Inefficient or insufficient camouflage.

(4) Lack of patience when pinned down.

(5) Noise or movement or reflection of equipment.

(6) Failure to have plans if surprised by the enemy.

(7) Failure to read signs of enemy presence.

(8) Failure to check and recheck course.

(9) Failure to stop, look, and listen frequently.

(10) Neglecting safety measures when crossing roads, fences, and streams.

(11) Leaving tracking signs behind.

(12) Underestimating time required to cover distance under varying conditions.

e. Evaders should understand progress on the ground is measured in stopover points reached. Speed and distance are of secondary importance. Evaders should not let failure to meet a precise schedule inhibit their use of a plan.

17-16. Movement Techniques Which Limit the Potential for Detection of an Evader (Single).

a. Evaders should constantly be on the lookout for signs of enemy presence. They should look for signs of passage of groups, such as crushed grass, broken branches, footprints, and cigarette butts or other discarded trash. These may reveal identity, size, direction of travel, and time of passage (Figure 17-38).

(1) Workers in fields and normal activities in villages may indicate absence of the enemy.

(2) The absence of children in a village is an indication they may have been hidden to protect them from action which may be about to take place.

(3) The absence of young men in a village is an indication the village may be controlled by the enemy.

b. A knowledge of enemy signaling devices is very helpful. Those listed below are examples of

Figure 17-38. Signs of Passage.

some used by communist guerrillas in Southeast Asia.

(1) A farm cart moving at night shows one lantern to indicate no government troops are close behind.

(2) A worker in the fields stops to put on or take off a shirt. Either act can signal the approach of government troops. This is relayed by other informers.

(3) A villager fishing in a rice paddy holds a fishing pole out straight to signal all clear; up at an angle to signal the troops are approaching.

c. The times evaders choose to travel are as critical as the routes they select. If possible, evaders should try to make use of the cover of darkness. Darkness provides concealment and in some cases there is also less enemy traffic. If, out of military necessity, the enemy is active during the hours of darkness, evaders may then find it wise to move in the early morning or late afternoon. Night movement is slower, more demanding, and more detailed than daylight movement; but it can be done. The alternative to night movement might be capture, imprisonment, and death. Evaders should then consider traveling under the cover of darkness first. However, if the enemy knows the evaders' position, or if the other factors dictate (terrain, vegetation, navigation consideration, etc.), other choices may have to be made. If travel is to be done during darkness, the terrain to be traversed should be observed during daylight if possible. While observing the area to be traveled, evaders should give attention to areas offering possible concealment as well as the location of obstacles they may encounter on their route. If the evader has a map, a detailed study of it should be made. However, it should be remembered that natural terrain features change with time of day and time of year. Certain features (ditches, roads, burned-off areas, etc.) may not be on the evaders' map if they are new. Such pretravel reconnoitering of an area will give evaders a head start on knowing how to adapt their travel movements and camouflage necessary from point to point.

d. Evaders should try to memorize the routes they will take and the compass headings to their destinations. This information should not be written down on the map or on other pieces of material.

e. If traveling through hilly country that provides cover, the military crest should be used as it may be the safest route. An evader's route should avoid game trails and human paths on the tops of ridges. The chance of encountering the enemy is greater on the tops of ridges. The risk of silhouetting during both day and night is also increased (Figure 17-39).

(1) When it is necessary to cross the skyline at a high point in the terrain, an evader should crawl to it and approach the crest slowly using all natural concealment possible. How the skyline is to be crossed depends on whether it is likely the skyline at that point is under hostile observation. Evaders may never be certain any area is not under observation. When a choice of position is possible, the skyline

Figure 17-39. Military Crest.

is to be crossed at a point of irregular shapes such as rocks, debris, bushes, fence lines, etc.

(2) Another important point about moving along the military crest is that it is easier to evade the enemy on the side of the ridge than it is to lose sight of the enemy on the top of a ridge or in the valley below. Evasion along the side of a hill will afford a better chance for evaders to reach sites which are suitable for air recovery.

f. Evaders should move slowly, stopping and listening every few paces. Additionally, they should not make noise and should take advantage of all cover to avoid revealing themselves. If spotted, they should leave the area quickly by moving in a zigzag route to their goal if at all possible. If the enemy finds evidence of the evaders' passage, their route may help confuse the pursuer as to direction and goal. Background noise can be either a help or hindrance to those who are trying to move quietly—both the evaders and the enemy. Sudden bird and animal cries, or their absence, may alert evaders to the presence of the enemy, but those same signals can also warn the enemy of an approaching or fleeing evader. Sudden movement of birds or animals is also something to look out for.

g. The following are some techniques of limiting or concealing evidence of travel. Evaders should:

(1) Avoid disturbing any vegetation above knee level. Evaders should not grab at or break off branches, leaves, or tall grass.

(2) Glide gently through tall grass or brush. Avoid using thrashing movements. A walking stick may be used to part the vegetation in front, and then it can be used behind to push the vegetation back to its original position. The best time to move is when the wind is blowing the grass.

(3) Realize that grabbing small trees or brush as handholds may scuff the bark at eye level. The movement of trees can be spotted very easily from a distance. In snow country it may mark a path of snowless vegetation that can be spotted when tracks cannot be.

(4) Select firm footing and place the feet lightly but squarely on the surface avoiding the following:

(a) Overturning duff, rocks, and sticks.

(b) Scuffing bark on logs and sticks.

(c) Making noise by breaking sticks.

(d) Slipping, which may make the noise of falling.

(e) Mangling of low grass and bushes that would normally spring back.

h. Evaders can mask their tracks in snow by:

(1) Using a zigzag route from one point of concealment to the next and, when possible, placing the unavoidable tracks in the

shadows of trees, brush, and along downed logs and snowdrifts.

(2) Restricting movement before and during snowfall so tracks will fill in. This may be the only safe time to cross roads, frozen rivers, etc.

(3) Traveling during and after windy periods when wind blows clumps of snow from trees, creating a pockmarked surface which may hide footprints.

(4) Remembering that snowshoes leave bigger tracks, but they are shallower tracks which are more likely to fill in during heavy snowfall. Branches or bough snowshoes make less discernible prints.

i. Evaders' tracks in sand, dust, or loose soil should be avoided or else marked by:

(1) Using a zigzag route, making use of shadows, rocks, or vegetation to walk on to mask or prevent tracks.

(2) Wrapping cloth material loosely about the feet, this makes tracks less obvious.

j. By moving before or during wet or windy weather, evaders may find that their tracks are obliterated or worn down by the elements. Along roadways, evaders should be particularly cautious about leaving their tracks in the soft soil to the side of the road. They should step on sticks, rocks, or vegetation. They shouldn't leave tracks unless there are already tracks of natives on the road and their tracks can be made to look like the existing ones (small bare feet, tire sandals, or enemy footgear). Rolling across the road is a method of avoiding tracks. Walking in wheel ruts with the toes parallel to the road will help conceal tracks. If the road surface is dry, sand, dust, or soil tracks may be "brushed" out to make them look old or will help the wind erase them more quickly. This must be done lightly, however, so the tracks do not look as if they have been deliberately swept over. Mud will retain footprints unless the mud is shallow and there is a heavy rain. Evaders should try to go around these areas.

k. Many principles and techniques which work for the individual are also appropriate for use in groups of evaders. While it is not true that the only safe way to evade is individually, there is a certain danger in moving with a group.

(1) It is generally not advisable to travel in a group larger than three. If possible, the senior person should divide the group into pairs. Paired individuals must be compatible since any disagreement may prove fatal during the evasion process. Group travel can be advantageous because supplementary assistance is available in case of injury, in defense against hostile elements, in travel over rough terrain, and it provides moral support.

(2) In group travel, movement becomes critical. Movement attracts attention.

(3) The intervals and distances between individuals in the group should be made according to the terrain and the time of day. Intervals will probably be greater during the day. The natural but extremely dangerous tendency to "bunch up" is to be avoided when traveling in a group (Figure 17-40).

(4) Under favorable conditions it is possible for the enemy to see 100 yards into open woods. If the undergrowth is light, the route must be farther into the woods, and the interval between evaders must be greater. Added consideration must be taken in deciding whether to travel at night or during the day. The leader will direct and guide the group to and from the best positions. All communications within the group should be made with silent signals only. Security in group evasion is of paramount importance. All members should stay alert. Security posts, lookouts, or guards should be designated for periods of rest or stopping.

(5) Various formations are available for use by the evading group during periods of travel. The group must be flexible and able to adapt to changes in the conditions of the situation. The type of formation may also change with the route. In choosing a formation, the following points should be considered:

(a) Group control and intercommunications.

(b) Security.

(c) Terrain.

(d) Speed in movement.

Figure 17-40. Group Travel.

(e) Visibility.

(f) Weather.

(g) Enemy placement.

(h) The need for dispersion.

(i) Flexibility of change in speed and direction of travel.

(6) A "formation" is merely the formal arrangement of individuals within a group. This formal arrangement is designed to give the greatest dispersion consistent with adequate intercommunication, ease and speed of movement, and flexibility to change direction and speed of travel at a moment's notice; that is, close control. Any arrangement which provides the above advantages is satisfactory. The Army, which constantly moves groups of men of various sizes on various missions, has found the squad file, squad column, and squad line to be satisfactory.

(a) In a squad file, the personnel are arranged in a single file, or one directly behind the other, at different distances. The distance will vary depending on need for security, terrain, visibility, group control, etc. It is primarily used when moving over terrain which is so restrictive that the squad cannot adopt any other formation. It is also used when visibility is poor and squad control becomes difficult. When people are in a squad file, it is easy to control the group and provide maximum observation of the flanks. This is a fast way to travel, especially in the snow.

-1. However, there are disadvantages to this type of movement. A major one is that this formation is visually eye-catching. All the noise of the group is also concentrated in one place. This type of formation is easily defined and infiltrated. Depending on how many people are in the group, the area they walk through can become very packed down and easily detected by the enemy.

-2. If the group in the squad file is small, some members may have to double up on jobs; however, the "point" or lookout (scout) in the lead should only perform that job (Figure 17-41).

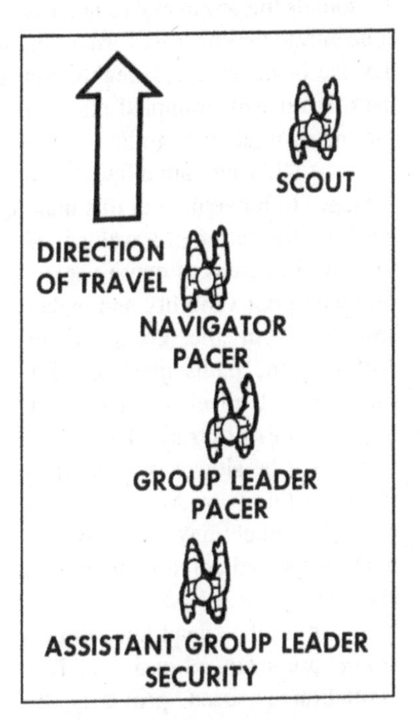

Figure 17-41. File Formation.

(b) In the squad column, personnel are arranged in two files. The personnel are more closely controlled and yet maneuverable in any direction. There is greater dispersion with reasonable control for all-round security. It is used when terrain and visibility are less restrictive because it provides the best means of moving armed personnel into dispersed all-round security. It is easy to change into either the file or the line.

-1. There are many advantages to this type of formation, a major one being a greater dispersal of personnel. Visually, this way of moving is less eye-catching, and the sounds the group makes are less concentrated. With this formation, the rate of travel is reasonable, yet there is no well trampled corridor for enemy trackers to follow.

-2. There are a few disadvantages to movement in this manner. This formation of travel is hard to control in areas of dense vegetation with poor visibility and makes straying from the group likely. Although the paths may be faint, this mode of travel will also form a large number of trails. The rate of travel will be slower overall when traveling this way. An example of how personnel may be dispersed using the squad column methods is shown in figure 17-42.

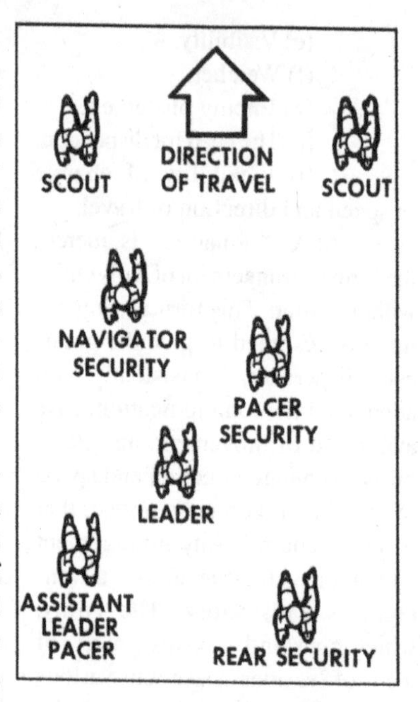

Figure 17-42. Squad Formation.

(c) In the squad line, all personnel are arranged in a line. This formation is used primarily by the Army as an assault formation because it is best for short, fast movements.

-1. The advantage of this formation is that it is the quickest way to cross such obstacles as roads, fences, and small open spaces. It provides for tight control of individual movement while providing security for short-distance moves.

-2. The disadvantages are that there are extreme communication and control problems. Some personnel in this formation may be forced to traverse undesirable rough terrain in contrast to the other two formations. Figure 17-43 illustrates the organization of personnel in this type formation. The speed at

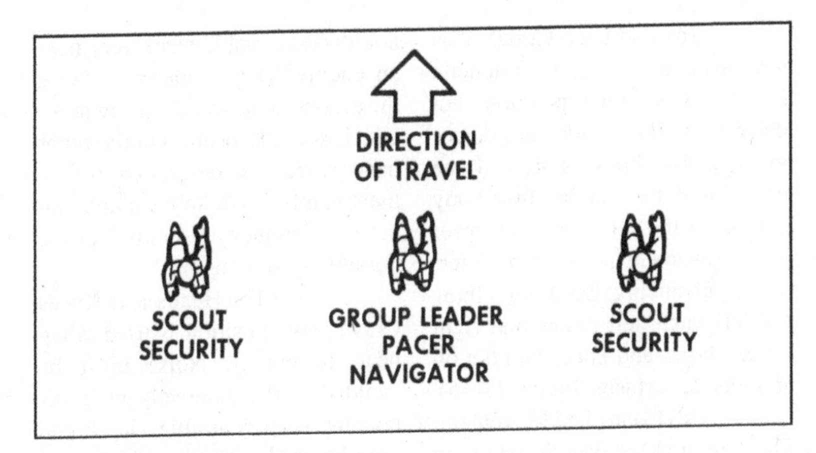

Figure 17-43. Squad Line.

which these formations will progress will vary with terrain, light, cover, enemy presence, health of personnel, etc.

1. Regardless of the formation used, the evader should pay particular attention to the technique used to travel. Knowing how to walk or crawl may make the difference between success and failure.

(1) Correct walking techniques can provide safety and security to the evader. Solid footing can be maintained by keeping the weight totally on one foot when stepping, raising the other foot high to clear brush, grass, or other natural obstacles, and gently letting the foot down, toe first. Feel with the toe to pick a good spot—solid and free of noisy materials. Lower the heel only after finding a solid spot with the toe. Shift the weight and balance in front of the lowered foot and continue. Take short steps to avoid losing balance. When vision is impaired at night, a wand probe, or staff, should be used. By moving these aids in a figure-eight motion from near the ground to head height, obstructions may be felt.

(2) Another method of movement is by crawling. Crawling is useful when a low silhouette is required. There are times when evaders must move with their bodies close to the ground to avoid being seen and to penetrate some obstacles. There are three ways to do this: the low crawl, the high crawl, and the hands-and-knees position. Evaders should use the method best suited to the condition of visibility, ground cover, concealment available, and speed required.

(a) The Low Crawl. This can be done either on the stomach or back, depending on the requirement. The body is kept flat and movement is made by moving the arms and legs over the ground (Figure 17-44).

(b) The High Crawl. This is a position of higher silhouette than the low crawl position, but lower than the hands and knees position. The body is free of the ground with the weight of the body resting on the forearms and lower legs. Movement is made by alternately advancing the right elbow and left knee, left elbow and right knee, elbows and knees laid flat on their inside surfaces (Figure 17-45).

(c) Controlled Movement. The low crawl and high crawl are not always suitable when very near an enemy. They sometimes cause the evader to make a shuffling noise which is easily heard. On the other hand, carefully controlled movement can be made to be silent, and these techniques present the lowest possible silhouettes.

(d) The Hands-and-Knees Crawl. This position is used when near an enemy. Noise must be avoided, and a relatively high silhouette is permissible. It should only be used when there is enough

Figure 17-44. Low Crawl.

Figure 17-45. High Crawl.

ground cover and concealment to hide the higher silhouette involved (Figure 17-46).

Figure 17-46. Hands-and-Knees Crawl.

m. If evaders are moving as a group and are forced to disperse, being able to account for everyone after regrouping is important. Likely locations for rallying points are selected during map study or reconnaissance.

(1) Selecting a Rallying Point. The group leader must:

(a) Always select an initial rallying point. If a suitable area for this point is not found during map study or reconnaissance, the leader can select it by grid coordinates or in relation to terrain features.

(b) Select likely locations for rallying points en route.

(c) Plan for the selection and designation of additional rallying points en route as the patrol reaches suitable locations.

(d) Plan for the selection of rallying points on both near and far sides of danger areas which cannot be bypassed, such as trails and streams. This may be done by planning that, if good locations are not available, rallying points will be designated in relation to the danger area; for example, "...50 yards this side of the trail," or "...50 yards beyond the stream."

(2) Use of Rallying Points. If dispersed by enemy activity or through accidental separation, each evader in the group should be prepared to evade, individually, to the regrouping (rallying) point to arrive at a predesignated time. If this is not possible, the individual will become a "lone evader." The group should not make any effort to locate someone not reaching the rally point. The group should formulate a new plan with new rallying points and clear the area.

(a) Rallying points should be changed or updated as they are passed. Points should not be directly on the line (route) of travel. By selecting points offline, the job of searchers (trackers) is made more difficult and the chance of being "headed off" or "blind stalked" by the enemy is reduced.

(b) If the group is dispersed between rallying points en route, the group rallies at the last rallying point or at the next selected rallying point. The group leader announces the decision at each rallying point as to which point the group will rally.

(3) Actions at Rallying Points. Actions to be taken at rallying points must be planned in detail. Plans for actions at the initial rallying point and rallying points en route must provide for the continuation of the group as long as there is a reasonable chance to evade as a group. An example of a plan would be for the group to wait for a specified period, after which the senior person present will determine actions to be taken based on personnel and equipment present. Even during movement phases, it is important to be able to check on the presence and status of all group members. A low toned, actual head count starting at the rear of the formation might be one way to do this; hand signals is another. One reason for keeping in touch with everyone is to establish new plans or adjust old ones while moving.

17-17. Barriers to Evasion Movement:

a. Obstacles. Throughout the evasion episode, many obstacles may be encountered which may impede evaders or influence the selection of travel routes. These barriers or obstacles can be divided into natural ones, such as rivers or mountains, and human ones, such as border guards or manmade fences or roads. Some of these obstacles may be helpful while others might be a hinderance.

(1) Rivers and Streams. When crossing rivers and streams, bridges and ferries can seldom be used since the enemy normally establishes checkpoints at these locations. This leaves a choice of fording, swimming, crossing by boat, or using some improvised method (Figure 17-47).

(2) Mountains. In mountainous areas, survival may be the primary concern. It may be neces-

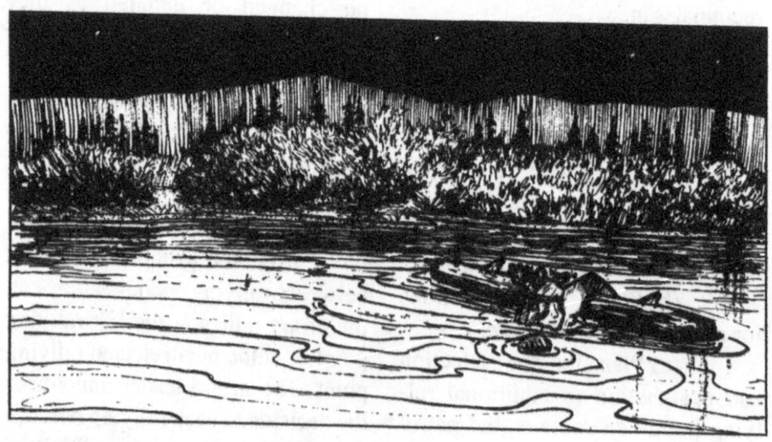

Figure 17-47. River Crossing.

sary for evaders to remain in one location for an extended period of time, perhaps even waiting for the coming of spring before attempting travel. Many mountainous areas, however, are havens which afford cover, water, food, and low population densities. Also, the chances of receiving assistance from people in areas where homes and farms are separated by great distances are more likely. When traveling in mountainous regions, evaders should not forget to use the military crest if concealment is available. In plains areas, evaders should use depressions, drainages, or other low spots to conceal their movements. Route selection should be planned with the utmost care to avoid unnecessary delays caused by cliffs, large bodies of water, and flat areas.

(3) Vegetation. Some swamps, drainage areas, and thickets may be too thick for evaders to penetrate, and may require that a detour or alternate route be used. If the vegetation can be moved through, evaders should take care not to leave evidence of passage by disturbing the growth.

(4) Weather. Weather can sometimes be used to screen evaders from the enemy. Certain weather conditions mask the noise made by traveling. Moving during a rainstorm may erase the footprints left by an evader; but after the rain the soft soil will leave definite signs of passage. Thunder may mask the sounds evaders make, but lightning may cause them to be seen. Snowstorms may be used to cover evaders' signs and sounds, but once the storm is over, evaders must use extreme care not to leave a trail.

b. Artificial Obstacles. Evaders may also encounter a wide variety of artificial obstacles while traveling within enemy territory or when attempting to leave a controlled area. As a general rule, evaders should not attempt to penetrate these obstacles if they can be bypassed. If an analysis of the situation reveals the obstacles cannot be bypassed, evaders must be skilled in the methods and techniques for dealing with specific artificial barriers to evasion. If possible, move to a less fortified (controlled) area or find a better damaged area in the barrier.

(1) Evaders may encounter trip wires. These wires may be attached to pyrotechnics, booby traps, sensor devices, mines, etc. These wires are normally thin, olive green (or other colors that blend with the environment), strong, and extremely difficult to see. A supple piece of wood can be improvised and used as a wand to detect these wires (Figure 17-48).

(a) A tripwire may be set up to be from 1 inch to a number of feet above the ground and to extend any number of feet from the device to which it is attached.

(b) The pressure necessary to activate a sensing device, mine,

Figure 17-48. Trip Wires.

pyrotechnic device, or other trap-associated device to which the wire may be connected can vary from a few ounces to several pounds. This means that the evader must be very careful when attempting to determine the presence of these devices.

(c) Once a tripwire has been detected, evaders should move around the wire if possible. If not possible, they should go either over or under it. They should not tamper with or cut the wire. If one device is discovered, be alert for "backup" devices.

(d) A number of devices activated by tripwires have a combination pressure-release arming mechanism. Cutting the wire or releasing the tension in the wire may activate the device. Some devices are electrically activated when there is a change in the current flowing through the attached wire—either because the wire is cut, in some way grounded, or otherwise altered. Evaders should take extreme care to avoid touching tripwires, but if contact is made, they must try to avoid sufficient pressure for activation.

(2) Illumination flares may also present a problem to evaders. These, of course, can be activated by evaders themselves by a tripwire, by the enemy in the form of electronically activated ground flares, or by flares dropped from an overflying aircraft. Other overhead flares may be fired by mortars, rifles, artillery, and hand projectors.

(a) The illumination flares may burn as bright as 20,000 candlepower and illuminate up to a 300-foot radius in case of a ground flare, or a much larger area in the case of an overhead flare which is lofted or dropped and burns high above the ground.

(b) If evaders hear the launching burst of an overhead flare, they should, if possible, get down while it is rising and remain motionless. If caught in the light of a flare when they blend well with the background, they should freeze in position and not move until the flare goes out. The shadow of a tree will provide some protection. If caught in an open area, they may elect to crouch low or lie on the ground and, as a general rule, should not move after the area is illuminated.

(c) However, if they are caught in the light of a ground flare and their position is such that the risk of remaining is greater than that of moving, they should move quickly out of the area. If within a series of obstacles or an obstacle system, evaders must remember running can be extremely dangerous because of the obstacles in the area and the fact that movement, especially fast movement, catches the eyes of an observer. If it is determined they cannot quickly move out of the area because of possible serious injuries due to existing obstacles or because they may be observed by the enemy, evaders should drop to the ground and conceal themselves as much as possible.

(d) Evaders should remember the light of a flare (either ground or overhead bursts) is temporarily blinding and the eyes should be covered to conserve night vision.

(e) If caught by a flare when actually penetrating an obstacle such as a concertina wire, evaders should get as low as possible, stay still, and cover their eyes. The light of a flare can act to an evader's advantage because the searching enemy will lose its night vision which may add to the evader's chance for success in departing the area after the flare has burned out. The light of a flare also creates very dark shadows which, under some circumstances, can afford good concealment from enemy observation.

(3) Various types of chain and wire fences may hinder the progress of evaders who are moving to the safety of friendly areas.

(a) For indications of electrical fences, evaders should watch for dead animals, insulators, flashes from wires during storms, and short circuits causing sparks (Figure 17-49). A quick simple test can be conducted to determine if a wire is electrified. This test is made

Figure 17-49. Electrified Fences.

by carefully approaching the wire holding a stem of grass or a small, damp stick on the wire. If the wire is charged, a mild shock will result but will not cause injury.

(b) Evaders should use a wand to check for booby traps between strands of multi-strand barbed wire. Generally, they should penetrate the fence under the wire closest to the ground with the body parallel or perpendicular to the wire, depending on circumstances (Figure 17-50). They should lie flat on their backs both to project the lowest possible silhouette and to provide good visibility of the wire against the sky. The probe can sometimes be used to lift the wire. If the lowest wire is close to the ground and is tight, evaders may have to modify their approach to the problem.

(c) The apron fence is penetrated in the same manner as any multi-strand fence—one wire at a time. Evaders should check the area between wires before proceeding.

(d) Concertina wire is penetrated with the body perpendicular to the wire using a probe to lift the wire if it is not secured to the ground (Figure 17-51). If the wire is secured to the ground, the evader can crawl between the loops. If two loops are not separated enough, they may be tied apart using shoe laces, string, suspension line, or strips of cloth. After passing through, the ties should be removed for future use and to erase evidence of travel.

(e) Chain link fences should be avoided completely if at all possible. These fences are usually found only in highly sensitive zones. This means the area is probably more highly guarded and patrolled than other areas. There also may be other traps or devices

Figure 17-50. Penetrating Wire.

Figure 17-51. Penetrating Concertina Wire.

installed. The fence may also be electrified. If the fence must be penetrated, the evader should go under it if possible (Figure 17-52). If digging is required, the soil should be placed on the opposite side so it can be re-placed to remove evidence. Climbing the fence is recommended only as a last resort.

(f) Evaders may encounter rail and split-rail fences while evading or escaping. The fences are penetrated by going under or between lower rails if possible. If not, evaders should go over at the lowest point, projecting as low a silhouette as possible (Figure 17-53). They should check between the rails and on the other side of the fence to detect tripwires or booby traps. Firmness of the ground should be checked on both sides of the fence. The body should be parallel to the fence before penetration.

(4) Raked or plowed areas may be found in areas of both low and high density security. If such an area is encountered, evaders should roll across the area, after making sure it is not a mine field, to avoid leaving footprints; or they may side-step, walk backwards, or brush out footprints. Any of the above may be

Figure 17-52. Penetrating Chain Link Fence.

Figure 17-53. Penetrating Log Fence.

Figure 17-54. Common Barriers.

done when it is a requirement not to leave clear-cut evidence of the direction of movement.

(5) Roads are common barriers to evasion and escape. When roads are encountered, evaders should closely observe the road from concealment to determine enemy travel patterns (Figure 17-54). Crossing from

points offering best concealment such as bushes, shadows, etc. is best (Figure 17-55). Evaders should cross at straight stretches of road in open country and on the inside of curves in hilly or wooded areas. This allows the evader to see in both directions so the chances of being spotted or surprised in the open is minimized. Avoid leaving tracks both in the road and on the shoulder of the road.

(6) Culverts and drains offer excellent means of crossing a road unobserved (Figure17-56).

Figure 17-55. Crossing Roads,

Figure 17-56. Crossing at Culverts.

Figure 17-57. Crawling Over Railroad Tracks.

(7) Railroad tracks often lie in the path of evaders. If so, evaders should use the same procedures for observation as for roads. If it is determined that tracks are patrolled, a check should be made for booby traps and tripwires between tracks. Aligning the body parallel to the tracks with face down next to the first track, evaders should carefully move across the first track in a semi-pushup motion, repeating for the second track and sub-sequent sets of tracks (Figure 17-57). If there is a third rail, they should avoid touching it as it could be electrified. Sound detectors can also be attached to the tracks and can reveal any crossing if a track is touched. If determined from observation that the tracks are not patrolled, evaders should cross in a normal walking or hands-and-knees manner, attempting to attract as little attention as possible. Evaders should try to keep their hands and feet on the railroad ties to prevent leaving foot or hand prints in the adjacent soil or gravel.

(8) Deep ditches (such as tank traps or natural drainages) may be obstacles with which evaders must deal. Ditches should be entered feet first to avoid injury to the head or upper torso should there be large rocks, barbed wire, or other hazards at the bottom (Figure 17-58). Using a wand to detect tripwires and booby traps in the ditch and on the sides is highly recommended. Maintaining a low silhouette upon exiting the ditch is imperative.

(9) Open terrain complicated by guard towers or walking patrols is a definite hazard to evaders. These areas should be avoided if possible. If it becomes necessary to traverse open terrain or come near guard towers, evaders should

stay low to the ground and, when possible, travel at night or during inclement weather. Use terrain masking since night vision devices may be used near border areas (Figure 17-59).

Figure 17-58. Climbing Down Cliff.

(10) The problem of crossing areas which have been contaminated as a result of enemy or friendly NBC operations may arise. Chemical contamination should be suspected when the following are observed (Figure 17-60).

(a) Shell craters with liquid in the bottom.

(b) Liquid droplets on vegetation.

(c) Water with "film" on the surface.

(d) Unexplained dead animals.

(e) Unseasonal discoloration of vegetation.

NOTE: Without protective clothing, mask, and accessories, evaders should bypass contaminated areas if possible.

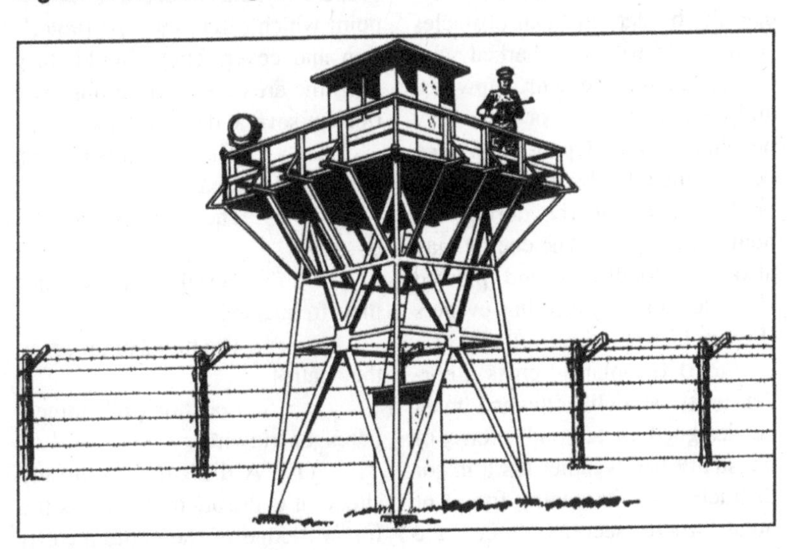

Figure 17-59. Open Terrain.

Figure 17-60. Contaminated Areas.

(11) CAUTION: Stay away from borders unless absolutely necessary. The crossing of one or more borders presents a major problem. These areas may be located in any type of terrain.

(a) In areas where there is no well-defined terrain feature to indicate the border, artificial obstacles such as electrified or barbed wire fences, augmented with tripwires, anti-personnel mines, or flares may be encountered. Open areas may be patrolled by humans or dogs, or both, particularly during the hours of darkness. The enemy may also use floodlights and plowed strips as aids to detecting evaders (Figure 17-61).

(b) The plan to cross a border must be deliberate and must be designed to take advantage of unusually bad weather (as a major distraction to the enemy force) or areas where security forces are overextended. These areas are usually found where there are natural obstacles.

(c) Crossings should be made at night, when possible, in battle-damaged areas. If it is necessary to cross during daylight hours, evaders should select a crossing point which offers the best protection and cover. They should then keep the area under close observation for several days to determine:

-1. The number of guards in the area.

-2. The manner of their posting.

-3. Aerial patrols and their frequency.

-4. The limits of the areas they patrol.

-5. Location of mines, flares, or tripwires.

(12) A difficult task in any situation is the attempt to cross the forward edge of the battle area. If

Figure 17-61. Border Crossings.

unable to determine the general direction to friendly lines, evaders should remain in position and observe the movement of enemy military forces or supplies, the noise and flashes of the battle area, or the orientation of enemy artillery. After arriving in the combat zone, evaders should select a concealed position from which as much of the battle area as possible may be observed. They can then select a route and critical terrain features on which they can guide when infiltrating back to friendly positions under the cover of darkness. Several alternate routes should be selected with care to avoid "easy" approaches to friendly lines which are more likely to be covered by friendly fire and enemy patrols. If in uniform, select exposure time during daylight hours and be close enough to be easily recognized by friendly troops.

(13) Evaders should also watch out for friendly patrols. If a patrol is spotted, evaders should remain in position and allow the patrol to approach them. When the patrol is close enough to recognize them, evaders should have a white cloth displayed before the patrol gets close enough to see the movement. A patrol may fire at any movement. Shouting or calling out to them jeopardizes both the patrol and the evader. Evaders should stand silently with hands over their head and legs apart so their silhouettes are not threatening. If evaders elect not to make contact with a patrol, they should, if possible, observe their route and approach friendly lines at approximately the same location. This may enable

them to avoid mine fields and booby traps. (NOTE: The practice of following any patrol is extremely dangerous. The last persons in line are charged with security, and anyone following them would be considered hostile and eliminated.)

(14) If unable to contact a friendly patrol, the only alternative for evaders may be to make a direct approach of front-line positions. This will require them to crawl through the enemy's forward position near forward friendly elements. This action should be done during the hours of darkness. Once near friendly lines, however, evaders should not attempt to make contact until there is sufficient light for them to be recognized.

(15) In dogs, the ability to perceive odors is much greater than that of humans.

(a) In this portion, the term dog is meant to describe only the animals specifically trained in the areas of patrolling, guarding, and searching. Since dogs are basically odor-seeking animals, anything developed to work against their odor-seeking capabilities is worthy of experimentation for survivors (evaders).

(b) A problem which must be considered is evaders will not be working solely against a dog, but against a dog handler as well. There is no simple, sure method of evading a dog. Some possible means which could be tried by evaders are:

-1. Dogs detect the fatty acids in dead-skin cells that humans shed by the thousands every day. Scenting dogs may be distracted by scattering an irritant such as pepper behind the evader or traveling across an asphalt road on a hot day.

-2. Dogs track better when the weather is humid and the air is still—there is less evaporation and dissipation of odor.

-3. If evaders know they are being followed by dogs, either from the landing site or as a result from an escape, they should try to use water to conceal their tracks and eliminate their scent.

-4. If there is a choice of terrain and it is possible to travel on a hard surface, evaders should do so rather than travel on soft ground.

-5. Evaders should always attempt to move downwind of a dog. This should be attempted when they are traveling in open country or penetrating obstacles such as dog guard posts or border areas. If penetration of obstacles or escape is planned (after careful location and study of guard and dog areas of responsibility and their methods), evaders should select a time for movement when noise will distract the dog to a point away from the planned maneuver.

-6. If evaders are physically capable, they should attempt to maintain the maximum distance possible from dogs. Moving fast through rugged terrain will slow and probably defeat the handlers of dogs. Here evaders must choose between making mistakes in travel techniques while evading or being caught by dogs if they don't move fast enough.

17-18. Evasion Aids:

a. Survival Kits. Personnel may sometimes find it practical to devise and carry compact personal survival kits to complement issued survival equipment. If E&E kits are provided, potential evaders should be familiar with them, their uses, and their limitations.

b. Maps. Any maps of the area in an evader's possession should not be marked. A marked map in enemy hands can lead to the compromise of people and locations where assistance was given. Evaders should be wary of even accidentally marking a map; for example, soiled fingers will mark a map just as plainly as a pencil.

c. Pointee-Talkee. The "pointee-talkee" is a language aid which contains selected phrases in English on one side of the page and foreign language translations on the other side. To use it, evaders determine the question and statement to be used in the English text and then point to its foreign language counterpart. In reply, the natives will point to the applicable phrase in their own language; evaders then read the English translation.

(1) The major limitation of the "pointee-talkee" is in trying to communicate with illiterates. In many countries the illiteracy rate can be astoundingly high, and personnel have to resort to pantomime and sign language which have been relatively effective in the past.

(2) "Pointee-talkee" phrases are presented under the following eight subheadings:

(a) Finding an interpreter.
(b) Courtesy phrases.
(c) Food and drink.
(d) Comfort and lodging.
(e) Communications.
(f) Injury.
(g) Hostile territory.
(h) Other military personnel.

Barter kits may be available in some commands. If not, crewmembers may elect to develop their own. Items for consideration should be selected from area studies. Some items to consider might be rings, watches, knives, local currency, coins, and lighters. Items should have no markings of personal significance or military value. Military items packed in kits may be considered if not essential to the evader. Flashing large amounts of cash or valuables can have negative results in a depressed, war-torn area. Show only small amounts and drive a hard bargain.

d. Other Evasion Aids. Information on other evasion aids and tools is available from unit intelligence officers.

e. Assisted Evasion. There may be people in a hostile nation or in an enemy-occupied country who are dissatisfied with existing conditions.

(1) History has revealed that in every major conflict there are groups of people in every country who will aid a representative of their government's enemies. The motivating force behind these groups may vary from purely monetary considerations to idealistic concepts of government reform. In many cases, their real objective will be the political advancement of their particular group. During WW II, many underground or resistance groups and movements were established in occupied countries. One of the major purposes or functions of these groups was the aiding of downed allied aircrew personnel. In most cases, the driving force behind these movements was patriotism and desire for political recognition for their cause.

(2) These circumstances favor active resistance movements. One of the functions of such movements may be the operation of escape and evasion (E&E) systems for the purpose of returning evaders to friendly territory.

(3) U.S. Special Forces may also organize and operate E&E mechanism.

(a) E&E Oifemizations. Assistance may range from that rendered by a sympathetic individual to elaborate E&E nets organized by local inhabitants. E&E organizations may be limited in nature, such as providing assistance to reach a national frontier, or they may be linked to larger organizations capable of returning to friendly control.

(b) Acts of Mercy. These are usually isolated events during which evaders may be provided food, shelter, or medical attention for a brief period of time. Local people may find an exhausted or incapacitated evader and provide that evader with limited sustenance. This type of assistance is frequently offered with reluctance or under fear of reprisal because an act of providing comfort to the enemy would mean punishment. Unless an evader is in immediate need of medical attention, acts of mercy may consist of only an offer of food followed by an urgent plea that the evader leave the area immediately. If an evader is physically able to depart with a reasonable chance of evading capture, he or she should do so. An evader should not insist on receiving additional aid other than what is offered by the person who renders assistance only through human impulses (Figure 17-62).

(4) Evaders must understand when dealing with any indigenous personnel while in enemy territory, their own actions will govern the treatment they will receive at the hands of these people. How evaders conduct themselves may also have much to do with getting back to their own forces should they fall

Figure17-62. Receiving Aid.

into the hands of irregulars friendly to their (the evaders) own cause. The following list of suggestions may be a useful guide in dealing with these people or groups.

(a) Evaders should understand that failure to cooperate or obey may result in death.

(b) Evaders should try to avoid making any promises they cannot personally keep.

(c) Evaders should remember that the four conditions called for by the rules of land warfare must be met before members of an irregular group can be recognized as qualifying for PW status in the event of capture. These same rules will also apply to evaders who are involved with these groups.

(d) Prisoners of war, according to the current Geneva Convention, are persons belonging to one of the following categories, who have been captured by the enemy. Members of other militias and volunteer corps, including those of organized resistance movements, belonging to a conflicting party and operating in any territory, even if this territory is occupied, provided that such groups fulfill the following four conditions of:

-1. Being commanded by a person responsible for subordinates.

-2. Having a fixed distinctive sign recognizable at a distance.

-3. Carrying arms openly.

-4. Conducting their operations in accordance with the laws and customs of war.

(5) If evaders join such a unit, the closest thing to a guarantee of treatment as a military person will be their uniforms if they are captured.

(6) If evaders have the opportunity to influence the group they are with, they should try to encourage them to abide by the four conditions mentioned.

(7) Evaders must avoid becoming associated in any way with atrocities these groups may commit against civilians, prisoners, or enemy soldiers.

(8) Evaders should not become involved in their political or religious discussions, take sides in their arguments, or become involved with the opposite sex.

(9) Evaders should show consideration for being allowed to share food and supplies. It may also be helpful if evaders understand and show interest in the assister's customs and habits.

(10) The overall best and safest course for evaders to follow is to exercise self-discipline, display military courtesy, and be polite, sincere, and honest. Such qualities are recognized by any group of people throughout the world. The impression left can influence the aid provided to future evaders.

f. E&E Lines. An E&E line is a system of one or more secret nets organized to contact, secure, and when possible, evacuate friendly personnel. Well-organized and supported lines normally can be expected to provide the following assistance:

(1) Temporary shelter, food, and equipment for the next phase of the journey.

(2) Clothing and credentials acceptable in the area to be traveled.

(3) Information concerning enemy security measures along the evasion route.

(4) Local currency and transportation.

(5) Medical treatment.

(6) Available native guides.

g. Conduct of E&E Lines. The success of an E&E organization depends almost entirely upon its security. The organization of a line includes much planning and work carried out under dangerous conditions. The security of the system often depends upon the evader's cooperation and working knowledge of how it functions, how it may be contacted, and what rules of personal conduct are expected of the evader. The following paragraphs summarize the major aspects of the operation of an E&E line.

(1) Contacting the Line. Premission briefings may inform evaders where to go and what actions to take to make contact with an E&E mechanism. After being picked up by an evasion and escape mechanism, evaders will be moved under the control of this mechanism to territory under friendly control or to a removal area, and arrangements may be made for air or sea rescue. The organizer of a line in friendly but enemy-occupied territory normally will have arranged a network of spotters who will be especially active when evaders are in the immediate area, but so will the enemy police and counterintelligence organizations. For this reason, certain precautions must be observed when making contact.

(2) Approach. Help may be refused by a person simply because he or she thinks someone else has seen the evaders approach to seek assistance. If captured with a local helper, an evader will become a prisoner, but the helper and perhaps an entire family may be more severely punished.

(3) Making Contact. Contacts with the natives are discouraged unless observation shows they are dissatisfied with the local governing authority, or previous intelligence has indicated

the populace is friendly. Evaders should proceed to, and remain in, the nearest SAFE area where arrangements for contact can be developed. If the E&E system is operating successfully, the spotter will know evaders are present and will search the immediate area, making frequent visits to designated contact points. Identification signs and countersigns, if used, may be included in the preoperational briefing. It is seldom advisable to seek first contact in a village or town. Strangers are conspicuous by day, and there may be curfews or other security measures during the hours of darkness. The time of contact should be at the end of the daylight period or shortly thereafter. Darkness will add to the chance of escape, if the contact proves to be unfriendly, and may be advantageous to the contact in providing further assistance.

(4) Procedure After Contact. If contact is made, evaders may be told to remain in the vicinity where spotted, or, more likely, they will be taken to a house or other structure used by the E&E net as a holding area. It must be decided at this time whether or not to trust the contact. If there is any doubt, plans should be made to leave at once. It is also possible that the house may not belong to the E&E organization but rather to someone who will look after evaders until arrangements can be made for the line to identify and accept them in E&E net.

(5) Establishing Identity. Verification of identity will be required before anyone is accepted as a bona fide evader. The constant danger facing the operators of an escape line is the penetration of the E&E system by enemy agents pretending to be evaders or escapees. Evaders should be prepared to furnish proof of identity or nationality. Since it may lead to later difficulties of identification, they should never give a false name—just their name, grade, service number, and date of birth. It is best for them to avoid talking as much as possible.

(6) Awaiting Movement on the Line. Delays can be expected while proceeding along the escape line. If the period of waiting is prolonged, frustration and impatience may become unbearable, leading to a desire to leave the holding area. This must not be done, because if seen by other people, the lives of the assisting personnel and the existence of the entire line itself may be endangered.

(g) Evaders should follow the orders of those assisting them. If kept indoors for any length of time they can keep fit by moderate physical exercise.

(h) The host should have a plan for rapid evacuation of the area if enemy personnel should raid the holding area; if not, evaders should have a personal plan to include measures for moving all traces of having occupied the area. If the net is being overrun and capture is imminent, evaders must be prepared to fend for themselves. The evader is the only one who knows more than one part of the net. The assisters may attempt to eliminate the security risk to the net.

(7) Traveling the Line. It would be a grave breach of faith and security for evaders to discuss, at any point on the line, the earlier stages of the journey. Evaders might be tested to see if they are trustworthy—they should discuss nothing of the net. For security reasons and to protect the compartmentalization of the line, no information should be revealed. It is also useless to ask where a line leads or how it will eventually reach friendly territory. Evaders should not try to learn or memorize names and addresses and, above all, they shouldn't put these facts or any other information in writing. Evaders should give the impression of having received no assistance from local inhabitants.

(8) Fellow Evaders. Caution is required in the case of fellow evaders on an escape line unless they are personally known. Even when it has been satisfactorily determined that another person is a genuine evader, no information should be given.

(9) Travel with Guides. If under escort, this fact should not be apparent to outsiders. In a public vehicle, for example, evaders should never talk to their guide or appear to be associated with the person in any way unless told to do so. This will lessen the possibility of both the evader and the guide being apprehended if one should arouse suspicion. It should always be possible for the guide to disown an evader if the guide gets into difficulty. When evaders are escorted, they should follow the guide at a safe distance, rather then walk right next to the person, unless instructions indicate the latter action is required.

(10) Speaking to Strangers. Evaders should never speak to a stranger if it can be avoided. As a last resort, they should pretend to be deaf and dumb or even halfwitted. This technique has often been successful. To discourage conversation in a public conveyance, they can also pretend to read or sleep.

(11) Personal Articles and Habits. Evaders should not produce articles in public which might show their national origin. This pertains to items such as pipes, cigarettes, tobacco, matches, fountain pens, pencils, and wristwatches. Evaders should also ensure their personal habits do not give them away; for example, they should not hum or whistle popular tunes or utter involuntary oaths. Again, in restaurants, imitating local customs in the use of knives and forks and other table manners is advisable. Study of the area before the mission may help evaders avoid making mistakes.

(12) Payment to Helpers. On an escape line, evaders should not offer to pay for board, lodging, or

other services rendered. These matters will be settled afterwards by those who are directing and financing the line. If in possession of escape kits or survival packs, evaders should keep them as reserves for emergencies. If they have no food reserve, they should try to build up a small stock in case they are forced to abandon the line.

(13) Evaders Conduct:

(a) Be polite by local standards.

(b) Be patient and diplomatic.

(c) Avoid causing jealousy. Disregard the sex of the assisters.

(d) Avoid discussions of a religious or political nature.

(e) Eat and drink if asked, but don't over indulge or become intoxicated.

(f) Don't take sides in arguments between assisters.

(g) Don't become inquisitive or question any instruction.

(h) Help with menial tasks as directed.

(i) Write or say nothing about the other people or places in the net.

(j) Don't be a burden; care for self as much as possible.

(k) Follow all instructions quickly and accurately.

17-19. Combat Recovery:

a. Air recovery is one of several means of transportation for downed crewmembers in their quest for their final goal—returning to their own lines and units. How recovery will be effected will depend on a number of factors, among them are the following:

(1) Terrain.

(2) Capability of the rescue vehicle.

(3) Condition of the survivor (evader).

(4) Enemy activity in the area.

(5) Weather conditions.

(6) High or priority mission.

b. Even though a maximum effort will be made to recover downed aircrew members, survivors can jeopardize the whole rescue operation and the lives of rescue personnel in a combat zone by not taking a responsible role in recovery operations. The responsibilities are many and varied but essential to a successful rescue mission. Evaders should recall that even though they may have little experience in participating in actual rescues, they must nonetheless be very proficient in their actions. Evaders should always remember other people are endangering their lives in an attempt to retrieve them.

c. There are two basic types of air-recovery vehicles— rotary wing and fixed-wing aircraft. The rotary wing type can extract evaders from remote areas. With air-to- air refueling, the range of these aircraft has increased; they are limited only by the endurance of the crewmembers who operate them.

d. Survivors must be proficient in the operation of all the survival equipment at their disposal. For example, they must be able to switch off the beeper on the radio to receive verbal instructions from the rescue commander.

e. The initial contact with rescue or other combat aircraft in the area must be done as directed by authorities in that theater of operations; for example, in Southeast Asia, a contact method of transmitting 15-seconds beeper, 15-seconds voice (call sign), and 15-seconds monitoring was used. The method an evader should use to establish contact with rescue will be briefed before the mission.

f. One important aspect of the rescue process is evader authentication by rescue personnel. Survivors must be able to authenticate their identity through the use of questions and answers, responding as directed before the missions. Authentication in a combat area changes rapidly to reduce the chances of compromising the rescue efforts. Survivors (evaders) must keep up with these changes so their rescue can be made without undue danger to rescue personnel.

g. The purpose of filling out and using a personal authenticator card is positive verification of an evader's identity which is essential before risking search and rescue aircraft or the lives of assisters. The purpose of the photographs, descriptions, fingerprints, etc., on the card is to ensure all possible means are used for the proper identification of personnel. Intelligence personnel are responsible for ensuring that personnel fill out the card completely and that they are aware of the purpose and use of the information. When filled in, the personal authenticator card becomes classified Confidential, and is reviewed at least semiannually by both the crewmember and intelligence personnel.

h. When selecting a site for possible evasion recovery, there is much for evaders to keep in mind. The area they choose could well decide the success or failure of the mission.

(1) The potential recovery site should be observed for 24 hours, if possible, for signs of:

(a) Enemy or civilian activity.

(b) Roads or trails.

(c) Farming signs.

(d) Orchards.

(e) Tree plantations.

(f) Domesticated animals or droppings.

(g) Buildings or encampment areas.

(2) The rescue site should be observable by aircraft but unobservable from surrounding terrain if possible. There should also be good hiding places around the area. The site should include several escape routes so evaders can avoid being trapped by the enemy if discovered. There should be a small open area

for both signaling and recovery. It would be beneficial if the surrounding terrain provided a masking effect for rescue forces in order to avoid enemy observation and fire.

i. The size of evasion ground-to-air signals should be as large as possible but must be concealed from people passing by. Evaders should remember the six-to-one ratio. The contrast these signals make with the surrounding vegetation should be seen from the air only. Any signal displays should be arranged so they can be removed at a moment's notice since enemy aircraft may also fly over the immediate area.

(1) All of the principles of regular (nontactical) signals should be followed by those building evasion signals. Crewmembers may be prebriefed as to the appearance of specially shaped signals (Figure 17-63).

(2) Evasion signals, like all others, must be maintained to be effective. At times, evaders may be preinstructed to set out their signals according to a prearranged time schedule.

j. Whatever signal devices are available to evaders, they must be able to use them (mirror, flare, gyro-jet, etc.) with proficiency. These signals, like those used in nontactical situations, are to be used either to gain the attention of friendly aircraft or rescuers, or when directed by rescue. Extreme care must be taken to minimize or eliminate chances of enemy elements spotting the signals. For example, the strobe light and mirror can be directed and aimed instead of being used in an indiscriminate manner.

(1) If evaders are in or on the water, they should use lights, flares, dye, whistles, etc., with extreme

Figure 17-63. Evasion Signals.

care as they are readily distinguishable over water at great distances.

(2) In addition to knowing how to use radios correctly to effect their rescue, evaders should also be familiar with the special points of evasion radio procedures; for example, ensuring the radio transmits continuously during a night recovery operation using the electronic locator finder (ELF) system.

(3) It is also the evaders' responsibility to communicate all signs of enemy activity, such as:

(a) Locating anti-aircraft emplacements.

(b) Identifying when they are firing.

(c) Assisting strikes by spotting hits (high, low, or on target).

(d) Determining effectiveness of hits.

(e) Notifying personnel of changes in small arms positions, etc.

(4) Downed evaders can normally expect to be hoisted to a helicopter by one of five methods: basket, Stokes litter, bell, horse collar, or forest penetrator. Other pickup devices which may be used are the McGuire Rig, Swiss Seat, Motley Rig, Stabo Rig, rope ladder, or rope. Another method of recovery which may be employed to extract evaders is the Surface-to-Air-Recovery (STAR) System. Evaders should remember whatever me pickup device, it should always be grounded before they grasp it.

(5) There is also fixed-wing capability of rescuing downed crewmembers. Evaders will be prebriefed as to which type rescue vehicle and systems to expect in their areas of operation. No matter which type is used, evaders must be capable of mounting and riding the rescue devices in a minimum amount of time, allowing the rescue craft to effect the rescue and depart quickly.

(6) As downed crewmembers, whether as a result of enemy action or of mechanical failure, it is important they all become familiar with the information that 1 day could prove instrumental in saving their lives. First, they must have prepared themselves to cope with the survival situation before an aircraft emergency. This can be done by including, as part of their preflight planning, a thorough inspection of survival equipment to determine its availability and serviceability. Second, they must realize there are many types of recovery vehicles available in the Air Force inventory, and a knowledge of all recovery techniques is a necessity. They should request detailed briefings from area or local rescue personnel who will explain the operational limitations and recovery potential for each. Third, evaders must be capable of fulfilling their part in the recovery operation. This can be done by knowing when, where, and how to initiate communications and how

to cooperate with the rescue aircraft crew.

(7) Other forms of assistance which may be available to evaders are Special Forces, Combat Control, RECON, SEALS, Riverine Operations, and Submarines. While aiding evaders is not the primary mission of these groups, as a secondary mission, they may travel in or near SAFE areas on their return journey to check for the presence of evaders.